Programming for the Series 60 Platform and Symbian OS

Programming for the Series 60 Platform and Symbian OS

DIGIA Inc.
Helsinki, Finland

JOHN WILEY & SONS, LTD

Published by John Wiley & Sons Ltd,
The Atrium, Southern Gate, Chichester,
West Sussex PO19 8SQ, England

National 01243 779777
International (+44) 1243 779777

e-mail (for orders and customer service enquiries): cs-books@wiley.co.uk
Visit our Home Page on http://www.wileyeurope.com or http://www.wiley.com

Library of Congress Cataloging-in-Publication Data

(applied for)

British Library Cataloguing in Publication Data

A catalogue record for this book is available from the British Library

ISBN 0-470-84948-7

Typeset in 10/12pt Optima by Laserwords Private Limited, Chennai, India
Printed and bound in Great Britain by Biddles Ltd, Guildford and King's Lynn
This book is printed on acid-free paper responsibly manufactured from sustainable forestry, for which at least two trees are planted for each one used for paper production.

Contents

Foreword by Nokia

Every day, millions of people around the world manage their mobile telephone communications with Nokia phones, without giving a second thought to the software inside their phones. This is not only a sign of good hardware design but also a sign of great user interface software and menu system design, often considered as one of the key success factors in Nokia's mobile phones.

For fixed communications people are used to browsing the Internet, sending and receiving email messages, or creating documents on their personal computers. Although smartphones are not intended to replicate the functionality of personal computers they introduce true mobility and always-on data connectivity to people who want to have access to normally desktop-bound resources anywhere they go.

How, then, is it possible to design highly functional yet simple-to-use smartphone software that combines both voice communications and data connectivity applications? Early on, our user interface software design team concluded what the key factors for a successful smartphone were. It had to be as easy to use as a mobile phone. It had to integrate voice communication and mobile capabilities of a phone and key applications, such as multimedia messaging, scheduling, and browsing, into an attractive product package. We believe that years of experience with mobile phones, user interface development, and listening to our customers have paid off in the Series 60 Platform.

The Series 60 Platform is a mobile terminal software platform built on Symbian OS. The goal was to design an open, standards-based platform for one-hand operated smartphones with color screen and keypad that features an intuitive user interface and key applications. Symbian OS was selected because it has been designed from the beginning for mobile phones and provides a stable platform with low power consumption and advanced memory management for a wide range of mobile phones, applications, and services. It is no coincidence that, besides Nokia, all major manufacturers in the mobile phone industry have invested in Symbian.

Technical capabilities are important, but the attainable market size is crucial when considering a platform for application development. The first product built on the Series 60 Platform available in shops was the Nokia 7650. Siemens IC Mobile and Matsushita Communication were the first companies that licensed Series 60 for integrating it in their own phones.

These companies are also contributing to the future development of the platform, thereby demonstrating how the open, standards-based product concept works for the benefit of the entire emerging smartphone market. Software developers, operators, and IT vendors can rely on this platform, which is both shared and supported by the industry.

Forum Nokia welcomes you to the world of smartphones and to the world of Series 60 software. It is great to see you join the pioneers of mobile application developers!

Jouko Häyrynen
Vice President, Forum Nokia
Nokia Plc.

Foreword by Digia

In the mobile industry there are a number of factors linked to the success of smartphones – starting with the hardware architecture and ending up with the applications. When a service or an application is installed on a smartphone, it must provide a positive and optimized user experience.

Cross-platform standards, interoperability, captivating applications, high bandwidth networks, the robust Symbian OS, and suitable multimedia middleware are the platforms on which the next generation smartphones able to meet their full potential, leading to excellent consumer acceptance and market growth.

As the market moves from simple voice devices to complex voice and data devices, new technology and new concepts in mobile software, such as the Series 60 Platform, are being introduced. Together with its customers and partners, Digia Inc. is playing its part in building a standard environment for interoperable smartphones.

Digia Inc. has developed its strategy to contribute to market growth. The company is a global Finnish mobile software supplier of Personal Communications Technologies and Solutions for smartphones, and has been cooperating with Nokia on Series 60 Platform development for several years. Digia will offer the required integration and customization services to any licensee of the Series 60 Platform.

Digia has proven its capability by developing major parts of Symbian OS phone projects, helping to reduce the time-to-market span significantly for these new phones. Digia technology expertise and consultancy cover Symbian OS phone architecture from base porting to user interface customization, helping its customers to create, optimize, and differentiate their products.

This book is a landmark for the Symbian economy in wrapping Digia's expertise in the chapters of a single book. I would like to show my gratitude to Digia customers and business partners who have contributed in the making of this excellent book. A sincere thank you goes to everybody at Digia who has put their expertise in the chapters, thus making this book possible.

Jari Mielonen
Chief Executive Officer
Digia Inc.

About the Authors

Tino Pyssysalo was the main author and project manager responsible for creating this book with the help of the co-authors listed below. Tino holds an MSc (Eng., with honors) and licentiate of technology (with honors) majoring in theoretical computer science, graduating from Helsinki University of Technology. He has done research in the areas of data communication and telecommunications software for several years by formally specifying, verifying, and designing communication protocols. Tino worked for five years as a senior teaching assistant and professor at the University of Oulu, teaching and researching mobile virtual reality in which area he is currently completing his PhD thesis (doctor of technology). He has authored and co-authored more than 30 scientific papers and articles. His current position in Digia is as senior software specialist, and his main duties include providing an expert's view on software development projects and giving advanced technical training in Symbian OS.

Terje Bergström holds an MSc (Eng.) in computer engineering from Helsinki University of Technology. He majored in embedded systems, with usability studies as a minor subject. He joined Digia a year ago and has been working as a software engineer specializing in client–server software in a major Series 60 project. Previously, Terje worked for several years for Nokia Networks, where he developed software tools for internal use.

Jürgen Bocklage has a (Dipl., diploma in engineering degree (full honors)) in European computer science from the University of Applied Sciences, Osnabrück, Germany, and a BSc degree (Eng.) in computer science at the Espoo-Vantaa Institute of Technologies. Jürgen joined Digia in 1999. He has extensive knowledge on Symbian OS C++ and Java technologies. His previous writing experience includes contributing a chapter to the Jonathan Alin book *Wireless Java for Symbian Devices* (John Wiley, 2001).

Pawel Defée has a degree in mathematics and software engineering from Tampere University of Technology. He has over five years of cumulative experience in designing and implementing Symbian OS software for several successful products at Nokia, Symbian, and Digia, where he is currently employed as a software specialist. His areas of expertise include designing user interface frameworks and designing application software and system-level components.

Patrik Granholm has been working on numerous Symbian OS projects at Digia since the beginning of 2000. In his current position as a team manager, Patrik is responsible for a messaging competence team. Previously, he has worked on developing object-oriented systems, such as accounting software for Emce Solution Partner Ltd.

Juuso Huttunen has experience in various software development tasks, from implementation to architectural design. He has gained both C++ and Java development experience in Symbian OS through several demanding projects. Currently, he is working in Digia's Lappeenranta office, as a software engineer in the Java competence team.

Ville Kärkkäinen has an MSc in computer science, graduating from the University of Helsinki. In Digia, his current position is as a team manager. Ville is responsible for the competence development of the Communications and IP team in Digia's Helsinki office. He is also working as a project manager in a large project, concerning a Series 60 smartphone.

Matti Kilponen is a bachelor of visual and media arts, graduating from Lahti Polytechnic, Institute of Design. He has several years of design experience in multimedia and user interaction, and his current position in Digia is as an interaction designer. His previous career includes positions as a project manager and information designer at Razorfish, specializing in mobile content. His user-centered design expertise in Digia covers the Nokia Series 60 user interface style application design, user interface specification writing, interaction design, and early design verification. He specializes in the interface design of applications for the Nokia Series 60 Platform.

Timo Kinnunen has an MSc (Eng.) in industrial engineering and management, graduating from the University of Oulu, Finland. He has several years experience in interaction design for mobile phone user interfaces. Timo's current position in Digia is as a senior interaction designer in which he leads Digia's User Experience Group. His previous career includes user interface and concept designing in Nokia. Timo has also authored scientific articles on concept design and has contributed to a number of pending patents on user interaction. His user-centered design expertise covers user needs research, concept design, interaction design, and usability engineering, focusing especially on the Symbian OS user interfaces, such as Nokia Communicator, Nokia Series 60, and UIQ.

Tomi Koskinen has a BSc (Eng.), graduating from Lahti Polytechnic Institute. He has been involved with Symbian OS in large-scale Series 60 projects, for over three years. During this time he has accumulated a thorough understanding of the platform and the available user interface components.

Mika Matela has an MSc (Eng.) in information technology, graduating from the University of Lappeenranta. Currently, he is working at Digia as a software engineer. Previously, he worked as a designer at Kone Co. in Hyvinkää R&D center. Mika is experienced in various areas of product creation and, most recently, he has been involved in developing messaging solutions in Symbian OS.

Ilkka Otsala is a student at Helsinki Polytechnic, where he is studying for a bachelor's degree in software engineering. Ilkka's current position in Digia is as a software engineer. Previously, he worked at Nokia, where his main area of responsibility was user interface software. During his employment at Digia, Ilkka has gained a more thorough understanding of user interface software engineering on Symbian OS, especially the Series 60 Platform.

Antti Partanen has an MSc (Eng.) in computer engineering, graduating from the University of Oulu. His current position in Digia is as a software specialist. Antti's previous positions include working as a software design engineer in Nexim Ltd, as a researcher at the University of Oulu, and as a software engineer in Tellabs Inc. He has experience and a thorough knowledge of many areas of Symbian OS. His main areas of expertise are communication protocols, system and communication architecture, and embedded systems.

Timo Puronen has worked at Digia for over two years, and currently his position is as a software specialist. He has been involved in designing and developing messaging and communications software for the Series 60 Platform. He has studied at the Department of Computer Science at the University of Helsinki, and he is currently preparing his master's thesis on utilizing reusable software component frameworks in Symbian OS system software.

Jere Seppälä has a BSc (Eng.) from the Department of Telecommunication Technology, graduating from the Swedish Institute of Technology. He is currently a technology manager responsible for competence development at Digia's Helsinki office. He was previously a project manager at Digia and responsible for large software projects concerning a Series 60 smartphone. Before his career at Digia, he was employed by Ericsson. Where, he was involved in software production, for five years, within different network standards, such as GSM, PDC, PCS, GPRS, and WCDMA.

Juha Siivola has an MSc (Eng.) in computer science, from Lappeenranta University of Technology. In 1994, Juha joined Nokia and worked as part of the team that pioneered the first revolutionary smartphone: Nokia Communicator 9000. As connectivity and interoperability became more and more important in mobile phones, he focused on the management of connectivity and subcontracting for the Nokia 9110 product. In February 1999, Juha joined Symbian, where he contributed to the definition of the Symbian Connect 6.0 release, later used in the Nokia 9210. In February 2001 he moved back to Finland and joined Digia where he works as head of product and technology management.

Saiki Tanabe has a BSc (Eng.) in software engineering, from Helsinki Polytechnic Institute. The objective of his graduate study was to analyze how to initialize an Intel processor in an object-oriented operating system (Symbian OS). He has been working at Digia for two years as a software engineer, designing and implementing application programs for Symbian OS. His main area of expertise is system level user interface design, and he belongs to the user interface competence team at Digia's Helsinki office.

Jukka Tarhonen has worked at Digia in several positions, and his current position is as a team manager. Jukka is responsible for the Java competence team in Digia's Lappeenranta office. His has a thorough knowledge of Java in Symbian phones as well as of Symbian implementation of the K virtual machine (KVM). His key expertise includes developing extension APIs for KVM (i.e. device-specific Java MIDP extensions in the Series 60 Platform).

Tommi Teräsvirta has worked at Digia since early 2000. He has several years experience in training, consulting, and development with Symbian OS, starting from EPOC Release 3. As the training manager, he is responsible for all Digia's Symbian OS related training. His duties also include technical consultancy in various software projects. He has lectured and delivered technical training sessions for hundreds of current and future Symbian OS professionals.

Tuukka Turunen has an MSc (Eng.) in computer engineering from the University of Oulu and is a licentiate of technology in embedded systems, also from the University of Oulu. His current position at Digia is as a program manager. Tuukka's previous career included working as an R&D manager responsible for all product development projects in Nexim Ltd, as a senior assistant in computer engineering at the University of Oulu, and as a software engineer at Nokia. He has authored over a dozen scientific publications and conference presentations in the areas of communication protocols and virtual reality, as well as having contributed to several technical magazines. His main areas of technical expertise are communication protocols and embedded systems, and recently he has become increasingly interested in user experience design.

Tommi Välimäki holds a BSc (Eng.) from Helsinki Polytechnic, Stadia (graduating in 1992, at which time the school was called Helsinki Technology School). Tommi's area of specialization was telecommunications engineering, and his graduating thesis was on microprocessor's real time operating system. Tommi joined Nokia in 1993 and started his current career in Digia in the beginning of 2000. He has several years experience in software testing, development, training, and research. His current position within Digia is as a senior testing engineer for Symbian OS and Series 60 environment.

Acknowledgements

This is the first book ever written by Digia, and it was quite an experience for all the people involved. Despite the fact that we had lengthy experience in writing the technical documentation as part of the Symbian OS and Series 60 Platform software development projects, this book taught us that book writing is something else.

The idea of the book came up within Digia at the end of 2001, as we saw a need to make life just a bit easier for the thousands of developers eager to learn more about programming for this exciting platform. It has been an honor for us to put a fraction of our knowledge and experience on paper and share it with readers.

Being a professional software company, we could not figure out a better way of managing the writing of this book than establishing a project with a project manager, team, and steering group and creating detailed time, resource, and budget estimates. Our admiration of people who have written a whole book increased tremendously during the project. Also, having a large number of people involved in the creation of the book made the project management even more challenging. We hope that readers appreciate the extra insight that comes from the fact that the various chapters have been written by experts in those areas.

This was a true team effort and the result of many people working closely together. The authors volunteered to do the writing, which in practice meant that many of them ended up spending their evenings, nights, and weekends working on this book.

We would like to take this opportunity to thank all who contributed to the book and supported us during the project. If this book is useful to you, remember that these people had a great deal to do with that.

First of all, we would like to express our gratitude to the numerous people at Nokia and Symbian with whom we have worked throughout the past years. Without the support and input from both Nokia and Symbian, writing this book would not have been possible. Thanks for trusting us and giving the opportunity to work in this exciting area.

From the publisher John Wiley & Sons, we would especially like to thank our publishing editor Karen Mosman, for the lessons learned and for the valuable pieces of advice on how books should be written – thanks for your patience and good cooperation! Thanks to Jo Gorski for her efforts in promoting the book, and Geoff Farrell for making this book available for corporate buyers. Additionally, big thanks goes to the rest of the team at John Wiley & Sons, including the copy editors, artists, and production people who worked hard on this project.

Several people outside Digia have reviewed and commented the manuscript, and with their comments the outcome has been significantly improved. We are very grateful to all the individuals who have commented either parts of or the entire content of the draft of the book. Special thanks are due to external reviewers Robert van der Pool and Vladimir Minenko from Siemens, Sudeesh Pingili and Richard Baker from Panasonic, John Roe from Symbian, Rob Charlton from Intuwave, as well as Olli Koskinen, Matti Kakkori, and Mika Muurinen from Nokia, for their thorough and detailed review. If you have ever technically reviewed a proof copy, you know how much work this is. These people went an extra mile and obviously read in their own time, and for this we thank them profusely.

We would also like to thank all the people who made the book more enjoyable to read. Among them are Marketta Räihälä, who made the English more understandable, Martti Lepistö, who improved the graphical representation of the images in the book, and Ville Koli, who did the cover design of the book.

Several of Digia's professionals have contributed as authors to the writing of this book, and even more have helped by giving ideas, reviewing, supporting, and commenting. Especially helpful comments in the formal review were received from: Jari Hakulinen, Jani Harju, Tuomas Harju, Kari Hautamäki, Kimmo Hoikka, Eeva Kangas, Ossi Kauranen, Jari Kiuru, Henri Lauronen, Petri Lehmus, Markku Luukkainen, Pasi Meri, Tomi Meri, Ville Nore, Jari Penttinen, Heikki Pora, Sami Rosendahl, and Tapio Viitanen. The source code examples were reviewed by Kimmo Hoikka and Petri Poikolainen, for which the authors wish to acknowledge them.

Special thanks are also due to Pekka Leinonen for organizing the formal reviews to all the chapters, and to Tino Pyssysalo who, in addition of being the principal author, was the project manager of the book creation project. Tino did an outstanding job of managing and coordinating all the various pieces of the project. The rest of us really appreciate all your efforts!

The authors wish to acknowledge all the members of the steering group of this project: Minna Falck, Jari Kiuru, Pekka Leinonen, Inka Luotola, Tino Pyssysalo, Juha Siivola, and Tuukka Turunen, as well as the Digia management team and all our other colleagues for all their support and efforts that made this book possible.

On the personal side of things, we want to thank our families and friends. Thanks for understanding and thanks for being there.

And finally, we would like to express our gratitude to all the other people who are not listed here by name but who helped us in 'running the marathon' this book presented: THANK YOU!

1

Introduction to the Series 60 Platform

The market drive towards advanced mobile services such as multimedia messaging and wireless access to the Internet sets high requirements for the terminals and networks. New devices (and networks) must be able to handle still and moving images, music, allow access to the Internet for messaging and browsing, and provide a diverse set of applications in addition to operating as a mobile phone.

As 2.5G and 3G networks emerge, the focus is turned to the terminal – what kind of capabilities are required by these new services and how can they be fulfilled? This book presents one significant platform for smartphones – devices that are designed to meet the increasing requirements of the new services – Series 60. It is a platform made by Nokia on Symbian OS (Operating System), which in turn is an operating system designed for mobile devices.

Owing to these requirements of the new services, the task of making the software for wireless devices is becoming increasingly complex. As new protocols, communication interfaces, services, and applications are integrated, the required R&D effort grows rapidly. This increases the development cost significantly and sets high demands on interoperability testing. Symbian was formed in 1998 by Ericsson, Motorola, Nokia, and Psion to provide a common standard and to enable the mass marketing of a new era of wireless devices. Matsushita (better known by its brand name, Panasonic) joined Symbian in 1999, in January 2002 the Sony Ericsson joint venture took a share of Symbian, and, in April 2002, Siemens also joined Symbian as a shareholder (www.symbian.com).

The cost of the hardware components (i.e. bill of materials) used for making a mobile phone is one of the most important factors in making competitive devices, but the cost of developing and maintaining the increasing number of features of the software also becomes more significant as the world evolves towards wireless multimedia. When the mobile phone shipping volumes are low, the price of the hardware components, especially the most expensive ones, such as the color display, is high. When components are made in high volumes, the price of a single component is reduced. New requirements are set for the operating system by the need to open the devices for developing third-party applications in order to have a more diverse set of applications to attract more users and to increase volumes.

A typical smartphone is attractive for users looking for more features compared with the basic mobile phone. Also, for those seeking better personalization, a smartphone is likely to be a natural choice. A smartphone is, as its name states, a phone, but with capabilities such as color screen, advanced messaging, calendar, browser, e-mail, synchronization with other devices, and, installation of applications.

From the device manufacturer perspective, it is vital that the software is stable and reliable – especially if it is possible for the user to install new applications. A serious software or hardware defect may force a manufacturer to call back all the mobile devices. The costs of such an operation are very significant even when compared with development costs. From the software point of view, the right selection of software platform will reduce the risk of failure.

Series 60 Platform is particularly appealing for the terminal manufacturers, as it allows Symbian OS with multimedia capabilities to be used in devices that are only slightly more expensive than the devices equipped with a proprietary operating system. Series 60 Platform provides the benefit of open interfaces, efficient power management, and advanced multimedia capabilities. From the application developer perspective, Series 60 is also very interesting, as the compact size and reasonable price of the devices leads to large market penetration, thus making the application development worthwhile as well.

This book is written by the software professionals of Digia – a company that has been heavily involved with Symbian OS and Series 60 Platform and application development since early 1999. During that time, knowledge of the platform has been accumulated in demanding, time-critical projects and in developing products for Symbian OS.

This book is a handbook of software development for the Series 60 Platform, written by software professionals for software professionals – and for individuals who want to become experts on Series 60 software development. The approach of this book is to build on top of the previous publications on Symbian OS and to provide a guide for the art of developing software for the Series 60 Platform. The intended audience is anyone interested in learning more about Symbian OS and the Series 60 Platform. It is specially beneficial for software developers who can use the code examples for illustrating the key aspects of Series 60 software development.

1.1 Operating Systems for Smartphones

Series 60 Platform is targeted at smartphones. But what makes a smartphone smart? There is no single characteristic that defines a mobile device as a smartphone. However, it is clear that the devices in the smartphone category possess certain capabilities. In this section, the most typical characteristics are presented, and the requirements of smartphones are discussed.

A device having the size and form factor of a normal phone, while providing, a graphics-capable color screen, value-adding applications such as messaging tools (e.g. e-mail, advanced calendar, and contacts book) and the ability to install new applications is categorized as a smartphone. Typically, smartphones are aimed at buyers who are seeking a replacement phone and who are willing to pay a little more for additional features. Examples of smartphones are shown in Figure 1.1.

Figure 1.1 Examples of smartphones: a smartphone looks like a normal cellular phone and typically provides a color screen capable of displaying graphics and several value-adding applications

In addition to features easily noticed by the user, there are other types of requirements for smartphones. As they are embedded systems (like all mobile phones), behavior accepted for typical personal computers cannot be tolerated. The operating system of the smartphone needs to be very reliable and stable – system crashing and rebooting are most undesired features. It must also fit into a very small amount of memory and use the resources sparingly. A real-time nature with predetermined response times is required of the parts that deal with the cellular networks, and it is preferred that the entire system provides fast response times. The overall quality of the operating system and the software platform of the smartphone needs to be very good, as the cost of replacing the software with a new version is high.

Smartphones are devices that are always on and typically run for weeks or months without restarting. Actually, many users turn their devices off only when traveling by plane – in all other situations the silent mode is sufficient. Another issue is the nature of data and storage media used with these devices. As the users store important personal data – such as their itineraries and precious memories – loss of data simply cannot be tolerated. These set stringent requirements for the memory management in the smartphones, and a smartphone operating system must be robust and support design principles that allow other software to be reliable. Robustness of the operating system is one of the key criteria to be considered when selecting the platform for smartphones. Especially important is the performance in error conditions, and it is vital that the user data and system integrity are not compromised in any situation.

Key requirements for hardware components are a small form factor, high tolerance of abuse, and extremely low power consumption (and low manufacturing and integration costs). The power consumption is particularly vital, as the amount of energy needed is stored in the battery, and most users do not want to charge their devices daily. In addition to selecting hardware components suitable for achieving low power consumption, the software has to operate in such a way that energy is conserved to the extent possible in all situations – while providing the response times and performance needed by the user or the network.

The operating system of a smartphone is the most critical software component as it depicts the nature of software development and operating principles. The most important requirements are multi-tasking (with multi-threading), real-time operation of the cellular software, effective power management, small size of the operating system itself, as well as the applications built on it, ease of developing new functionality, reusability, modularity, connectivity (i.e. interoperation with other devices and external data storage), and robustness.

Based on the choices of the world's top mobile phone manufacturers with the largest market share (Nokia, Motorola, Samsung, Sony Ericsson and Siemens holding almost 80% of the market), the most significant alternatives for extending smartphone functionality in a phone are either Symbian OS or the manufacturer's proprietary operating system. Although it is possible to select some of the operating systems used in personal digital assistants, such as Palm OS and Pocket PC (or its smartphone variant called Smartphone 2002) or to take an open-source approach with Linux, these solutions tend to raise problems, such as lack of power, excessive build cost, and fragmentation.

The hardware platform – especially the selected processor architecture – influences the suitability of the different operating systems for the device. For example, Symbian OS is currently used mainly with the ARM architecture and instruction set. However, it is possible to use Symbian OS with virtually any hardware platform by fitting the operating system to the new platform (called base porting). It also means that for the same type of hardware – actually, even for exactly the same hardware – it is possible to fit many of these smartphone operating systems. In practice, the cost related to creating and supporting the smartphone software (and also the fact that third-party applications would then be incompatible) makes an industry-standard operating system a more practical approach.

Power management is a usability issue to some extent. Although it is very important to use effective power-saving modes whenever possible, it cannot happen at the user's expense. This means that the response times need to be short even when the system is in sleep mode. Reaction to, for example, the user pressing a button, or the network signaling an incoming call, has to be immediate. Also, the boot sequence needs to be short and preferably allow user interaction even when some services are not yet invoked.

A typical design choice supporting modularity and allowing the robustness needed in consumer devices is the **microkernel** approach. Only very little code (the microkernel) runs in the privileged mode while other system components are built on top of it to provide a modular and extendable system. Additional benefits of the microkernel approach are increased security and robustness. As only a small part of the operating system runs in privileged mode, the likelihood of errors causing total crashes is minimized.

Generally, in software platforms the modularity and architecture define the extendibility and lifecycle of the platform. In operating systems, the modular approach is most 'future-proof' as it allows new functionality to be added in the form of additional modules and functionality to be removed that is no longer needed. Modularity can also benefit memory consumption, as applications can rely on the same software components as do the system services, thus reducing the total memory consumption by resource reuse.

Considering all the issues presented above, analyzing them against the cases presented for new smartphones and selecting the best operating systems sounds like a straightforward task for the terminal manufacturer. In many ways it is, but additional challenges arise in providing the software that 'sits on top' of the core operating system. Matters such as availability of standard communication protocols, the interoperability of the system, interfaces and tools allowing third-party software development, the capability to support common development and content creation methods, an attractive and customizable user interface allowing good usability, support, and, naturally, the cost need to be carefully balanced as well.

This section has covered the requirements of a smartphone operating system from the perspective of the terminal manufacturer. In the next two sections, an approach answering the demands – namely, Symbian OS and Series 60 Platform – is presented. Although this section has looked at the choices of creating smartphones from the viewpoint of the device manufacturer, the aspects covered affect the application developers as well. Comparison of different operating system choices has been kept to a minimum in order to give a general view on the topic.

1.2 Symbian OS

Symbian was formed from Psion Software by Nokia, Motorola, Psion, and Ericsson in June 1998. In 1999 Matsushita (Panasonic) and in April 2002 Siemens joined Symbian as shareholders. From the very beginning, the goal of Symbian was to develop an operating system and software platform for advanced, data-enabled mobile phones. For this purpose, the EPOC operating system developed by Psion formed a solid foundation. It was a modular 32-bit multi-tasking operating system designed for mobile devices.

1.2.1 Roots of Psion's EPOC Operating System

Psion developed personal organizers – small mobile devices with extremely long operating times (i.e. small power consumption) and an impressive feature set (all the essential office tools) in an attractive package – for which they had developed the EPOC operating system. It was not their first operating system, and it was developed in the mid-1990s with a strong object-oriented approach and C++ programming language. It actually turned out to be so good that it was a suitable choice for the mobile phone manufacturers for the operating system of the future communication devices. A detailed history of Psion and the EPOC operating system can be found in (Tasker *et al.*, 2000).

When Symbian was formed, the operating system was further developed to suit a variety of mobile phones. A lot of effort has been put into developing the communication protocols, user interfaces, and other parts of the system. The modular microkernel-based architecture of the EPOC operating system allowed this and, after EPOC release 5 (last version, used mainly in Psion 5), the operating system was renamed 'Symbian OS'.

Symbian is a joint venture between leading mobile phone manufacturers formed to develop a common operating system suitable for mobile communication devices. The operating principle is quite simple: Symbian develops and licenses Symbian OS containing the **base** (microkernel and device drivers), **middleware** (system servers, such as the window server), a large set of **communications protocols**, and a **test user interface** for the application engines of the operating system. Licensees develop the user interfaces to suit their purposes, and they also have the ability to license their user interface and application set on top of Symbian OS to other Symbian licensees – as Nokia has done with Series 60.

The terms of licensing Symbian OS are equal, and there are (at the time of writing) 10 licensees in the list, containing (in addition to the owners Ericsson, Matsushita, Motorola, Nokia, Psion, Siemens, and Sony Ericsson) companies such as Fujitsu, Kenwood, and Sanyo. Symbian develops new versions of the operating system, but the licensees can make their own alterations to the look and feel of the system, to the applications, and to the development tools. This is important in order for the licensees to be able to adapt their devices to their brand and also to have a variety of different types of devices to address different market segments.

For the developer community, Symbian offers technical support and guidelines for developing software. This is particularly helpful to the smaller licensees who do not have the resources to set up their own developer network and support areas.

Symbian provides training directly and through its training partners. For the licensees, Symbian offers support via its Professional Services department, and more support and work on a pay-per-service basis is available from the Symbian Competence Centers – companies specialized in helping the Symbian licensees in developing their terminals.

1.2.2 Operating System Structure

The core of the Symbian OS consists of the base (microkernel and device drivers), middleware (system servers, security, and application framework), and communications (telephony, messaging, and personal area networking). This core remains common to different devices supporting Symbian OS. Naturally, when Symbian OS is fitted to a new hardware the base needs to be changed (**base porting**), but this does not affect the upper layers.

When two devices have different user interface libraries providing an entirely different look and feel they still contain the common code from Symbian OS. This provides interoperability, simplifying developing applications for several different devices and decreasing the costs of making different kind of devices. In this section, the principles of Symbian OS architecture are covered to provide grounds for understanding the concepts. Different user-interface styles are also presented. The structure of Symbian OS v6.1 for Series 60 Platform is shown in Figure 1.2.

Application engines
Phone book, calendar, photo album, notepad, to-do, pinboard

Installed applications
Games, self-developed applications, utility software

Messaging
SMS, MMS, e-mail, fax

MIDP

Java
KVM

Application framework
GUI framework (Avkon, standard Eikon, Uikon), application launching, and graphics

Personal area networking
Bluetooth, infrared

Multimedia
Images, sounds, graphics

Communications infrastructure
TCP/IP stack, HTTP, WAP stack

Security
Cryptography, software

Telephony
HSCSD, GPRS

Base
User library, kernel, device drivers

Legend:

GPRS	General Packet Radio Service
GUI	Graphical User Interface
HSCSD	High Speed Circuit Switched Data
HTTP	Hyper Text Transfer Protocol
KVM	K Virtual Machine
MMS	Multimedia Messaging Service
SMS	Short Message Service
TCP/IP	Transmission Control Protocol / Internet Protocol
WAP	Wireless Application Protocol

Figure 1.2 Structure of Symbian OS v6.1 for Series 60 Platform; for abbreviations, see the Glossary

Symbian continuously develops Symbian OS to contain the protocols and features most needed by the licensees. The first Symbian-released software was Symbian OS v5, a unicode version of EPOC Release 5.0 from Psion. Symbian OS v5 was used in one family of smartphones – the Ericsson R380 series (R380s, R380e, R380 World). The first actual open platform release was Symbian OS v6.0 in spring 2000, used in Nokia's 9210, 9210c, 9290, and 9210i Communicators. Symbian OS v6.1 was shipped January 2001 and is used as a base technology for the Series 60 Platform (e.g. in the Nokia 7650 and 3650 imaging phones). Symbian OS v7.0 was released in spring 2002 and is common to several upcoming devices, such as Sony Ericsson P800.

As new versions of Symbian OS become available it is possible for the licensees to develop new product on top of these. The development lifecycle for a new mobile communication product typically lasts 2–3 years (depending on the amount of resources, experience, and the product in question). A variant of an existing product can be made significantly faster (in even less than a year).

It is not feasible to change the operating system used in a products each time a new version is developed. For example, the Nokia 7650 uses the Series 60 Platform built over Symbian OS v6.1. It is technically possible to change the version of the underlying Symbian OS to a new one. Whether it will be done depends on the benefits gained with the new version compared with the costs of implementing the change. Unlike in the desktop computing world, a new version of the operating system is not fitted into the existing hardware but into devices that are made after the release of the new operating system version.

1.2.3 Generic Technology

Most parts of the generic technology remain unchanged between different devices utilizing Symbian OS. The architecture of the system is modular, and it is designed with a good object-oriented approach. Most of the operation is based on a client–server model to allow all applications to use the services provided by the system, as well as other applications. This provides a very flexible system, allowing secure and robust resource allocation. It also saves significantly in the binary size and development effort of the applications, as the most used functionality is provided by the platform. Generic technology also contains a security framework that provides certificate management and cryptography modules (Mery D., 2001).

The base is the bottom layer of the operating system. It consists of the microkernel, device drivers, and user library. The microkernel is run directly in the processor in privileged mode. It is responsible for power management, memory management, and owns device drivers. The device drivers are the hardware–software interface layer needed for accessing, for example, the display and audio devices of the terminal as well as the communication channels. The user library in turn provides many of the functionalities such as error handling, cleanup framework, hardware abstraction layer, client–server architecture, as well as context (i.e. process and thread), and memory management, used by virtually all programs.

Application framework in turn is a set of reusable libraries providing means for handling text, graphics, and sound. As smartphones are sold globally, it is vital that internationalization and localization are well supported in the Symbian OS. **Internationalization** (i.e. customization of applications, presentation, and content to support local units, standards, practices, and traditions) is handled mainly by the application framework. **Localization** (i.e. translation of the text strings to different languages and character sets) is done separately for each application.

One important part of the application framework is the window server, which also provides keyboard and pointer support (for devices with a touch screen). The graphics framework is part of the application framework and contains parts that are common to all terminals as well as parts that are specific to the user interface library or manufacturer (depending on the amount of customization performed).

Communication architecture contains the infrastructure needed for communications and protocol stacks of the most needed communication protocols. One of the key strengths of Symbian OS is the provided set of communication methods and their tested interoperability. In addition to the protocols provided with the system, the licensee and the application developers are able to create support for additional protocols in the form of new protocol modules or by building the needed protocol into an application.

1.2.4 User Interface Styles

As Symbian was formed, the founding members each had their own plans for making the new era of mobile terminals. The most significant differences were found in the user interaction, and thus the concept of reference design was introduced. The idea is that most of the system remains unchanged even though the terminals are totally different. The original three reference designs were Crystal, Quartz, and Pearl. Crystal and Quartz are communicator-type devices, Crystal having 640 × 200 pixel horizontal half-VGA display and Quartz offering 240 × 320 pixel quarter-VGA in portrait orientation (and a touch screen with stylus). Pearl is the smartphone reference design, of which Series 60 Platform is the most significant example.

With the introduction of Symbian OS v7.0, Quartz has been renamed UIQ and is provided in slim (208 × 320) and normal (240 × 320) versions by UIQ Technology. Crystal is now called Series 80 by Nokia. The current set of user interface styles is presented in Figure 1.3. In addition to these, all licensees are able to provide their own user interface style or to modify the existing styles.

As the display is one of the most expensive components of a smartphone, the Series 60 Platform is quite attractive to terminal manufacturers with its 176 × 208 pixel screen resolution. UIQ provides a slightly larger display, with pen-based operation, valuable in many types of applications; Series 80 has the largest screen size. At the time of writing, terminals using it are the Nokia 9210, 9210c, 9290,

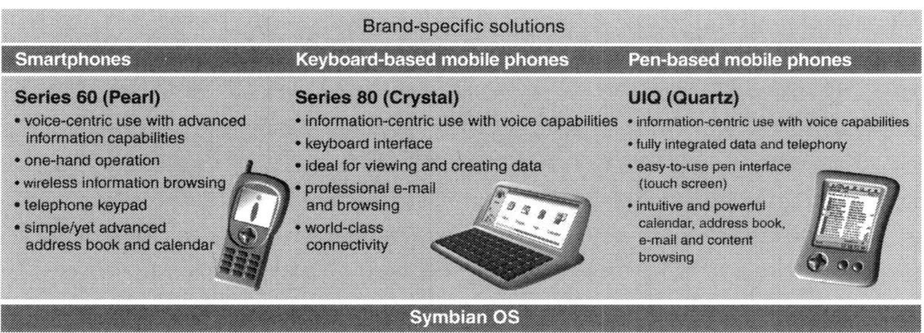

Figure 1.3 Symbian OS: Series 80, UIQ, with its two widths, and Series 60 are currently the most important user interface styles

and 9210i communicators; Nokia has not made any announcements on licensing Series 80. The Sony Ericsson P800 in turn uses the slim version of UIQ, Nokia 7650 and 3650 use the Series 60 Platform.

Each user interface style allows customization of the look and feel of the user interface for each licensee and device. When Nokia, for example, licenses Series 60 Platform to some other manufacturer, the user interface is likely to look somewhat different. This is natural, as the physical shape of the smartphone is different. Simple customization of the graphical user interface may consist of changing the shape and colors of the graphics (bitmaps), changing menu text and structure, and the addition of new components, e.g. applications. The purpose of customization is not only to tie the smartphone tighter to the licensee brand but also to make the device more usable for the intended user group.

1.2.5 Application Development

Symbian OS supports application development with C++ for building native applications and with Java for building Java applications and midlets (i.e. Java MIDP applications). C++ development is done with Symbian's extensions and differs somewhat from, for example, C++ development for Microsoft Windows. The biggest difference from the development with standard C++ is the unique approach to exception handling, which was not supported by the compilers available in the early phase; it is also better suited to its purpose than is the generic approach.

Development with Java can be somewhat easier than C++ programming language, and the developer community for Java is larger. The performance issues and available interfaces restrict Java development to certain types of applications. For these, the decision to develop with Java for Symbian OS is a sound choice.

The development tools are available from tool partners, including Metroworks, Borland, and Appforge. Currently, Series 60 uses

Microsoft's Visual Studio in the C++ Software Development Kit (SDK). The development is done on ordinary desktop computers, and the SDK provides good development support and allows applications to be tested without the target hardware (to some extent). With the development tools, the compilation for the target device can also be performed, and installation packets (Symbian Installation System – SIS files) can be created. Although development with the SDK allows testing of the application to some extent, it is also important to install the application into the target device to test the operation fully.

Testing the software in the target device requires a big effort as the tester needs to enter all the commands manually to test the functions of the software. It is especially time-consuming to perform tests several times (e.g. it takes a great amount of time to test memory allocation operation 10 000 consecutive times to see whether it always succeeds). There are tools available for automating the testing in the target device and for providing a better quality final software. This type of tool significantly reduces the effort needed for performing thorough tests.

Some other important aspects of developing applications for the smartphone are interaction design, user experience, security, and localization for different languages. Usability and interaction design of software developed for smartphones is demanding, as the user interface is very small, allowing only the most important information to be included. Also, the possible user base is very diverse, and help files or user manuals should not be needed at all – users of smartphones do not necessarily have experience with, for instance, computers. Security is provided by the operating system to some extent, but it is important not to compromise the system (in the sense of being vulnerable to attacks) through the developed applications installed by the user.

Localization of the application for different languages – including those with different character sets – is supported quite well by Symbian OS. This is a very important feature, as developing a new smartphone is such an investment that in most cases it requires global sales in order to be profitable. The platform is localizable for several countries (in terms of units, languages, and character set), and the third-party applications can also easily be translated to several different language versions.

1.3 Series 60: Smartphone Platform

This book is written about programming for Nokia's Series 60 Platform running on top of Symbian OS. This section contains an overview of the key features and application set of the Series 60 Platform. Some details are also presented to highlight the possibilities of development for this platform.

Why should some device manufacturer license Series 60 Platform to use on top of Symbian OS, and why does Nokia want to license this platform to its fierce competitors? The answer to the first question is relatively easy, as Series 60 Platform and Symbian OS provide a user experience and application set unlike any other smartphone platform in a very competitive package. An example of the applications view and other screenshots of the Series 60 Platform are shown in Figure 1.4.

Figure 1.4 Nokia's Series 60 Platform: screenshots; Series 60 Platform provides a graphical user interface and a large set of applications

It takes great effort from a mobile phone manufacturer to create a comparable user interface and application set – even if they start building this on top of Symbian OS. Actually, this partially answers the second question – to be able to make the developer base larger, and thus to create more content, Nokia wants to license its smartphone platform to all manufacturers. This in turn is important, as a large amount of available content and a large number of devices from different manufacturers lead to an increased number of users (and thus terminal volume), which reduces the cost of the components needed in the device.

At the time of writing Siemens, Samsung, and Matsushita (Panasonic) have announced they are licensing the Series 60 Platform. In addition to Nokia, they are likely to be the first device manufacturers to offer smartphones based on the Series 60 Platform.

For a Series 60 licensee the platform forms a solid base for the much faster creation of smartphone devices than can be achieved by other approaches. In addition to accelerated time-to-market, competitive advantage is gained as the needed hardware components for building the smartphones are available at a reasonable price. The supported display resolution of 176 × 208 pixels allows the creation of reasonably priced smartphones. It is also easier to enter the smartphone market with Series 60 Platform as there is already an established developer base. Series 60 terminals are binary-compatible with each other; that is, the application developed for Nokia's Series 60 terminals are interoperable with other Series 60 terminals (with the exception of unsupported features, such as trying to take a picture with a device without a camera).

The Series 60 Platform 1.0 (built on Symbian OS v6.1) provides the communication technologies needed in smartphones such as: e-mail (with POP3, IMAP4, SMTP), WAP 1.2.1 stack, SyncML 1.0.1, MMS, Bluetooth, GPRS, and, naturally, all the other necessary protocols supported by Symbian OS v6.1 (see Glossary for terms and abbreviations). The list of applications provided with the platform includes a phonebook, calendar, notepad, photo album, clock, calculator, composer (for creating ringing tones), e-mail client, SMS and MMS clients, as well as the telephone application. All these features are encapsulated into a well-designed graphical user interface with tested usability (Nokia, 2001c).

It is possible to create a smartphone with just these reference applications, but the licensees are likely to add new features and applications (such as games) to the devices. Naturally, the new licensees are likely to customize the platform simply by creating different types of graphics and color schemes, but they may also use advanced techniques such as altering the default applications. For this purpose, the development is, naturally, done with C++. Application development in Java for Series 60 is done with the **Mobile Information Device Profile** (MIDP). This allows a certain degree of interoperability with MIDP applications, such as simple games that will run on any MIDP-enabled device, if the capabilities are suitable.

1.4 Contents of this Book

This book is organized in four parts:

1. Software Engineering on the Series 60 Platform,
2. Graphics, audio, and User Interfaces,
3. Communications and Networking, and
4. Programming in Java.

Most of the text is targeted at software developers who have some experience in programming with Symbian OS, although we have included a few chapters for those developers starting from the beginning.

For beginners, a good starting point is provided by Chapters 4, and 5, which give a description of the software build process and build tools on the Series 60 Platform as well an introduction to Series 60 software development. In addition, the essentials of Symbian OS programming and development support are included in these chapters.

More experienced Symbian OS developers may jump directly to any part that seems to provide interesting information. It is recommended that chapters of each part are read in the given order, because each chapter builds at least partially on the previous chapter. Every part begins with an introductory chapter, which presents an overview of the topic. These opening chapters also give background information beyond the Series 60 Platform and are useful to anyone who wants to gain an overview of some technology supported in Symbian OS.

1.4.1 Contents of Part 1

The software engineering principles on the Series 60 Platform are walked through in Part 1. Chapter 2 gives a high-level overview of the platform, including the features of the user interface, supported communications technologies, and preinstalled applications. This chapter should be read by anyone who wants to know what the Series 60 Platform actually is and how it is related to Symbian OS.

In the design of Symbian OS and user interface libraries provided by the Series 60 Platform, the same design patterns have frequently been applied. In Chapter 3, model–view–controller, observer, client–server, adapter, microkernel, and state design patterns in addition to some idioms relating to memory management are described. Chapter 3 helps in understanding how design patterns have influenced the application framework of the operating system and how these design patterns are applied in application development to make software components more robust, reusable, and portable. As an example, the most frequently used design patterns are applied in the design of a simple rock–paper–scissors game.

Chapters 4 and 5 are targeted mainly at programmers who have no earlier experience in Symbian OS. In these chapters, the software build process and the use of build tools are described. Development support can be found in the web links provided. Owing to the limited space we cannot describe all the details of Symbian OS, but we do describe the basics of memory management, exception handling, multi-tasking, client–server framework, and descriptors to help the reader to understand the code examples given in the rest of the book. We use code fragments of the rock–paper–scissors game designed in Chapter 3 to emphasize the characteristics of Series 60 Platform and Symbian OS programming.

The quality of smartphones and applications is determined by their usability. That is why it is essential to take usability issues such as user interface and interaction design into account when designing software for smartphones. Chapter 6 describes the user interaction design process, starting from user needs and ending with usability verification. By reading Chapter 6, developers will learn how to design their applications for usability.

Part 1 ends with a chapter presenting software testing for smartphones (Chapter 7). The scarce memory resources, the robustness requirements of applications, and other special characteristics of smartphones increase the importance of ensuring good quality applications. Each software developer should know the testing techniques and strategies as well as the possibilities and limitations of the available testing tools.

1.4.2 Contents of Part 2

A thorough description of user interface (UI) programming, including graphics and audio, is given in Part 2. Chapter 8 gives an overview of the basic concepts – such as the GUI application framework (the Uikon framework and its Avkon extensions), views, controls, observers, menus, dialogs, and list boxes – you need to know to understand the rest of the chapters in this part. The Uikon framework provides the basic functionality of GUI applications, such as application launching and user event handling. It consists of two logical parts – a generic part called Uikon, and user interface style-specific parts, that determine the look and feel of GUI applications. In Series 60, the user interface style-specific parts are an implementation of standard Eikon and Avkon.

Chapter 9 contains a description of the structure of the Series 60 screen and introduces the important concept of panes. The three panes taking the whole display area are: status pane, main pane, and control pane, and their subpanes. In Chapter 9 the use of these panes is explained, with code examples.

List boxes are a common way to present data on the Series 60 Platform. There are several list types provided by the platform. The application developer may choose to use plain, grid, or hierarchical views with single or double items, single or multiselection, etc. A thorough description of all issues relating to lists and list types, with code examples, are given in Chapter 10.

In Chapter 11 the rest of the user interface components of the Series 60 Platform are described. Applications user interfaces are built from the described components, the set of which includes dialogs, notifications, editors, and queries. The use of these components is explained with numerous code examples. Chapter 11 is essential for all GUI application developers.

Views are pages that perform a specific task. The view architecture is a mechanism for switching between the views and to send messages between views in different applications. For example, a user may store received contact information in the messaging application by switching to a view of the phonebook application, without starting the phone application by hand. Chapter 12 describes how views are defined and used on the Series 60 Platform.

Graphics is not the only part of the user interface; with respect to user experience, audio may be as important. Chapter 13 contains a description of how to use different audio formats to record and play back audio data on the Series 60 Platform. Again, we have included code examples to clarify the use of audio.

By following the conventions and recommendations from Symbian, the porting of an application between user interface styles should require only the porting of the user interface part. Chapter 14 gives developer information on, how to customize the look and feel of user interface components and how to port the Series 60 Platform to new devices. Chapter 14 also includes a description of application porting from previous user interface styles to Series 60.

1.4.3 Contents of Part 3

Part 3 concentrates on networking and communications. Chapter 15 is an introductory chapter in which we describe supported communication techniques and also background to information to provide an understanding of how different protocols work and what kind of characteristics they have. An important part for all developers is the description of communications architecture of Symbian OS. This helps in understanding how to use any supported communications service.

Chapter 16 gives examples, of how communications services, described in Chapter 15, are used in applications. The description follows the structure of the communications architecture, presented in Chapter 15. Use of programming interfaces of any of the Symbian OS communications servers (Serial communications server, Socket server, and Telephony server) is described, with code examples. An example of the use of these programming interfaces is how different communications technologies may be used to transfer the protocol messages of the rock–paper–scissors game. Chapter 16 is useful to anyone who is going to use communications services in applications.

Messaging is another way of communication and the subject of Chapter 17. The messaging architecture and messaging examples (with source code) using e-mail, short messages, multimedia messages, and smart messages are described.

Connectivity issues related to file management, remote printing, and synchronization of data between the smartphone and other devices are described in Chapter 18. In addition, SyncML language for data synchronization is described in this chapter.

1.4.4 Contents of Part 4

Java programming on the Series 60 Platform is the topic of Part 4. Java 2 Micro Edition platform, with its CLDC (Connected Limited Device Configuration) and MIDP definitions for smartphones are described. The possibilities and limitations of the configuration are discussed in Chapter 19. In addition, a discussion of Java MIDP applications, called midlets, form part of the content of Chapter 19, which should be read by anyone who is considering developing software in Java for the Series 60 Platform and needs to know what are CLDC and MIDP and what kind of limitations they set.

Midlet and Nokia user interface classes, support of networking, and installation of midlets over the air are discussed in Chapter 20. The use of each class is described in detail, and an example project and code examples (time-tracker application) are used to show how the classes are used. Any midlet developer should find this information useful in the development of his or her own midlets.

The source code examples used throughout the book, as well as both example applications (the rock–paper–scissors game developed in C++, and the time-tracker application developed with Java MIDP), and related graphics can be downloaded from the web pages of this book at www.digia.com/books.

Part 1

Software Engineering on the Series 60 Platform

2

Overview of the Series 60 Platform

The Series 60 Platform is designed for one-hand operated smartphones. It is based on Symbian OS and completes it with a graphical user interface (UI) library called **Avkon**. The Avkon library has been designed to enable the implementation of intuitive and easy-to-use applications. According to the design goals of the platform, using a smartphone should not be more difficult than using an ordinary mobile phone. In addition to the UI library, the platform contains a set of reference applications providing a public API (application programming interface) to the services of these applications. The reference applications help software developers to use a proper UI style.

This chapter provides a more detailed overview of the Series 60 Platform than presented in the previous chapter. We describe the physical characteristics of the user interface, the set of built-in applications, and the supported communications technologies to give the reader an idea of the possibilities and limitations of the platform. The software examples are found in the following chapters, where the essential design patterns useful in application development are described.

2.1 User Interface

The user interface of the smartphone is one factor with which the device manufacturer may differentiate its products from those of other vendors. Differentiation is the key word for surviving and succeeding in the global mobile phone markets, and thus each device manufacturer intends to provide a unique and competitive user interface, satisfying customer needs. Licensees can customize the Series 60 Platform user interface according to their own look-and-feel requirements. However, all the programming interfaces to the third-party developer will be the same in all Series 60 phones. The platform supports a screen size of 176 (width) × 208 (height) pixels by default. In future versions, the vertical resolution should be modifiable. Phones based on the platform should have a display supporting 4096 colors or more. There is no need for a pen or full keyboard, which enables phones to be used by one hand. However, it is possible also to use an external keyboard if the device manufacturer wants to build the required hardware support for it (Nokia, 2001g, pages 8–9).

The Series 60 platform user interface is designed to make use of the smartphone as easy and intuitive as use of a mobile phone. At the highest level, the application user interfaces may be considered as consisting of two parts: browsing elements and detailed views. By browsing, users make selections (e.g. an entry from a list). After the selection, the corresponding detailed view of the entry displays the data on the screen.

For data input, the platform specifies a 12-key ITU-T (International Telecommunication Union – Telecommunication Standardization Sector) numeric keypad (0–9, #, *) with additional function keys. Function keys include two softkeys (left and right softkeys having a textual label on the bottom of the screen), five-way navigation key (up, down, left, right, and select), an application launch-and-swap (application) key (bringing up the application shell), and send and end keys for call handling (Nokia, 2001c). An illustration of the keypad is shown in Figure 2.1.

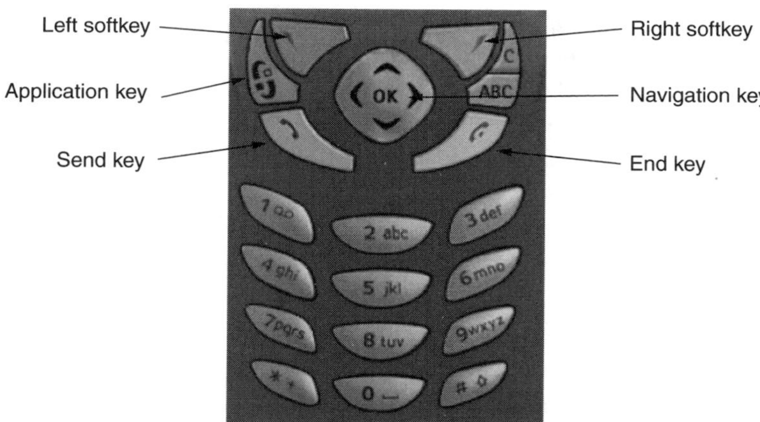

Figure 2.1 A keypad: keys supported by the Series 60 Platform

The left softkey is typically labeled Options, because it is used to open the options menu. In other states, Select, OK, and Yes are used to select items in menus or lists and to give a positive reply to a confirmation query. The right softkey is used for returning to the previous state and thus is usually labeled Back. It is also used for exiting

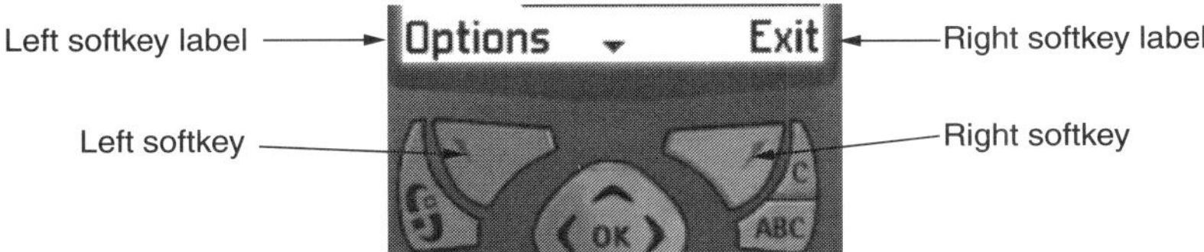

Figure 2.2 Softkeys and their labels

an application, canceling a procedure, or to give negative replies, in which cases the Exit, Cancel, and No labels are commonly used. Typical labels are shown in Figure 2.2. Software developers define the labels in application resource files.

The keypad also specifies a clear and an alpha toggle key for improving the text input. The alpha toggle or edit key is the only key that may be pressed down simultaneously with any of the navigation keys, in rare cases, for which both hands are required. One implementation of the Series 60 user interface is provided by Nokia 7650 phone, shown in Figure 2.3. Manufacturers can extend their products based on Series 60 to have other functional keys or even an external keyboard. Extra keys may be used to control the hardware or applications, for example in speech recognition or volume control.

Figure 2.3 Nokia 7650 phone

Keys may allow interaction, when they are pressed down, held down, or released. Typically, the interaction occurs when the key is pressed down, but the behavior of the edit key is different, because it may be pressed simultaneously with other keys. Based on Nokia's specification (Nokia, 2001g, pages 18–19), a short key press lasts less than 0.8 seconds. A long key press may cause another interaction in addition to the key press, after a duration of 0.8 seconds. A long key press may also perform a key repeat, in which the same function associated with the key is repeated at a user-specified frequency.

2.1.1 User Interface Software and Components

The graphical user interface (GUI) framework provides launching and UI functionalities for applications. It consists of three parts to make porting between the UI styles easier. The Uikon framework is the same in all UI styles, but each style has its own implementation of standard Eikon and extended functionality, which is called Avkon in the Series 60 Platform. Standard Eikon and Avkon are implemented by Nokia, while Symbian provides the Uikon framework. Uikon and the underlying libraries (application architecture and control environment) it uses provide a large part of UI functionality (launching an application, handling inputs, and drawing graphics) in an abstract form. The standard Eikon and Avkon make UI functionality concrete and supply a look-and-feel for Series 60. GUI applications are implemented by use of base classes provided in Uikon and Avkon, as shown in Figure 2.4. A more detailed description of the GUI framework is given in Chapter 8.

The Avkon library provides not only application architecture base classes from which each GUI application is derived but also

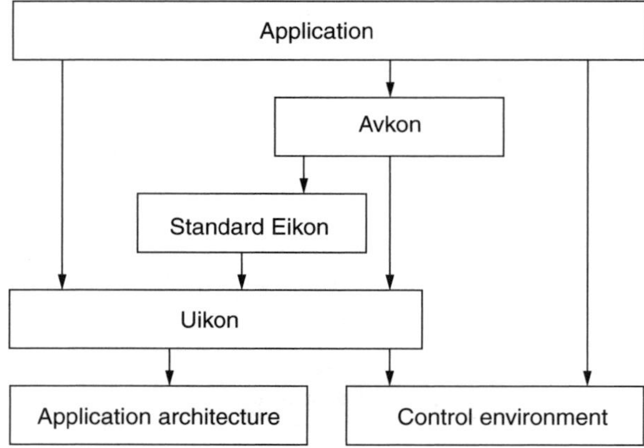

Figure 2.4 Graphical user interface framework

Avkon-specific controls to be used in the applications. Controls are GUI components, displayed on the screen. Commonly used controls include menus, dialogs, notifications, list boxes, confirmation queries, and edit windows.

Menus are opened with the left softkey, labeled Options. The contents of the menu is context-specific and may be dynamically changed in the application. For example, a communication application may hide receive and transfer menu items, until the connection to the destination has been established. Menus may also contain submenus. For example, the user may choose to send a document to the receiver, using either Bluetooth, infrared or short message service (SMS), as shown in Figure 2.5.

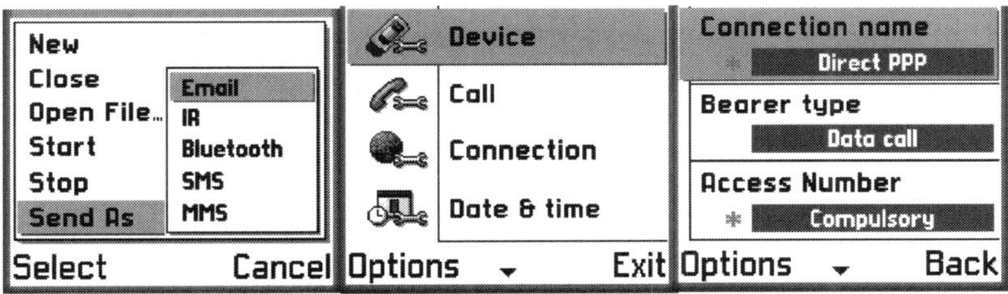

Figure 2.5 Examples of user interface components: menu (left), a single list with icons (middle), and a dialog box (right)

Dialogs are pop-up windows that are used in numerous cases to query some information with the user. The information may be a password, file name, any editable text, time, date, telephone, and so on. A simple dialog may ask the user to confirm a procedure before execution, but more complicated dialogs contain several lines, each having a label and a control. The control may be an edit window allowing the user to edit the text in the dialog.

Avkon notifications replace information and query windows of the previous UI styles. Notes are dialogs that may disappear without user interaction. They are used to notify the user about exceptional situations, when running the software.

List box is not a single UI component but contains a number of list and grid types. Lists contain items, which are browsed vertically, whereas grids are two-dimensional lists, allowing navigation in both the horizontal and the vertical direction. Lists may allow looping from the end of the list back to the beginning of the list. In addition, there are a single and double column, and single and multiple selection lists. A detailed description of the list types is given in Chapter 10; the rest of the UI components and examples of their usage are given in Chapter 11.

2.1.2 User Interface Structure

The GUI of a smartphone is shown on the screen. The screen has one or more windows either filling up the entire screen or part of it, in which case the window is termed 'temporary'. Each application runs in a window of its own and, in addition to that, applications may use temporary windows (i.e. pop-up windows) if required. However, the application window is not used directly for display but as a parent for various subcomponents, called 'panes'. Each pane may contain subpanes, which may contain subsubpanes, and so forth, until at the lowest level, the subcomponents are called elements. In Figure 2.6, an example of three panes (status, main, and control) occupying the window are shown. An application may display its data on the main pane. The status pane is used for showing status information, such as battery charge status or received signal strength. The control pane displays the labels of the softkeys and occupies the bottom of the screen.

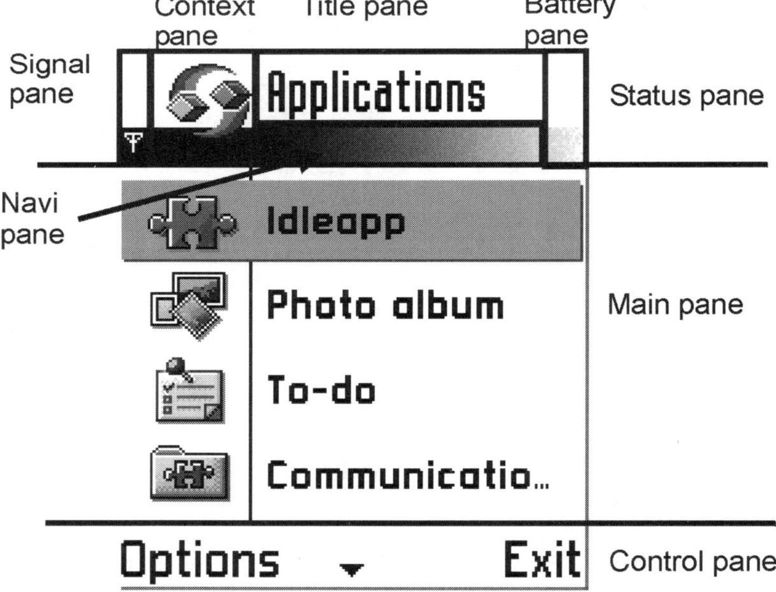

Figure 2.6 A screen with a window having three panes

2.2 Communications Support

The Series 60 Platform has been designed for voice services, data communications, and content browsing, and hence a wide set of communications technologies is supported. Connectivity allows short-range connections between a smartphone and other devices to synchronize data or to transfer files. Short-range connections themselves enable multiparty data transfer or, more interestingly, multiplayer games. Cellular networks may be used for Internet access, messaging, and voice communications.

Application designers, utilizing supported communications services, require standards for ensuring interoperability between any devices. The Series 60 Platform uses global protocol standards for communication. For instance, in remote connectivity purposes the Series 60 Platform introduces the SyncML protocol. The protocol is used for data synchronization between applications across multiple platforms and for standardized device management functions. However, in the first release of the Series 60 Platform, device management is not yet supported. An example of data synchronization is the exchange of vCard and vCalendar objects between a smartphone and a desktop.

SyncML also provides an alternative for proprietary device management implementations, which may be difficult to use by the end-user. The protocol allows the use of several protocol bindings in the exchange of synchronization or management data. Three bindings are supported in Symbian OS: HTTP (hyper text transfer protocol), WSP [wireless application protocol (WAP) Session Protocol], and OBEX (object exchange).

2.2.1 Communications Technologies

Communications support is based on standardized communication technologies and protocols. For short-range connections IrDA (infrared data association) and Bluetooth are the most often used protocols. Protocols of the TCP/IP (transmission control protocol/internet protocol) suite are used for networking over a CSD (circuit-switched data) or GPRS (General Packet Radio Service) connection. Several messaging protocols are supported from simple short messages to e-mail and multimedia messages. For browsing both HTTP and WAP. However, the current version of the Series 60 Platform does not provide a programming interface to HTTP stack. The WAP protocol is used for delivering MMS (multimedia messaging service) messages in addition to browsing.

TCP/IP suite is provided for networking and it enables usage of other protocols, such as SMTP (simple mail transfer protocol), POP3 (Post Office Protocol 3), and IMAP4 (Internet message access protocol 4). SMTP is a protocol used for sending messages (i.e. for uploading e-mail messages from a client application to an e-mail server). The same protocol is used to deliver messages between servers so that the server used by the receiver will be reached. The receiver may download the message from the server by using either the POP3 or IMAP4 protocol. Several other applications may use protocols [TCP, UDP (user datagram protocol) or IP] from the TCP/IP suite as a bearer.

To create an IP (UDP or TCP) connection to an Internet address, a connection to an Internet access point (IAP) must be established. For that purpose, a physical layer connection is created using either CSD or GPRS. PPP (point-to-point protocol) is used to transfer IP packets over a serial physical link.

E-mail messages may be transmitted between the smartphone and network servers without an IP connection by using, for instance, the SMS service. The third messaging service supported by the Series 60 Platform is MMS. The MMS allows for the inclusion of sound, images, and other multimedia content in the messages.

For short-range connectivity, there are several protocols that can be used. RS-232 is supported by the Series 60 Platform, although Nokia 7650 does not have an RS-232 port. Wireless connectivity protocols include IrDA and Bluetooth protocols, the latter using ISM (industrial, scientific, medical) radio frequencies. The detailed description of all protocols and their usage in own applications is provided in Part 3 of this book. Table 2.1 contains a list of supported communications technologies in the Series 60 Platform and their essential features.

In addition to applications based on the standard technologies, new services and middleware may be provided. Examples include mobile portals, payment solutions, location-based solutions, advertising, and virtual communities. All messaging or data transfer protocols are implemented by using the same framework provided by Symbian OS. New protocols and the support of new message types may be implemented with plug-in modules, dynamically extending the capabilities of the phone.

2.2.2 Messaging

Messaging provides an ideal way of communication when no immediate response from the receiver is required. Messages may be stored in the server to wait for the receiver to download or be pushed directly to the receiver, if online. E-mail messages are an example of the former so-called **pull** model. SMS messages are delivered using the **push** model.

The success of SMS messaging has shown how beneficial it may be if the same standard has a worldwide support. Now the development is changing from SMS to MMS, with sound and image capabilities, creating a new dimension for chatting, entertainment, and person-to-person messaging. In addition to SMS and MMS, the importance of e-mail in

Table 2.1 Supported communications technologies

Technology	Description	Primary APIs	Speed	Comment
Bluetooth	Short-range radio – frequency communication	RFCOMM L2CAP SDP BT security manager BT serial API	Up to 1 Mbps	Raw speed
E-mail	E-mail messaging	IMAP4 POP3 SMTP		
GPRS	Packet radio service for GSM cellular network		Up to 107.2 kbps in one direction	operator and terminal dependent
GSM	Global system for mobile		9600–14 400 bps	
HSCSD	High-speed circuit-switched data		Up to 57.6 kbps	
IrDA	Infrared data association protocol stack	IrTinyTP IrMUX	Up to 4 Mbps	
MMS	Multimedia messaging service	MMS		Minimum maximum message size 30 kB
SMS	Short message service	SMS	160 characters in one message	
SyncML	Synchronization protocol		Content protocol	
TCP/IP	transmission control protocol/internet protocol suite	TCP UDP IP	Depends on the physical bearer	
WAP	Wireless application protocol	WSP WDP	Depends on the bearer	

Note: for abbreviations and definitions of application programming interfaces (APIs) see the Glossary.

corporate messaging should not be underestimated. E-mail messages may be uploaded and downloaded to the corporate server using the standard protocols and, if required, in a safe way by encrypting the content by, for example, using a secure socket layer.

The Series 60 Platform provides an API for sending and receiving SMS, MMS, and e-mail messages. E-mail messages are transferred using the SMTP, IMAP4, or POP3 protocols. In addition to existing protocols and their APIs, software developers may implement their own plug-in modules for specific message types and messaging protocols. The messaging architecture takes care of checking, if the required messaging protocol is supported. If it is, the architecture takes care of loading the required plug-in modules for transferring, editing, and storing messages as well as for notifying messaging events. The plug-in architecture is not restricted to messaging; any protocols may be implemented using similar principles.

Smart Messaging

Smart messaging is messaging directed to applications residing on the phone. Nokia has defined an open standard called Nokia Smart Messaging (NSM; the specification can be found at www.forum.nokia.com). Symbian has defined an extension to NSM, named BIO messaging (bearer-independent object). The transport service for BIO messages can be, for instance, Bluetooth, SMS, or infrared. Smart messages are used to synchronize data for PIM (personal information management) applications or to transfer configuration settings and data for other applications. Data for PIM applications include vCard and vCalendar objects, which are used to synchronize phone book and calendar application entries. Configuration settings may be used to configure, for example, the WAP service, based on Nokia and Ericsson WAP configuration specification. In addition to that, smart messages may be used to deliver ringing tones, operator logos, and picture messages, which are all NSM-specific. Each message type has its own MIME (multipurpose internet mail extension) content type or port number by which it is recognized. In Symbian OS, the identification of each message type requires a BIO information file and, depending on the bearer, also the MIME type. The contents of this file and examples of smart messaging are provided in Chapter 17.

Personal Information Management Functions

PIM applications handle the user's personal information. As all communications in the Series 60 Platform, also the personal data interchange (PDI) is based on industrial standards. This allows the PDI to take place between a number of different devices running different operating systems. The two main technologies supported – vCard and vCalendar – are developed by the Versit Consortium and are now maintained by the Internet Mail Consortium (IMC). Electronic business cards or vCards may carry names, addresses, telephone numbers, e-mail addresses, Internet URLs (universal resource locators), and graphics for photographs, company logos, or audio clips. The vCalendar data format is used for the PDI of calendar and scheduling information between different platforms. Typical exchanged items are events and to-do entries. However, to-do entries cannot be delivered inside vCalendar objects in the Series 60 Platform. Example items between PIM applications will be given in Chapter 18.

2.2.3 Browsing

Browsing of arbitrary content is difficult in a handheld device because of its limited memory and small display. The WAP protocol family provides functionality similar to the World Wide Web but is designed to accommodate smartphones and other handheld devices, having only a low bandwidth connection to the content server. WAP may be used for news browsing, gaming, online auctions, or chatting over several different bearers, such as SMS and GPRS. In the future, one of the most common ways to use WAP may be the transfer of MMS messages. WAP over GPRS provides fast connection times compared with circuit-switched data connection and, although the data is delivered in a packet-switched manner, there is sufficient bandwidth for WML (wireless markup language) content. HTTP stack is implemented in the Series 60 Platform, but the HTTP APIs are not available for C++ application developers.

The Series 60 Platform supports WAP 1.2.1, which has many useful features for smartphones. In addition to normal content downloading, WAP settings may be transferred or content may be pushed into the terminal. WTAI (wireless telephony application interface) specification is also supported for standard telephony-specific extensions, such as call control features, address book, and phonebook services. Nokia has made a mobile Internet toolkit product available for application designers and content publishers to create WML content. The toolkit is available at www.forum.nokia.com.

2.2.4 Example of Using Communications Technologies in the Delivery of Multimedia Messages

The use of the MMS service in the client application shows how several protocols can be involved in one communication service. The protocols that the MMS uses are depicted in Figure 2.7. On the network side, the SMTP protocol can also be used to deliver multimedia messages between MMS servers (WAP Forum, 2001). In addition to the protocols, Figure 2.7 shows Symbian OS communications servers participating in the delivering of the MMS message.

The MMS application can use client and server message type modules (MTM), which both provide an API. Access to the modules takes place through the message server, which is a standard Symbian OS server. The MMS client MTM provides services for message creation,

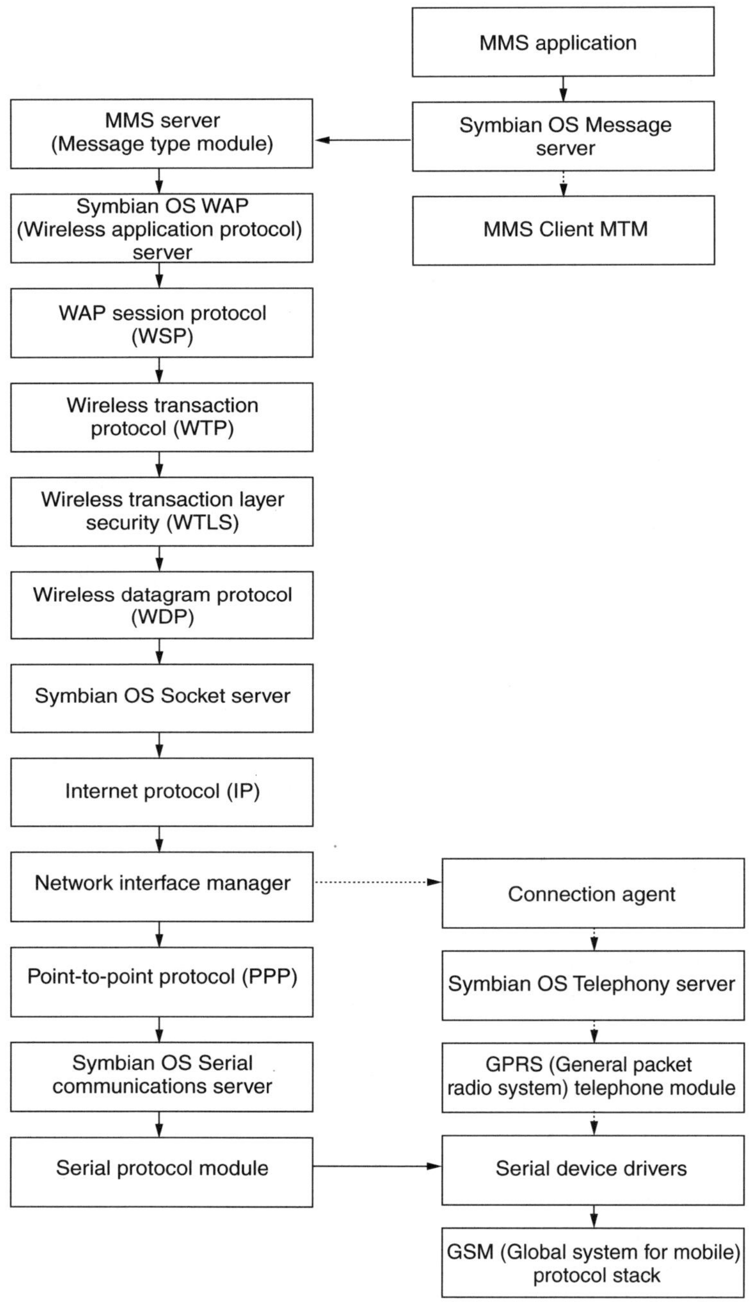

Figure 2.7 Protocols involved in the multimedia message service (MMS). Control is changed to Symbian OS Telephony server

storing, and for header manipulation. The body is created in the MMS editor. The main function of the MMS server MTM is to send the message.

The MMS service uses the WAP stack to send and receive messages. The exact functionality will be explained in Chapter 17. The protocols in the WAP stack are accessed through the Symbian OS WAP server. This server is not a WAP server, providing WAP services, but an internal server sharing resources with WAP protocols.

WAP datagrams are delivered over the IP protocol. The IP connection is created at an Internet access point using GPRS or CSD. Because IP has no knowledge how to route WAP datagrams, before the connection has been established it asks the network interface manager for a route. The manager uses a connection agent to create a route using the services provided by Symbian OS telephony server. The dashed line in Figure 2.7 indicates that the control is changed to the Telephony server, although data are not going to be delivered through these components. When there is a physical connection, the telephone module loans a port to the network interface manager, which then gives the port to the PPP protocol. Access to the GSM signaling stack takes place through an internal modem, the use of which requires the services of the serial communications server and device drivers.

2.3 Applications

Series 60 Platform provides not only ready-to-run applications but also a few reference applications. These provide a public API to get access to the application services. For example, the API of the photo album application provides easy access to the image find service. A list of available applications are shown in Table 2.2 (Nokia, 2001c).

In addition to the existing applications, users may download applications or develop new applications themselves. The rest of this book deals with software development for the Series 60 Platform.

2.4 Summary

The Series 60 Platform is designed for one-hand operated smartphones. All navigation, browsing, and selection functions of the user interface can be made with one's thumb. The platform contains a UI library called Avkon, which is designed to allow the implementation of usable and intuitive applications. The platform also contains a set of reference applications providing a public API, which may be used to access the services of the applications. The platform supports a

Table 2.2 Ready-to-run applications in the Series 60 Platform

Application	Description
Phonebook	Contact database integrated with messaging and other applications; supports vCard data format for visitor card exchange and synchronization
Calendar	Scheduling application; supports vCalendar data format
To-Do	A list of To-Do entries can be downloaded into this to-do list
Notepad	For text entries and editing
Photo Album	Store for images
Pinboard	Application management tool for the multi-tasking environment
Clock, Calculator, Unit Converter	World clock, business calculator, and a selection of entertainment items
Composer	Allows users to compose new tunes
E-mail, SMS, and MMS	Client software for messaging applications
Telephony applications	Voice recorder, telephone settings, call logs and message indicators, user profiles, call forwarding (divert), speed dialing, and voice dialing and voice tags
Application installation	For installing new software via the PC connectivity suite or over the network
Synchronization	SyncML 1.0.1 synchronization engine; supports data transfer over mobile networks [wireless application protocol (WAP)], Bluetooth, and infrared data association (IrDA); supports vCalendar 1.0 and vCard 2.1 data formats
Security	Security settings and software certificate management

Note: MMS, multimedia message service; SMS, short message service.

wide range of standardized communications technologies for data synchronization, content browsing, messaging, and networking. The next chapter describes some useful design patterns that may be used in the application development. Also, the basic application framework classes used in every Series 60 application are explained.

3

Design Patterns for Application Development

Design patterns describe solutions to common software problems. They help software developers to apply good design principles and best practices of software development. Design patterns when applied properly improve the quality of the software with respect to, for example, reusability and maintainability. Design patterns are reusable software building blocks. They provide not only a common language for the designers in some problem area but also class and collaboration diagrams to be applied as well as documentation, implementation, and testing specifications. Design patterns, however, have consequences, which must be taken into account before applying them.

This chapter covers the most common and useful design patterns applicable in Symbian OS. Examples of using Model–View–Controller, Observer, Client–Server, Adapter, Microkernel, and State design patterns are given. Some patterns are applied in the design of a simple rock–paper–scissors game. In addition to this, coding idioms characteristic of Symbian OS programming are provided. To understand class and the other diagrams used, the reader should have a basic knowledge of the UML (unified modeling language). Readers who are not familiar with UML may read, for example, the UML user guide written by its original designers (Booch *et al.*, 1999).

3.1 Design Patterns in Software Development

There are several process models for software development, depending on the application area, experience of the development team, and characteristics of the developed software. Regardless of the applied process, there are usually four phases to be identified: gathering of requirements, analysis, design, and implementation, which also includes the testing phase.

During the gathering of requirements, the software developer tries to specify all customer and end-user requirements of the software. The requirement specification contains a description on the information and information flow, functionality, behavior, and validation criteria (Pressman, 1999) of the requirements. UML provides several tools for specifying and further analyzing the requirements. Information and information flows may be described using **object diagrams** and **sequence diagrams**. **Use cases** encapsulate functionality; the **sequence diagrams** describe different scenarios of this. Behavior description contains the handling of the user events and commands. UML state charts may be used in the behavior description. (Douglass, 2001). Validation criteria are used later in the software process to test whether the implemented software meets the specified requirements. This phase is called validation testing, which is one of the subjects of Chapter 7.

The software developer analyzes each requirement and performs further investigation to recognize which classes may be used to encapsulate each object and what kind of relationships there are between the classes. The development process continues with the design phase, where the developer investigates how each requirement should be implemented. The design starts at the architectural level, specifying the system and its external interfaces. The system is then decomposed into subsystems. Functional decomposition divides complicated functions into a set of less complicated functions which are easier to design. More importantly, decomposition also helps in recognizing existing reusable subsystems. Design patterns are used in the design of the subsystems if a subsystem can be recognized as encapsulating a common software problem.

One problem in the design process is how to handle changes in requirements. Change that takes place during the design process naturally also affects the design and often violates the original design. In addition to this, developers are often not motivated to check all effects caused by changes in the design. A good design should have high cohesion and low coupling. High cohesion means that a subsystem includes only those functions that are closely related to each other; low coupling means that interactions between subsystems are kept to a minimum. So, a change should affect only one subsystem, leaving others unchanged.

One way to try to increase quality in software design is to apply design patterns during the design. They are commonly used in object-oriented software development to capture solutions to common software problems. Excellent tutorials to design patterns are those by Gamma *et al.* (2000) and Buschmann *et al.* (1996). As the whole Series 60 Platform is object-oriented there are many patterns applied in the implementation of the platform itself.

3.1.1 Design Pattern Categorization

The patterns are categorized to make the finding of a suitable pattern easier. The categorization may be done with respect to scale, including architectural patterns, design patterns, and idioms. **Architectural patterns** define high-level relationships and structures and are modeled with component diagrams. **Idioms** provide implementation of a specific pattern in a specific environment. They may, for example, implement a mechanism to avoid memory leaks in software. **Design patterns** are scaled between architectural patterns and idioms. They are modeled with class diagrams, representing the roles and relationships of each class in the model.

Another categorization of design patterns is based on purpose. This categorization includes creational, structural, and behavioral patterns. Creational patterns encapsulate instantiation of concrete classes. Structural patterns compose larger structures from classes and objects, and behavioral patterns describe object communication patterns. Some design pattern categories and patterns belonging to categories are shown in Table 3.1, categorized by scale (architectural, design, or coding) and purpose.

Table 3.1 Design pattern categories

Purpose	Architectural patterns	Design patterns	Coding Idioms
Creational	Abstract factory	Factory method Singleton	Two-phase construction Cleanup stack
Structural	Microkernel	Class adapter	Thin template
	Model–View–Controller	Object adapter	
Behavioral	Active objects	Template method Observer	Exception handling Iterator

3.2 Design Patterns in Symbian OS

All three architectural design patterns shown in Table 3.1 are applied in the design of Symbian OS itself. Symbian OS as a whole follows the **microkernel** which can easily be extended and adapted to the changing requirements and new environments because of its high modularity. The microkernel separates the core functionality from extended and user-specific parts. The size of the kernel core in Symbian OS is about 5% of the size of the whole operating system, which ranges from 500 kB to 15 MB, depending on Java support and preinstalled applications. The separation of the core and the other components make the system highly modular, which improves the portability of

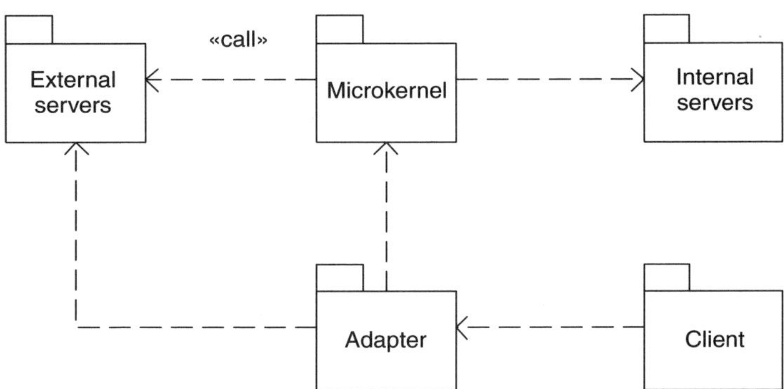

Figure 3.1 Microkernel architecture

the platform and makes upgrading and customizing the platform much easier than a monolith architecture. The microkernel architecture is presented in Figure 3.1.

The microkernel component in Figure 3.1 implements atomic services that are needed for all applications throughout the system. It maintains systemwide resources, such as memory, processes, and threads, and controls the access to them. It also provides the functionality for interprocess communication. Functions that cannot feasibly be included in the kernel because of their size or complexity are separated into internal servers.

Internal servers extend the functionalities of the core. They handle, for example, graphics and storage media and may have their own processes or shared libraries loaded inside the kernel.

External servers use the services of the microkernel and internal servers to provide services for the clients. They are heavily used in Symbian OS. External servers handle communications (Serial communications server, Socket server, Message server, and Telephony server), graphics (Window and Font and bitmap servers), audio (Media server), and storage media (File server). They are executed in their own processes. Each external server provides a client-side application programming interface (API) that encapsulates the interprocess communication between the client and the server. The adapter provides a transparent interface for clients, hiding, for example, the details of the communication. Adapters thus improve the changeability of the system.

3.2.1 Model–View–Controller Pattern

The Model–View–Controller (MVC) design pattern provides the ability to vary different software components separately. The pattern improves the software robustness and reusability. The MVC pattern helps the software designer to fulfill the object-oriented design principles, such as the Open Closed Principle (OCP; Martin, 1996). The idea of the OCP is that developers must decide early on in the analysis or design phase which parts of the system will be expanded later and which will stay fixed. According to the OCP principle, the design is extended by adding new code and classes by inheriting existing base classes rather than modifying the existing ones. Completed and tested code is declared closed, and it is never modified (only errors will be fixed). The MVC paradigm satisfying these principles is well-known and the main idea in the pattern is simple, as can be seen in Figure 3.2.

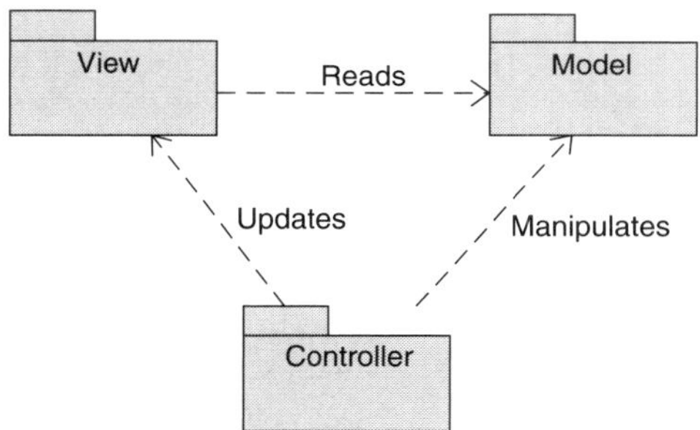

Figure 3.2 The Model–View–Controller paradigm

The MVC can be thought of as a low-level pattern and as an architectural, high-level, pattern. Architectural patterns can be implemented both on the operating-system level and on the application level. An example of an operating-system level architectural pattern is the layered user interface (Application Architecture–Control Environment–Uikon–Avkon) in Symbian OS. Architectural patterns provide the higher-level abstractions and guide how the applications should be built within a specific environment. The architectural patterns normally advise which lower-level patterns can be applied during the program development.

The MVC paradigm divides the application into three modules:

- Model: contains and manipulates the data in the program;
- View: defines how the data of the model is presented to the user; the view forwards received commands and requests to the controller;
- Controller: defines how the user interface reacts to received commands and requests.

MVC divides the application into modules that are logically decoupled from each other. By doing this, the components are more decoupled in the application. A decoupled component can be modified or removed without making changes to the other components. For example, changes to the Controller does not necessarily require changes to the View.

Other design patterns can also be applied in MVC, such as composite, observer, factory method, and decorator patterns (Gamma *et al.*, 2000). The **composite pattern** is useful when complex views are needed. This pattern enables the handling of individual objects and compositions of objects uniformly. The **observer pattern** defines how the observers can be notified and updated when the state of the subject changes. The **factory method** defines an interface for creating an object. The factory method is usually used in Symbian OS in the creation of a document of an application.

Use of Model–View–Controller in the Series 60 Platform

Series 60 applications follow the MVC paradigm because Symbian OS advices to use it. An application consists of an engine, user interface and view component. Avkon application is a DLL (dynamic link library), which is started by apprun.exe. Avkon is part of the GUI framework, which provides concrete controls and standard dialogs. Apprun executable loads and launches the application. At the beginning of program execution the framework expects the application to return an object (derived from `CAknApplication`). After application creation the framework calls the application class method `CreateDocumentL()`. This method creates and instantiates the document object (derived from `CAknDocument`). Then the document object creates an instance of AppUI class (base class `CAknAppUi`). AppUI is responsible for the event handling. AppUI also instantiates the AppView class, which displays the document data to the user.

The engine contains the data in the same way as the model in the MVC paradigm. The engine normally contains most of the application logic and handles the data manipulation algorithms. The document in Symbian OS has a reference to the engine and should be considered as part of controller, because the document is merely a container for the control. In the same sense, the application can be considered as a part of controller. In Figure 3.3 AppView acts as a view and AppUI acts as controller. It processes requests and commands received and sends appropriate updates to the model and to the view. The model notifies the view when the model's state has changed. AppView and Avkon base classes are described in more detail in Chapter 8.

The MVC makes it possible to develop software components separately. Application logic in the engine can be changed without changing the view component. Also, the other way round, changes to

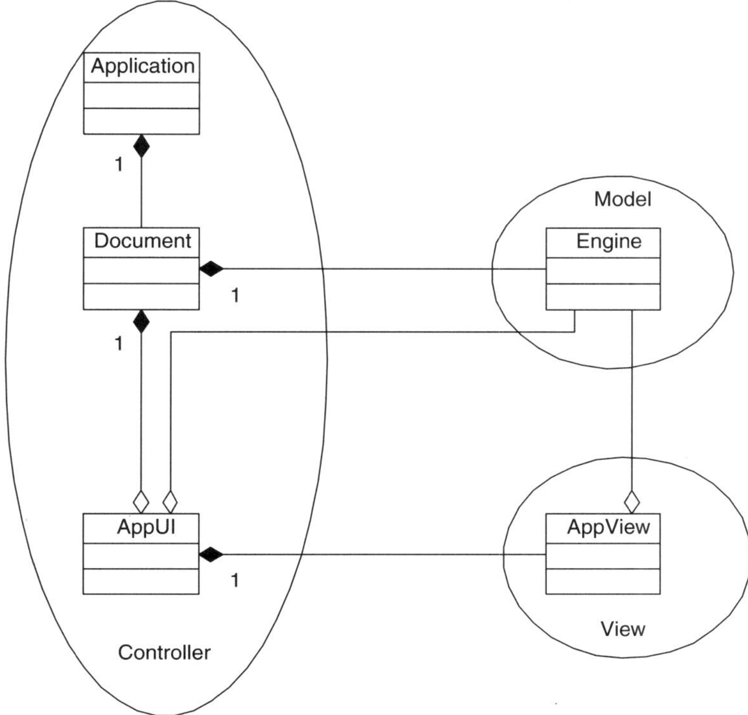

Figure 3.3 A simplified application architecture using MVC

the view do not necessarily cause changes to the engine. This is useful if there is a need to port the application to a different user interface (UI) style (Hoikka, 2001, pages 48–49).

3.2.2 Adapter Pattern

Adapter is a very common design pattern. It provides software developers with a way to adapt to changing requirements without changing the structure of the original design. That is why adapters are used more often in the implementation phase than in the design phase. The adapter design pattern is also called a 'wrapper' as it wraps some existing functionality into new environments. The adapter is essential in the design of the portable applications.

The adapter can be implemented either as a class or as an object adapter. However, a class adapter is not interesting in Symbian OS because it requires the use of multiple inheritance. Figure 3.4 shows the implementation of an object adapter. It uses aggregation rather than inheritance (used by the class adapter) to control the adaptee's

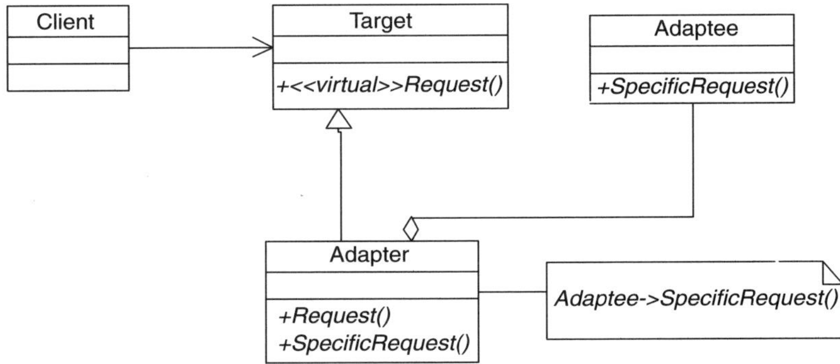

Figure 3.4 Object adapter

behavior. The client uses the adapter via interface, defined by the target. The adapter takes care of translating the client calls to the corresponding adaptee's methods. The adapter is transparent for the client.

The object adapter enables a single adapter class to be used with all adoptee type objects. The adapter can also add functionality to all adaptees at once. However, the object adapter has no way of overriding the adaptee's behavior.

3.2.3 Observer Pattern

The observer defines a one-to-many dependency between collaborating objects. The observer enables the partitioning of a system into observers that react when their subjects change state. It is used in many event-based systems to separate the event source from the event monitors. The observer design pattern is shown in Figure 3.5. The concrete observers attach themselves to the subject in a way that the subject knows which observers to notify when its state changes. The observers register themselves to the subjects in which they are interested. A subject may have multiple observers, and an observer may listen to several subjects. The update method may contain a flag, indicating which subject changed state in the case of a many-to-many relationship.

The problem with the standard observer pattern is that the type information gets lost if the subject or observer hierarchies are derived. A single observer does not know which subject subclass changed the state. There are, however, several special design patterns that solve the problem. Nevertheless, some of them are not applicable for Symbian OS environment because they require RTTI (run-time type indication) to work. RTTI can be seen as an extension to static (compile-time) type indication. It offers to the user information on

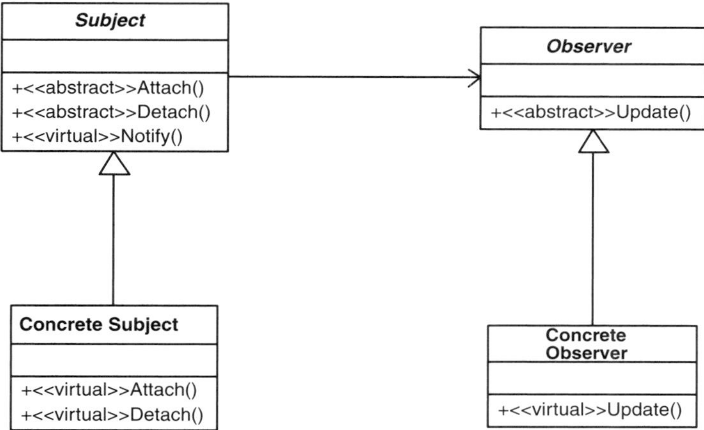

Figure 3.5 Observer design pattern

pointers and references, which is similar to the type information a compiler maintains during compiling.

3.2.4 State Pattern

The intent of the state pattern is to provide an entity with the possibility of altering its behavior when its internal state changes. The state pattern also lets us vary the state-based behavior by modifying the state transitions and to reuse the behavior separately. The state pattern is shown in Figure 3.6. The state pattern describes a context, a state machine that defines the entity that has a state-based behavior. It also provides a uniform interface for the states. The interface is abstract in most cases but can also contain some state-shared behavior or data. The transition from one state to another is usually defined inside the concrete states, but it can also be defined inside the state machine.

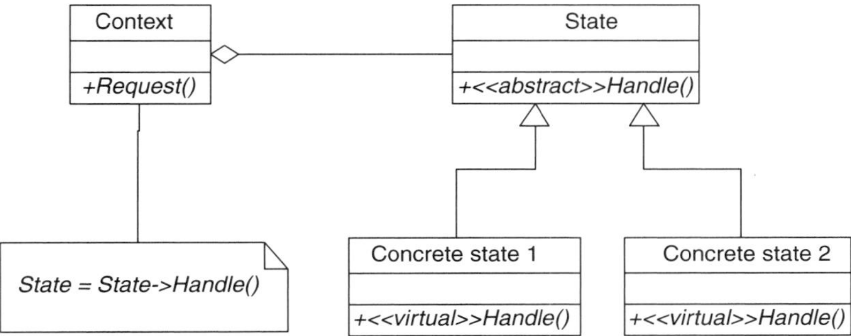

Figure 3.6 State pattern

Declaring it inside the states decreases the need to modify the context when state transitions are modified or new states are added.

The state change is invisible to the user; it has the same interface even though the inner functionality of the state machine changes. The state design pattern enables us to design very flexible and expandable solutions. The state transition can be defined by the state subclasses or by the state machine itself. The difficulty in implementing the state pattern arises when all the states use the same resources. The resources are often defined in the state machine, and the subclasses must get a reference to the owning class, which is not a very elegant solution.

3.2.5 Programming Idioms

Smartphones based on the Series 60 Platform may be run for months without rebooting. That is why it is important to take care that there are no memory leaks in the software. The leaks would eventually cause a shutdown of a system when no more memory could be allocated.

Memory may be leaked in cases where a pointer to an allocated memory area is lost. This happens when objects or memory are allocated in the heap, and pointers to allocated memory cells are stored in automatic variables in a function that may leave. The leaving mechanism is described in detail in Chapter 5. When the function leaves, all automatic variables stored in the stack are deleted, and pointers to memory cells are lost. To avoid this, Symbian OS provides a cleanup stack into which the pointers are pushed. The developer keeps the pointers in the cleanup stack as long as a possible memory leak may occur in case of leave. If the leave occurs, pointers are automatically pushed from the cleanup stack and the corresponding memory allocations are released.

The coding idioms of creating objects in the heap contain three concepts: non-public constructors, cleanup stack, and two-phase construction. First, the class should provide a static and public `NewL()` function for object creation. Users of the class cannot use normal constructors directly, because they are declared to be protected (allowing derivation from this class) or private (the class cannot be derived further). `NewL()` takes care of safe memory allocation, which avoids memory leaks.

`NewL()` encapsulates the use of the cleanup stack. First, the object should be created using an overloaded version of the new operator accepting an argument `ELeave`. This argument tells that the object creation leaves if memory cannot be allocated. Second, a pointer to a created object is pushed into a cleanup stack. Third, the second phase of the allocation is made. This includes the creation of any objects owned by the object just created. According to the idiom, this should be implemented in the function named `ConstructL()`. Finally, when it is safe (function leaves do not cause a lost of the pointer) the pointer is popped from the cleanup stack.

The code example below shows how the idiom is used to create an engine object of the rock–paper–scissors game. The rest of the design of the game is described below.

```
CRpsModel* CRpsModel::NewL()
    {
    CRpsModel* self = new (ELeave) CRpsModel;
    // Use of overloaded new
    CleanupStack::PushL(self);
    // Pointer pushed into the cleanup stack
    self->ConstructL();
    // Second phase construction
    CleanupStack::Pop(self);
    // Pointer popped from the cleanup stack
    return self;
    }
```

3.3 Design of the Rock–Paper–Scissors Game

To see how common design patterns are applied in software design, let us walk through a design of a simple Rock–Paper–Scissors game. The source code is available at www.digia.com/books. The same example is used in later chapters of this book, to show concretely the use of communications services, for example.

3.3.1 Requirements Gathering

The goal of the game is very simple. The challenger starts the game by selecting one of three items: rock, paper, or scissors. The rock may be put wrapped inside the paper, which may be cut into pieces with scissors, which can be damaged by the rock. If, for example, the challenger chooses the rock, the challenged player must choose the paper to win. If the challenged player chooses the scissors, the challenger wins. If both players are using the same items, neither of them gets the score and the result is draw. The number of moves may be configured in the software so that, for example, after each three moves the challenge is sent.

What are the requirements of this game? The challenger must be able to choose the item (e.g. using a GUI). In the Series 60 Platform an obvious key for selecting an item is the navigation key. The selection may be confirmed by pressing the key down. The selected item must be communicated to the other player (it should be possible to use any supported communication technology). For communication we must define the structure of the game PDUs (protocol data units; i.e. messages). The application can be launched from the shell but also when a smart message is received and directed to the application. Either player may be the challenger or the initiator of the game. The number of moves (selected items) should be configurable. After comparing the players' items in the moves, the result should be shown to the user.

Use Cases

Use cases are utilized to understand user's requirements (Douglass, 2001). They define the cases how the user may interact with the system. The case definition includes input and output parameters, consequences, relations to other use cases, exceptions, and timing constraints. There is a notation for use cases in UML, but cases may also be specified in text. The rock–paper–scissors (Rps) game may contain the following use cases.

Case A: the user wants to create a game challenge and send it to a certain recipient

- User launches the RpsGame application from the Menu.
- The user selects a sequence of moves using the navigation key and user interface. The sequence of rocks, papers or scissors is shown graphically.
- The user selects 'Challenge' from the options menu, which launches a smart message editor.
- The user fills in recipient number or address.
- The user selects Send, which causes the send of the message.

Another use case specifies the situation in which the challenge has been sent to the user, as described below.

Case B: an Rps smart message arrives

- The user opens the message, either from Inbox or Soft notification. This opens the Rps smart message viewer.
- User selects Accept from the Options list. This launches the Rps-Game applications.
- The user plays against the received challenge. Again a specific number of moves must be selected before the response is compared with the challenge.
- The user sends the result back to the other device.

3.3.2 Analysis

In the analysis phase the gathered requirements are studied further. First, we divide the UI part and game logic (engine, model) into their own components, as shown in Figure 3.7. The basic structure of the application follows the GUI framework, as in any Eikon (or Avkon) application. The separation of the user interface and the game logic makes it easier to test the application and use generic parts directly in other UI styles. The UI part takes care of the user events and application data, whereas the engine accesses the communications and the other system services, such as files, to send and receive messages between the players.

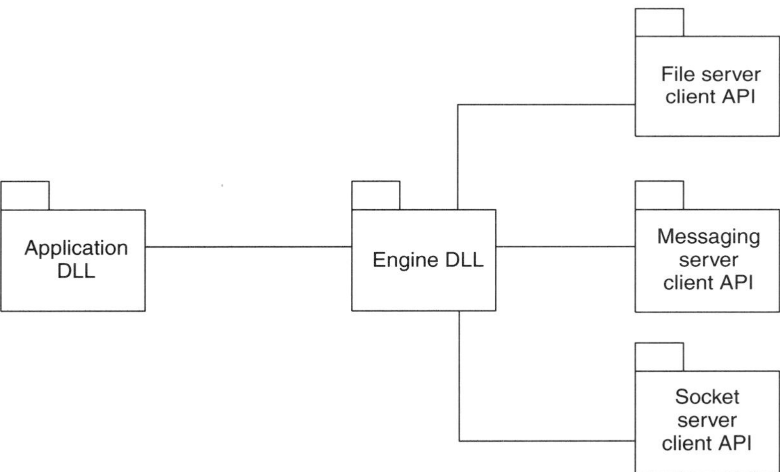

Figure 3.7 High-level architecture of the rock–paper–scissors game. Note: API, application programming interface; DLL, dynamic link library

3.3.3 Design

In the next phase, we consider what is required in the software so that it meets the set requirements (i.e. how it should be implemented). This is the place where design patterns are applied.

The design of the UI framework of the application is straightforward. The application architecture already determines which classes to use for which purpose (i.e. which has the role of the controller, which one is the view, etc.). There are four basic base classes (application, document, applications user interface, and container) and the fifth, which is independent of the application architecture and which implements the engine part. The application class contains the entry point of the application (i.e. the function used to start the application). The document is used to create the application user interface (App UI). In file-based applications, the document class creates a layer between the user interface, the engine, and the file in which the engine will be stored. The App UI takes care of handling the user events, and the container represents the user interface itself (i.e. all controls to be shown in the user interface). The basic classes and their relationships are shown in Figure 3.8.

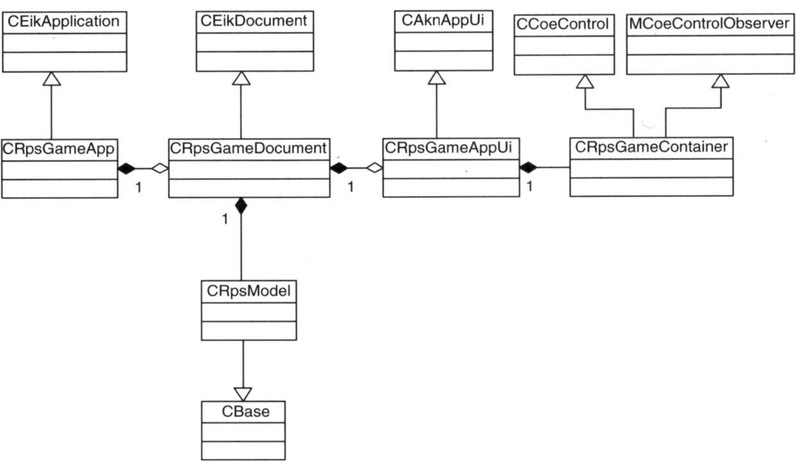

Figure 3.8 Initial class diagram of the rock–paper–scissors game

By studying the use cases it is possible to recognize situations in which we may apply design patterns. The user cannot, for example, send the challenge unless a certain amount of moves has been selected. The result cannot be calculated if either the challenge or the response is not available. Thus, in some **states** the application should respond to specific user requests only. More obviously, the states can be seen in communications. Because the state pattern increases the binary size of the application, we use just a single integer variable to indicate the state of the application. In the Series 60 Platform it is possible to, for example, hide some menu items in certain states. For example, in the game, the challenge menu item is hidden until the minimum number of moves has been selected.

Some classes need to know if there are changes in some other class. In this case it is reasonable to apply observers. They observe the state of the subject and notify the observer class about state changes. In Figure 3.8 we can see that `CRpsGameContainer` is derived from `CCoeControl` and `MCoeControlObsorver`. The former is the base class of all controls, which may be drawn on the screen. The latter provides a virtual `HandleEventL()` function. This function is called whenever the observed class calls `ReportEventL()` function. This enables the change of the view in the control in cases the view change in another control causes changes to this view as well. Observers are frequently used with asynchronous requests in the Series 60 Platform. They notify, for example, about message arrivals and service completions.

In Figure 3.9 we have applied also active object design pattern. Messaging, and communications in general, is highly asynchronous. There is no reason to use resources for waiting in applications, so the execution of waiting objects may be suspended until the request waited for has been served. In Figure 3.9 there are three communication classes derived from `CActive`, which is the base class for active objects. Each active object may wait for the completion of a single request at a time. That is why it is common to use three active objects: one requesting a connection, one requesting data transmission, and one requesting data. The details of `CSendAppUi` are not shown in the figure. This class provides another way of creating and sending messages in addition to communicating active objects.

The `CCommsEngine` class takes care of sending the message, when `CAsendAppUi` is not used. The message is created by the engine. It contains the moves of the user encapsulated in the `CRpsMoves` class. The moves (rock, paper, or scissors) are stored by the `CRpsGameContainer` class with respect to the selections of the user.

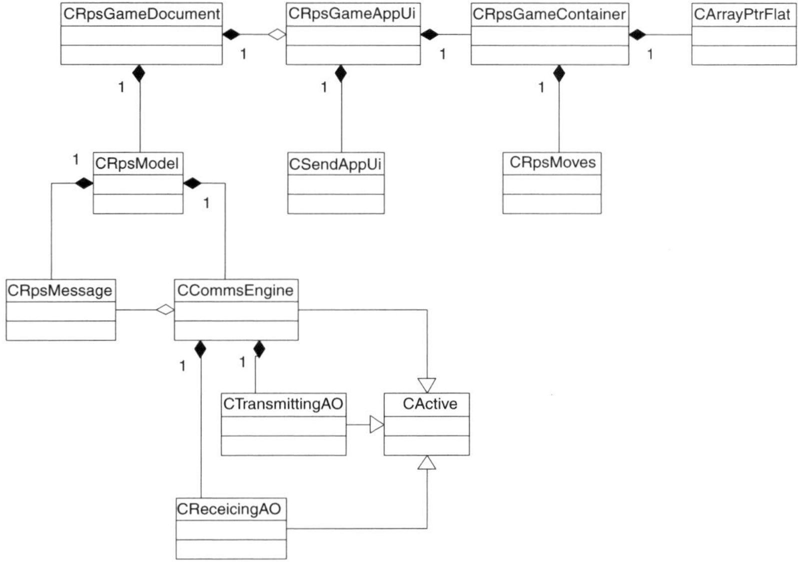

Figure 3.9 Skeleton design of the Rock–Paper–Scissors game

`CArrayPtrFlat` is used to store the bitmaps of the paper, rock, and scissors. The container (view) is capable of showing the bitmaps, because it is a control (i.e. derived from `CCoeControl` see Figure 3.8). All controls are typically drawn on the screen, although there are container controls, which just contain controls but are never drawn themselves. A large part of the source code of the game engine will be presented in Chapter 5.

3.4 Summary

Design patterns describe solutions to common software problems. To apply a pattern, the problem must be identified first. After that, the pattern provides the class and sequence diagrams, and, using these, existing components may be reused in the implementation. The most commonly used patterns in Symbian OS include MVC, microkernel, client–server, adapter, observer, and state patterns. Some of these have also influenced the architecture of the operating system itself to make it highly suitable for embedded systems, such as smartphones.

4

Software Development on the Series 60 Platform

After the design phase, which was the subject of the previous chapter, the software implementation starts. It is possible to use C++ or Java programming languages for software development on the Series 60 Platform. The development follows the same pattern regardless of the user interface (UI) style, although tools may differ. At least, there is a PC-hosted emulator providing a Microsoft Windows based implementation of the Symbian software platform. Microsoft's Visual C++ 6.0 (VC++) is used for source file editing, compiling, and building. Other UI styles may use other IDEs. For Java development, any Java integrated development environment (IDE) can be used (e.g. CodeWarrior). When using Visual C++, instead of building against Win32, the software is linked and built against a Symbian header and library files, which are included as a part of the SDK (software development kit). The binary built is then run in the emulator. In addition to the emulator, there are, naturally, building tools for target hardware and many utilities, making software development easier. For Java there is an own SDK – Nokia Series 60 MIDP (mobile information device profile) SDK for Symbian OS.

This chapter covers all the necessary development tools and instructions on how to build software for the emulator and for the target phone. We also show how to install software on the phone and how to take into account internationalization and localization issues. The end of this chapter contains a list of web links of the Symbian community. The links may be used for getting help and support for software development. This chapter gives enough information to understand how to build the code examples in later chapters. As an example, we build the rock–paper–scissors game and show which files have to be defined for build tools. All examples are available at www.digia.com/books.

4.1 Nokia Series 60 Platform Tools

On the Series 60 Platform, software development may take place either in C++ or in Java. C++ is the native programming language of Symbian OS, underlying the Series 60 Platform. All Series 60 Platform and Symbian OS application programming interface (API) calls are available for the C++ programmer. In addition, C++ gives the best performance and facilitates lower memory consumption. Since Java has been designed to run on any device, there are some compromises in its general functionality. A Java application cannot, for example, access all features of a device, such as the contact database or the calendar.

In spite of its restrictions, Java has become a popular programming language. The Series 60 Platform accommodates the J2ME (Java 2 Micro Edition) environment according to the MIDP of the connected limited device configuration (CLDC) definition. The CLDC is designed to be a small footprint configuration for devices with constrained resources. MIDP is a functional extension of the CLDC. It defines a profile for communication devices with restricted capabilities, such as smartphones. MIDP applications are called 'midlets'. The greatest benefit of using Java is its hardware independence, allowing developers to concentrate on the application development, not on the specific features of the hardware platform. In addition to that, there is a large development community providing support and help for application development. Java development is further discussed in Part 4 of this book.

4.1.1 Software Development Kits

The required Nokia Series 60 SDK is available at www.forum.nokia.com. SDKs supporting different programming languages or device families may be installed into one system, because files are installed into separate directories. If there are several SDKs installed, the one to be used is selected with the `EPOCROOT` environment variable, which should point to a Series 60 folder in the installation directory (e.g. `Symbian\6.1\Series60\`). Note that the drive letter must not be specified. By default, C drive is used, but there are no restrictions as long as the SDK and developed software are located in the same drive. An example of the directory overview is given in Figure 4.1. The existing folders depend on the installed SDKs.

The Series 60 directory contains four subdirectories: Series60Doc, for SDK documentation; Series60Ex, having examples on using the Avkon library and facilities of the platform; Series60Tools, containing an application wizard for creating minimal graphical user interface (GUI) applications; and Epoc32, containing files, initialization files, linked libraries, server executables, and prebuilt applications.

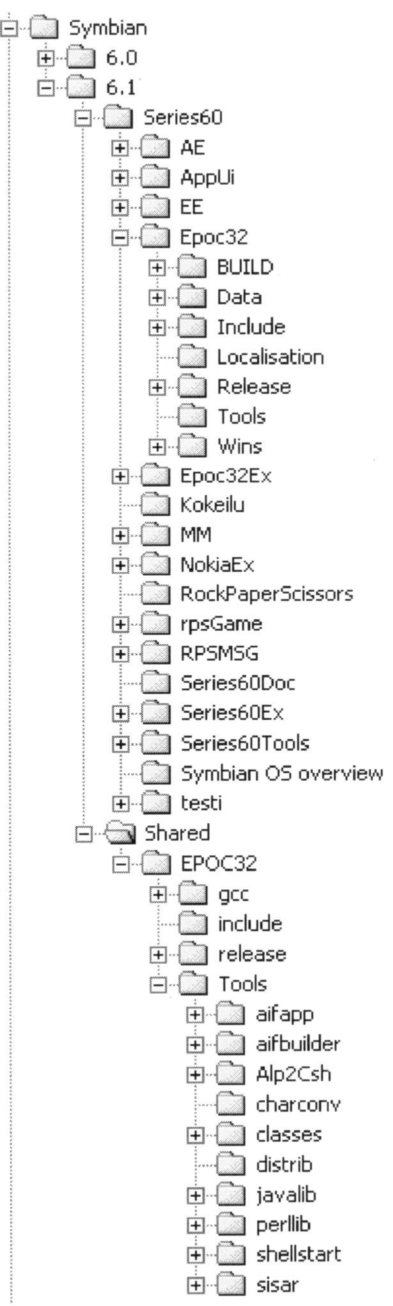

Figure 4.1 Installation directory overview

At the same directory level to which `EPOCROOT` points, there is another directory – Shared – containing tools for application development, testing, and deployment. The most often-used tools are emulator, SDK help, bitmap converter, resource compiler, AIF Builder (application information file builder), and application installer. One tool – Application wizard – is located in NokiaTools directory in `EPOCROOT`. There are also GCC (Gnu Compiler Collection) and linker tools for making builds for the target hardware.

4.1.2 AIF Builder

Each application program is run from a shell program, which may be activated by pressing the swap-and-launch application button in the phone. Before running the program, at least two files must exist: an application file having an extension .app, and an application information file having an extension .aif. The application file contains, at a minimum, a unique identifier (UID) to identify the program. Each file in the operating system contains unique identifiers, which consist of three 32-bit values and a checksum. A more detailed use of UIDs is given shortly.

The application information file may specify the icon for the application, the application caption in many languages, and capabilities, such as whether the application can be embedded or whether it is shown in the application launcher (i.e. shell). For a Java application, an additional text file having an extension .txt can be used. This file specifies the command-line arguments to the Java virtual machine when the application is launched.

AIF Builder helps to create the required files. It is independent of device family and programming language. AIF Icon Designer helps to create the bitmap for the application icon. The icon must be in a multi-bitmap format (.mbm), which is the specific format required by Symbian OS. Normal bitmaps may be converted into multi-bitmap format by using the bitmap converter tool bmconv. A screenshot of AIF Builder is shown in Figure 4.2.

4.1.3 Application Wizard

Application Wizard is an excellent tool for starting to learn the application architecture and the classes related to it. It provides a graphical means of creating a minimal GUI application. This minimal application may naturally be extended to develop more complicated applications. The tool creates four basic application architecture classes: `CXyzApp`, `CXyzDocument`, `CXyzAppUi`, and `CXyzContainer`, where Xyz is the name given to the application. The use of these classes and the

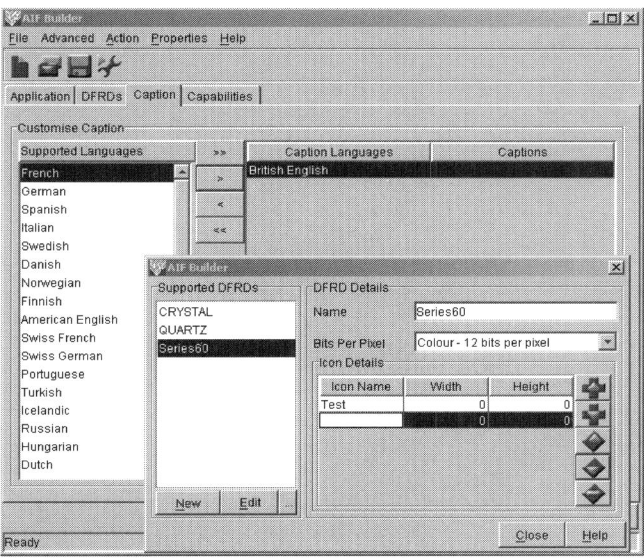

Figure 4.2 A view of the AIF Builder (application information file builder) showing language and user interface style selection

application framework will be discussed in Chapter 8. In addition to the application skeleton source code, files required to create workspace and makefiles are also created. Other tools, such as Menu Builder, may be used to develop the user interface further.

4.1.4 Build Tools

The tools delivered with the SDK make the building process simple. The environment neutral mmp project file of Symbian can be used to generate toolchain-specific project files for VC++ (emulator builds) and GCC (target builds). During the development, the project file manages all linking and building details. The project file also takes care that all outputs from the build and any required resources, such as resource files, icons, bitmaps, and sound files, are built into the appropriate location for debugging on the emulator. It is also possible to use the command line by using the same project file for creating required makefiles.

Makefiles are generated with the makmake tool. The developer may use makmake directly. However, more often a Perl script is used to call makmake. The bldmake project building utility is used to create a .bat file, which calls a Perl script, which in turn calls the makmake tool. The user has to provide an information file (having an extension .inf) for bldmake, which in the minimal form contains only the names of the project definition files (having an extension .mmp). These files may be generated automatically with Application Wizard or copied from example applications. The command (given on the command line) bldmake bldfiles generates abld.bat, which is invoked to call the Perl script calling makmake.

```
/* Example bld.inf file */
PRJ_MMPFILES
RPSGame.mmp     /* This is the UI part of the
  Rock-Paper-Scissors game */
RPSModel.mmp     /* Model or engine of the game */
PRJ_PLATFORMS

/* If you build for WINS, ARMS, ARM4 or THUMB, the
  platform does not need to be specified */
PRJ_EXPORTS
/* Specify any files you want to copy some destination
  (default \epoc32\include) */
PRJ_TESTMMPFILES
/* Specify any files needed for building test programs */
```

The project definition file defines the type of the build (e.g. application, server, library, device driver, protocol module). It is defined by targettype, but also by the UID number. The UID number has three 32-bit values, the first of which indicates whether the file contains data or is an executable program. The second UID defines the subtype. For data files, the subtype identifies the store (direct or permanent) containing the document associated with a specific application or an AIF, associated with the application. Stores are used to store streams, which are sequences of bytes. A complicated structure or object is externalized into several streams, as shown in Figure 4.3. The direct file store does not allow change of streams (overriding, deleting, replacing), but these functions are available for streams in the permanent file store.

There are three kinds of executable files in Symbian OS: normal executables (.exe), used mainly for servers, static interface libraries (.dll), used for shared libraries, and dynamic interface libraries, used in the implementation of plug-in software modules (their extension being library-specific). For servers, the second and third UIDs are not needed. For libraries, the second UID identifies whether the library is a shared library or provider library. In the case of a provider library, the UID also identifies the type of library (e.g. application, message-type module, device driver, protocol module, etc.). For example, the UID value 0x100039ce defines that this executable library is an application.

The third value uniquely identifies the application, document, or data file. The third UID of the rock–paper–scissors (RPS) game is 0x101F5BBD. For released software, the values in the range

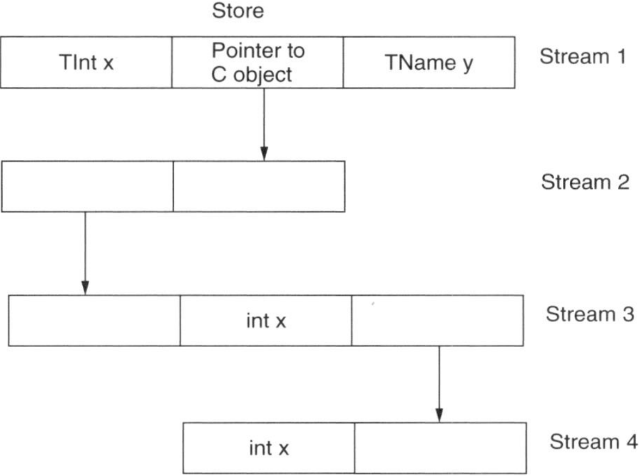

Figure 4.3 Streams and stores

0x01000000–0x0fffffff must not be used, since this range is reserved for testing purposes. Symbian manages the UID numbers, and developers may ask Symbian for UIDs for released software. This is done by sending an email to uid@symbiandevnet.com, the message specifying the number of required UIDs (typically, 10 at most).

An example project file is given below. The last line specifies, how the AIF resource file is compiled. The given line specifies that we compile a RPSGameAif.rss resource file, located in ..\aif, into an .aif file, RPSGame.aif, having an icon and corresponding mask. The icon is shown in Figure 4.4.

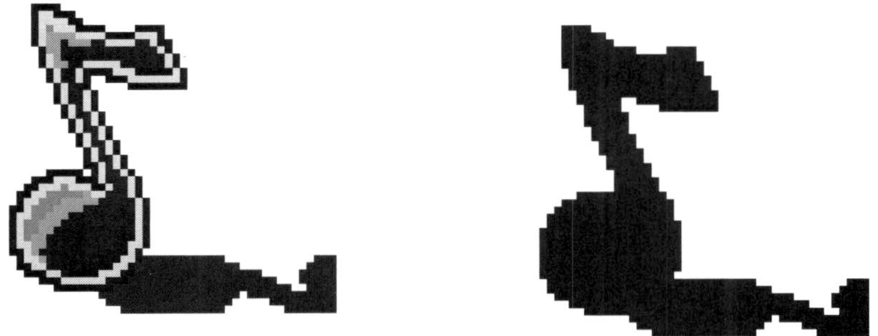

Figure 4.4 Icon of the rock–paper–scissors application and its mask

```
// MMP file of the RPS game
TARGET  RPSGame.app
TARGETTYPE  app
UID  0x100039CE 0x101F5BBD
TARGETPATH \system\apps\RPSGame
```

```
SOURCEPATH ..\src
SOURCE  RPSGameApp.cpp      // Only four basic GUI classes
SOURCE  RPSGameAppUi.cpp
SOURCE  RPSGameDocument.cpp
SOURCE  RPSGameContainer.cpp
RESOURCE ..\group\RPSGame.rss
LANG     SC      // Use the default language
USERINCLUDE . ..\inc ..\img
SYSTEMINCLUDE      . \epoc32\include
LIBRARY euser.lib apparc.lib cone.lib eikcore.lib
LIBRARY eikcoctl.lib avkon.lib fbscli.lib
LIBRARY etext.lib sendui.lib
LIBRARY estor.lib efsrv.lib rpsmodel.lib

START BITMAP rpsimages.mbm
TARGETPATH     \System\Apps\RpsGame
HEADER
SOURCEPATH     ..\img
#include      "..\img\images.txt"
END

// DEFFILE ?filename
AIF     RpsGame.aif ..\aif RPSGameaif.rss c12 icon44.bmp icon44m.bmp
```

The project file for the game engine library is much simpler. The second UID, 0x1000008D, reveals that the source files are built into a normal shared library:

```
ARGET  RPSModel.dll
TARGETTYPE  dll
UID  0x1000008d 0x101F5BBE
SOURCEPATH ..\src
SOURCE  RPSModel.cpp RPSMessage.cpp RPSModelDll.cpp
LANG     SC
USERINCLUDE . ..\inc
SYSTEMINCLUDE   . \epoc32\include

LIBRARY euser.lib estor.lib
```

The abld script is used to create required build files and build the application, including any resource files. The resource file defines all the resources of an application, such as text strings, menus, dialogs, and other customizable screen elements. These are managed outside the application source code in separate resource files, which makes, for example, the localization of the software easier.

To create build files for the emulator, i.e. WINS (Windows single process), the command **abld build wins** is used. In the same way, a target build is generated using the command **abld build armi**. Creating a workspace and project files for Visual C++, use command **abld makefile vc6**. The .dsw makefile is generated for applications which are going to be developed and debugged in the Visual C++ integrated development environment (IDE); this is a workspace file that can refer to one or more project files with the extension .dsp; the makmake generates a dsp file which is like a makefile used by the IDE.

For target hardware, three different executable binaries, or application binary interfaces (ABIs), may be generated: arm4, armi, and thumb. The plain binary for the ARM (Advanced Risk Machines) is arm4 build, which has a slight performance advantage over armi. However, arm4 does not interwork with thumb, which is the smallest build in terms of binary, because 16-bit rather than 32-bit instructions are used. The only binary that interworks with both thumb and arm4 is armi (arm interworking), and that is why this binary build should be used by third-party software providers. The relations between the binaries are shown in Figure 4.5 (Pora, 2002).

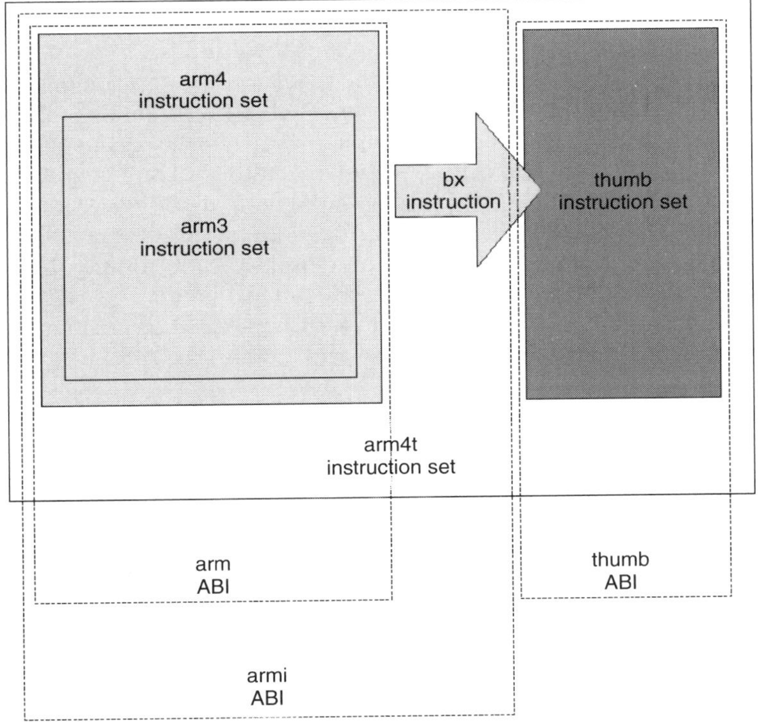

Figure 4.5 Thumb, arm4, armi, and their interrelationships; ABI, application binary interface

Note that there are necessarily several differences between emulator and target builds. One of the most obvious ones is that in Windows the software is run in a single process, whereas in the target several processes can be created. For example, in the target hardware a new process for each application is created, but in the emulator only a new thread for a new application is created when the application is launched. Since the processes are used for memory protection, there is no similar memory protection between applications and other software components in the emulator as in the target builds.

4.1.5 Emulator

Emulator is used to test and debug the software before deploying it in the target hardware. After building the application for WINS the emulator may be launched, and the application icon should appear on the display of the shell program. This can be done by pressing `Ctrl F5` in Visual C++. For the first time, Visual C++ prompts for the location of the executable to be run (this executable is the emulator). The debug emulator is located in the `%EPOCROOT%\epoc32\release\wins\udeb\epoc.exe` directory. After loading the emulator, the applications can be launched as shown in Figure 4.6, in which we start the RPS game.

The debugging is done in the same way as the debugging of any Windows application. It is possible to add breakpoints and step through code as normal. After testing throughout, the software may be uploaded to a real device. Since the binary format in ARM is different from Windows, the software must be rebuilt for armi by using **abld build armi urel** on the command line. The binary files may be uploaded directly using Symbian OS Connect (see Chapter 18) or creating first an installation package with the Sisar application.

4.1.6 Sisar

Sisar is an application for creating installation packages. An application typically consists of several files, such as executable, binaries, and resource files for supported languages. The Sisar application packages these files into a single installable .sis file.

4.2 Application Deployment

Executable and data files may be installed by storing them directly into the directories of the device. For applications consisting of several files, such as different executables, resource files for the support of different languages, and arbitrary data files in either textual or binary format, it is time-consuming and error-prone to install everything by hand. Sisar is a tool with which all files of an application may be packed into one package – a .sis file.

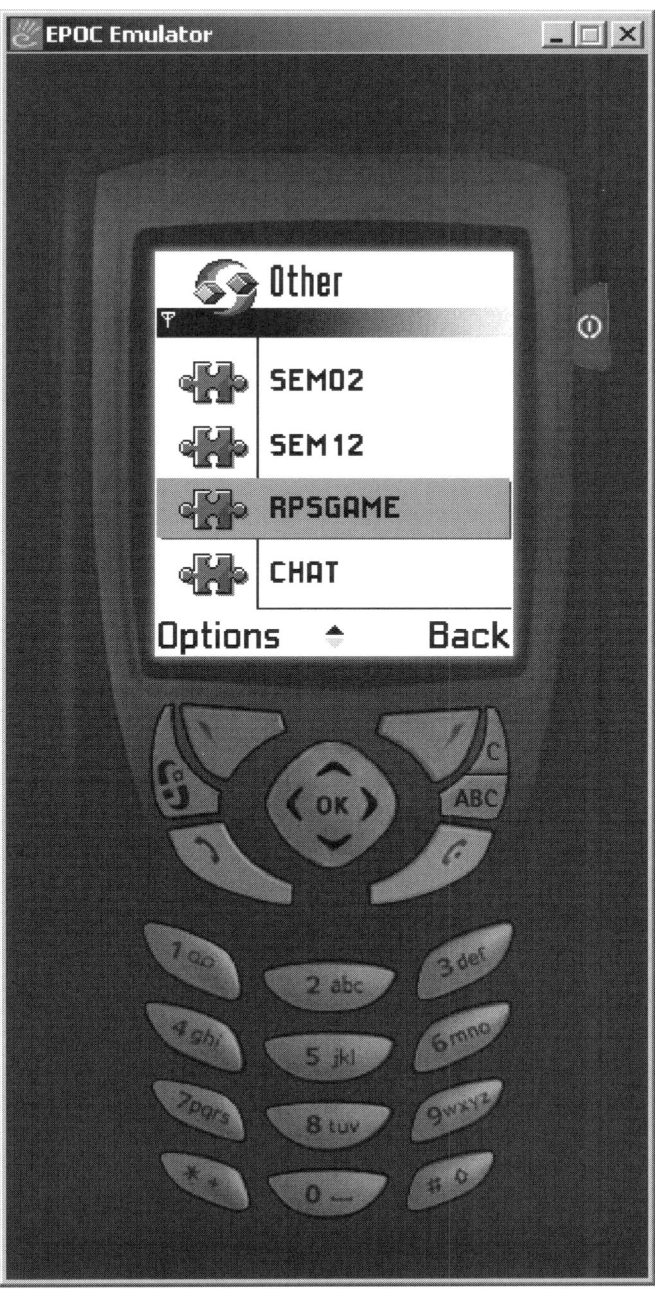

Figure 4.6 Series 60 Platform Emulator (version 0.3)

4.2.1 Package File Format

The .sis file creation is started by defining a sisar project into which the existing package file format files (.pkg) and AIF Builder project files (.aifb) may be imported. The package file format is defined in a textual file. The first line in the file specifies the languages, which are provided within the generated .sis file. The set of languages is specified using a two-character code, such as `&EN, FI`. In this example, English and Finnish are supported. The codes of supported languages can be found in SDK Help.

The package header should precede all other lines except the language line. An example of a package header is as follows:

```
#{"Rock-paper-scissors","Kivi-paperi-sakset"},(0x11111111),1,0,0,NC,
TYPE=SISAPP
```

First the name of the component is given in all supported languages, in our case in English and Finnish. All .sis files require a UID, which is given next. Although the installed files do not require a UID, the .sis file does require it.

The version is given using three numbers. In the example, 1,0,0 specifies version 1.0, with build number 0. All numbers may be in decimal or hexadecimal format. After the version number, there may be package options, which may be either SH (SHUTDOWNAPPS) or NC (NOCOMPRESS). The first options specify that applications should be closed during installation. By default, SH is unselected. By default, the .sis file is compressed, which may be changed by using the NC option.

Finally, the package type is defined. The package may contain an application SA (SISAPP) or any system component: SY or SYSSYSTEM (shared library); SO or SISOPTION (dynamically loaded component, code, or data); SG or SISCONFIG (configures an existing application); SP or SISPATCH (patches an existing component); SU or SISUPGRADE (upgrades an existing component).

All Nokia products based on the Series 60 Platform will use a built-in mechanism to check that the installed software is really Series 60 software. For this purpose, each Series 60 application should have an id-sequence identifying the software platform. The lack of the sequence results in a warning message. However, the user is still able to install the software. The id-sequence is the same for all Series 60 applications and is added into the .pkg file right after the header. The added line should be exactly like the one given below.

```
(0x101F6F88), 0, 0, 0, {"Series60ProductID"}
```

The next line (package signature) is optional and is included when the installation file is digitally signed:

```
*private-key-file-name,public-key-certificate-file-name
```

The signature is checked and the user may check the details of the certificate when the package is installed. The first file name specifies the private key file, from which the digital signature is created (see Chapter 15). The second refers to a file name which contains either the corresponding public key certificate or the certificate and its associated certificates, used to form a certificate chain. A third parameter may be used to specify a password to access the private key file, if it is encrypted.

Optional Lines

The rest of the file includes optional package lines. These are: condition blocks used to control the installation; option lines used to display a list of options to the user; language-dependent (only one is installed depending on the selected language) and language-independent files (installed regardless of the language); capability lines (used in the determination of the device or application capabilities); requisite components (components required by the installed component and present in the target); embedded .sis files (installed and also removed, when the main component is installed and removed); and comments (any lines preceded by a semicolon).

The condition blocks are used to control the installation. With them it is possible, for example, to use one installation file only to support any number of different Symbian OS phones. A condition in the condition block may be a condition primitive. There are NOT, AND, and OR operators available to combine individual conditions. A condition primitive may be a relation between an attribute, such as manufacturer, and a value. For example, manufacturer value 0 is Ericsson, as shown in the following example:

```
IF manufacturer=0 ; Ericsson
  "ericsson\rps.app"-"!:\System\Apps\rps\rps.app"
ELSEIF manufacturer=1 ; Motorola
  "motorola\rps.app"-"!:\System\Apps\rps\rps.app"
ELSEIF manufacturer=2 ; Nokia
  "nokia\rps.app"-"!:\System\Apps\rps\rps.app"
ELSE
  "unknown.txt"-"", FILETEXT, TEXTEXIT; Specifies that "unknown.txt" is
  ; shown in a dialog box and not installed on the target. TEXTEXIT
  ; specifies that Yes and No buttons are used in the dialog
ENDIF
```

See SDK Help to read the exact syntax and other attributes and their corresponding values for the conditions. A ('!') character allows the user to specify the drive letter.

The options line is used to ask the user to make selections affecting the installation. It is preferable to put these at the beginning of the package format file, so that options are asked at the beginning of the installation, although this is not required. Options are built-in attributes, that have either value 1 (selected) or value 0 (not selected). The value may be checked by using condition blocks, as above. For example, the following options could be used to ask the user if she or he wants to include Bluetooth and multi-player modules in the installation:

```
!({"Add Bluetooth Component","Lisää Bluetooth-komponentti"}, ; Option 1
{"Add Multi-player Component","Lisää monen pelaajan tuki"}) ; Option 2
```

Those files for which the selected language does not affect the installation are listed in the language-independent file lines. The line specifies the source and target directories for each installed file and, optionally, arguments that specify the type of file: FF or FILE (default value); FT or FILETEXT (text file shown in a dialog box during installation, but not installed into the target); FN or FILENULL (will be installed later); FM or FILEMIME (file is passed to an application, specified by MIME type); and FR or FILERUN (installed in the Symbian OS device). There are other arguments used to specify how the text dialog is cancelled, or when the installation file is run (used with FILERUN); for example, to open up a text document, describing the application:

```
"source\components\Math.dll"-"!:\system\libs\Math.dll"
```

Language-dependent files are, for instance, resource files, discussed further below. For example, for our RPS game, we could use the following line for language-dependent files:

```
{ "\source\apps\rps\group\rps.ruk"
  "\source\apps\rps\group\rps.rfi"}-"!:\system\apps\rps\rps.rsc"
```

The requisite components are specified with the UID, version number, and the name of the required component. The version must be at least the required one. The name does not have to match the component name. It is provided simply to give information to the user. Embedded .sis files may be added in almost the same way. The user has to specify the source .sis file and its UID. Look at the example below:

```
{0x10001111},1,0,0, {"My Math library"} ; Requisite component
@"YetAnotherComponent.sis",(0x10001112) ; Embedded .sis file
```

AIFB Files

In addition to package format files, sisar can import AIF Builder project files (.aifb). From these files, sisar reads the application UID and extra bar captions for the application. The .sis component will be set to support those languages specified in the .aifb file.

The .sis file is generated simply by using the GUI of the sisar. The Installation File Generator will generate the .sis file according to the guidelines of the imported package format and AIF Builder project files. The .sis files may then be delivered to the target device by using any supported communications media. The execution of the .sis file in the target causes the installation files to be copied into the correct locations.

4.3 Worldwide Localization Support

Symbian OS provides rather simple methods to make software products truly international. These same methods are also used in the Series 60 Platform. This section provides information for Series 60 licensees, other vendors, and third parties who want to localize their products to support multiple languages.

In Symbian OS, much of the information that defines the appearance, behavior, and functionality of an application is stored in a resource file – external to the main body of the program. By separating the application resources from its executable code provides many advantages, such as resources being loaded only when needed, resource files being compressed, and resource files being localized without recompiling the main program. Two types of resource files used in Symbian OS are resource files (.rss) and application information files (.aif). The difference between these file types is that application information files define the application behavior within the system context whereas ordinary resource files define the appearance and behavior within an application.

The applications can be localized simply by changing the resource file text associated with each menu item, task bar, or other control. Since changes to the text do not change the symbol information in the generated header file (.rsg) it is unnecessary to recompile the application to use the new file. Consequently, a resource file may be generated for each language supported, and the actual resource used is determined by the end-user at installation time. Compiled resource files for different languages are separated with the two-digit language codes in their file extensions (the .rsc extension becomes .r01 .r04 .r17, etc., depending on the language selection). Symbian has provided a list of recommended codes for a large number of languages; these should be used while localizing applications.

Symbian provides some additional tools to make the localization process even easier. For example, with Aif Tool, developers can create aif files easily, and the Rcomp (or epocrc) is used to compile resources.

For detailed examples and instructions of the tools, language codes, syntax, and use of resources, refer to Series 60 SDK.

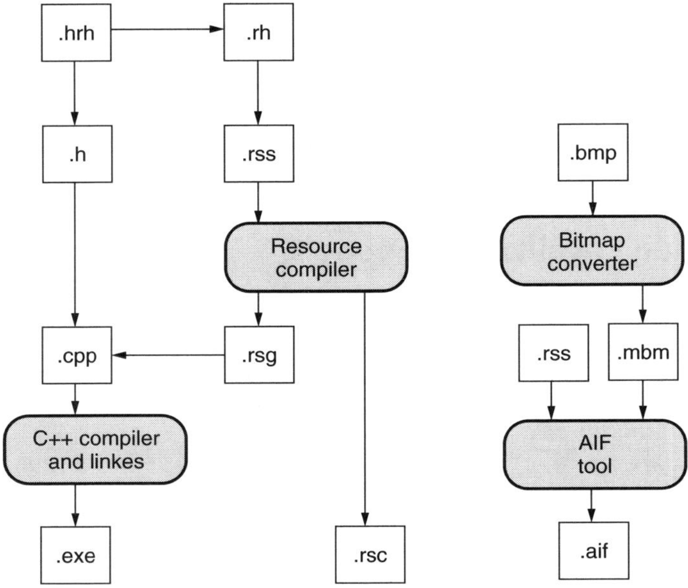

Figure 4.7 File types (white boxes) and tools (light-gray boxes) used in the resource compilation process; for a description of there files and tools, see the list in text

The purpose of the files and tools illustrated in Figure 4.7 is as follows:

- The C++ compiler and linker together take source files (.cpp) and produce output (.exe or .dll, or .app, or files of a similar extension).
- The resource compiler will convert a resource source file (.rss) into a compiled resource file (.rsc) containing the resources that the executable program will use. It also produces a generated resource header file (.rsg), which contains the symbolic IDs of these resources.
- The .rsg file is included in the .cpp file, so that the C++ compiler has access to the symbolic IDs of the resources that will be used.
- C++ class definitions are specified in a source header file (.h).
- Resource struct statements are specified in a resource header file (.rh). An application will include its own .rh file only if it requires its own struct statements.
- Flag values are defined in a file that must be available to both the C++ compiler and the resource compiler: the .hrh extension is

used for this, and .hrh files are typically included in the .h file that defines the classes (for C++), and the .rh file that defines the structs (for the resource compiler).

- The bitmap converter combines the original bitmap files (.bmp) into a multi-bitmap file (.mbm).
- The Aif builder uses .rss and .mbm files to build an application, information file (.aif). The .aif file specifies the icon, caption and embedding capabilities of an application.

4.4 Symbian Community and Development Support

This section provides useful information and a short introduction to the Symbian Community, support networks, and partner programs. Additionally, a number of developer resources, developer networks, and other references are presented as web links – thus providing more extensive support for development ideas and business opportunities for those developing software on the Series 60 Platform. This section is primarily targeted at independent software developers, third-party vendors, and for Series 60 licensees as a means of gaining quick access to other development resources and help not covered in this book.

The Symbian Community was formed in 1998 after the biggest manufacturers of handheld devices created an independent company called Symbian. The goal was nothing less than to create an open and worldwide standard operating system for use in portable devices such as smartphones and communicators. Thus, the Symbian Community was formed to meet the needs of providing development support for a number of different partners, developers, and vendors. Additionally, it was seen as important to share rapidly the knowledge and commitment with all big industry in order to make Symbian OS the standard operating system worldwide. This need was not least proven by competitors such as Palm OS, that had already shown the importance of the wide developer support network and by a large number of third parties who offered content (such as free and commercial applications). Symbian has built up or initiated with licensees a number of support activities within the community concept. Support activities for this concept are listed briefly below and are explained in more detail in the following chapters:

- licensee resources
- Symbian partner programs
- developer resources

- technology, industry, and training events
- development material

4.4.1 Licensee Resources

Symbian has created a special extranet for the licensees of Symbian OS. Also, licensees of the Series 60 Platform are able to access this extranet as Symbian OS is a part of the Series 60 Platform. It holds all current documentation relating to generic technology projects for all Symbian OS releases. The extranet includes, among other things, specifications, plans, schedules, progress reports, and any other useful documentation that licensees may need. Access to this database is obtained by licensing the Symbian OS. Current Symbian licensees and their was addresses are listed in Table 4.1

Table 4.1 Current Symbian licensees and their web addresses

Licensee	Web address
Nokia	www.nokia.com
Kenwood	www.kenwood.com
Siemens	www.siemens.com
Ericsson	www.ericsson.com
Psion	www.psion.com
Sanyo	www.sanyo.com
Matsushita (Panasonic)	www.panasonic.com
Motorola	www.motorola.com
Sony	www.sony.com
Sony Ericsson	www.sonyericsson.com
Fujitsu	www.fujitsu.com

4.4.2 Symbian Partner Programs

Symbian is allies with the providers of key technologies for Symbian OS and has created specific partner programs to clarify the role of each partner. There are seven partner programs: Symbian Competence Centers; Training Partners; Tools Partners; Semiconductor Partners; Technology Partners; Alliance Partners; and Network Operators. Each partner has an area of expertise that adds functionality to Symbian OS or knowledge to the Symbian Community. For example, Symbian Competence Centers, such as Digia Inc, provide, among other things, integration and customization services for Symbian OS phones. For current Symbian partner programs, go to www.symbian.com/partners.

4.4.3 Developer Resources

The list given in Table 4.2 gives addresses of developer networks and other websites providing additional support (which this book does not cover) primarily for independent software developers and vendors. Additionally, some marketplaces are listed for those wishing to sell or buy software that is based on Symbian OS.

Table 4.2 Developer resources

Resource	Web address
Symbian developer network	www.symbian.com/developer
Directories for resources related to Symbian OS	www.symbianpages.com
Forum Nokia	www.forum.nokia.com
Nokia marketplace	www.softwaremarket.nokia.com
Ericsson developer zone	www.ericsson.com/mobilityworld
Motorola developers	developers.motorola.com/developers
Digia services and support	www.digia.com
Java support	http://java.sun.com
All about Symbian OS version ER6	www.allabouter6.com
Tucows marketplace	www.tucows.com

4.4.4 Technology, Industry, and Training Events

Symbian arranges a number of events for licensees partners, and developers. For example, the Symbian World Expo Developer Conference is arranged yearly for developers, and the Symbian Partner Event is put on five partners. Symbian's Technical Training team and Symbian Training Partners (STP), such as Digia Inc, offer courses training engineers within the Symbian Community. Some resources and events are listed in Table 4.3.

Table 4.3 Symbian OS training resources and events

Resource or event	Web address
Symbian OS training	www.symbian.com/developer/training
Symbian OS STP	www.symbian.com/partners/part-train.html
Digia STP Training	www.digia.com/digiawww/home.nsf/pages/Training+Services
Symbian Events	www.symbian.com/news/news-events.html

Note: STP, Symbian Training Partners.

4.4.5 Development Material

In addition to this book, a lot of other material is available to support development related to Symbian OS in addition. Some of these and their web address are listed in Table 4.4.

Table 4.4 Interesting development material related to Symbian OS

Development material	Web address
Professional Symbian Programming book	www.symbian.com/books
Wireless Java for Symbian Devices book	www.symbian.com/books
Symbian OS Comms Programming book	www.symbian.com/books
Programming Psion Computers book	www.symbian.com/books
Code Warrior for Symbian Devices	www.metrowerks.com
Free-of-charge Series 60 SDK, tools, and documents	www.forum.nokia.com
Symbian Developer Network	www.symbian.com/developer

Note: SDK, Software Development Kit.

4.5 Summary

The Series 60 SDK provides a rich set of tools enabling an efficient means of to developing applications. In the build process, Perl scripts are used to create work spaces for the Visual C++ IDE and makefiles for GCC. After that, applications are built either in the IDE or on the command line. Other tools may be used for minimal application creation, bitmap conversion, AIF generation, and for preparation of an installation package. In the next chapter we start writing the source code for the application. In addition to examining the existing applications, the Symbian Community may be used to search for help during the development process.

5

Platform Architecture

The programming characteristics of the Series 60 Platform and the underlying Symbian OS originate mainly from Microkernel, Client–server, Model–View–Controller, and Observer design patterns in addition to idioms related to careful management of resources, such as memory. This chapter gives an overview of the platform architecture from the application engine programming point of view. We concentrate mainly on Symbian OS programming and leave the details of graphical user interface (GUI) programming to Chapter 8. Using the engine of the rock–paper–scissors (RPS) game as an example, we describe processes, threads, libraries, multi-tasking, the client–server framework, filing, memory management, and the use of descriptors.

5.1 System Structure

It is possible to implement three kinds of executables in Symbian OS: shared libraries, provider libraries, and server executables. The type is specified in the project file (.mmp), as shown in the previous chapter. In addition, there are differences in the implementation of the executables themselves. For example, all libraries must have an `E32Dll()` function as one entry point, but server executables require an `E32Main()` function and have no entry points.

Shared libraries have an extension .dll (dynamically loading link) in Symbian OS. Libraries may be dynamically loaded or unloaded, if there is no software component using the library. There is at a maximum one instance of a shared library in the memory at any moment in time. In case of previous software component have already loaded the library, its reference count is increased when another software component uses the same library. The reference number indicates the number of applications sharing the library. When the reference count is decreased to zero, there is no application using the library any more, and it may be unloaded.

The interface of the library is defined by entry points, which are members of a class declared with IMPORT_C macros in the header file and EXPORT_C macros in the implementation file. In the RPS game the engine has been implemented in its own shared library. Thus, there could be several GUI applications sharing the same engine code. The interface of the engine is defined as follows: The most important function is `AddDataFromReceivedMsgL()`, which is called from the document.

```
#include <e32base.h>
#include "RpsMessage.h"

class CRpsMessage;
class CRpsMoves;

// CLASS DECLARATION

/**
* Manages the game data. The protocol allows only one
* outstanding challenge.
*/
class CRpsModel : public CBase
      {
      public:
            IMPORT_C static CRpsModel* NewLC();
            IMPORT_C static CRpsModel* NewL();
            IMPORT_C ~CRpsModel();
      public:
            /**
            * Adds data of received msg to a model.
            * @return ETrue if the message was valid, and EFalse
            * if not.
            */
            IMPORT_C TBool AddDataFromReceivedMsgL
              ( RFileReadStream& aStream );

            /**
            * Checks, whether the player has a challenge message
            * @return iChallenge member variable
            */
            IMPORT_C TBool HasChallenge() const;
            IMPORT_C TBool HasResponse() const;

            /**
            * Sets the challenge member variable, if challenge msg
            * exists
            */
            IMPORT_C void SetChallenge( TBool aChallenge );
            IMPORT_C void SetResponse( TBool aResponse );

            /**
            * Returns challenge moves from the message
            * @return challenge moves (e.g. rock-rock-paper)
            */
```

```
        IMPORT_C const CRpsMoves& ChallengeMoves() const;
        IMPORT_C const CRpsMoves& ResponseMoves() const;
private:
        CRpsModel();
        void ConstructL();
private: // data
        CRpsMoves* iMove;
        CRpsMoves* iResponseMoves;
        CRpsMoves* iChallengeMoves;
        TBool iResponse;
        TBool iChallenge;
        CRpsMessage* iMessage;    // Received message
};
```

The function checks if there is any challenge message sent to this application. If there is, challenge moves are read to the memory and the player creates and sends a response message to the challenger. In case there is no challenge message, the player becomes the challenger of the game. Other functions are used to set and get the role of the player and return the moves in the challenge or in the response.

What is missing in the above example? There is no `E32Dll()` function. The function is implemented in a separate file together with our own `Panic()` function. The `Panic()` function requires two parameters: a category and a reason. Because the category is the same throughout the whole engine, we have implemented a function, that requires only the reason code:

```
#include <e32std.h>       // for GLDEF_C
#include "RpsMessage.h"      // Panic enums

GLDEF_C TInt E32Dll( TDllReason/* aReason*/ )
    {
    return (KErrNone);
    }

void Panic( TInt aPanic )
    {
    User::Panic( KRpsModelPanic, aPanic );
    }
```

Two other classes of the library (`CRpsMoves` and `CRpsMessage`) are declared below. These could be implemented in their own library to increase possible reusability even further, but in the RPS game they are part of the same library as the engine. First, we have defined global enumerations for message type and move type.

The main function of the `CRpsMoves` class is to store and handle individual moves. The type and owner of the move may be get and set with corresponding functions. `AddMoveL()` calls `AppendL()` to store a move in an array. We have used the `CArrayFixFlat` class, because stored elements (moves) have a fixed size. The **flat** array does not have as good a performance as **segmented** arrays in the case of frequent insertions and deletions of elements. We have made an assumption that a typical game consists of a few insertions and, at most, one deletion. The difference between flat and segmented arrays is that the flat array has only one contiguous buffer, including all array elements. A segmented array has several buffers in the memory segmented together.

`Count()` returns the number of moves in an array. This function is used, for example, in calculating the final score, when all elements of an array are gone through. Finally, elements may be deleted with a `Delete()` function.

The received message is stored into a file, which is read and written by the `CRpsMessage` class. It calls `CRpsMoves` to read the moves from the message. Reading and writing are handled by the `InternalizeL()` and `Externalize()` functions. Other classes take care of opening the file store and just pass a stream as a parameter to read and write functions. The code is as follows:

```
#include <e32base.h>
#include <s32file.h>
#include "RpsGameDocument.h"
//CONSTANTS
_LIT(KRpsModelPanic, "RpsModel.dll");

enum TRpsMessageType
    {
    EChallenge,
    EResponse,
    EResponseAndNewChallenge
    };

enum TRpsMoveType
    {
    ERock,
    EPaper,
    EScissors
    };

// FUNCTION PROTOTYPES
void Panic( TInt aPanic );

// FORWARD DECLARATIONS
class RReadStream;
class RWriteStream;

// CLASS DECLARATION
/**
* Contains an unlimited number of game moves, i.e. a Rock, Paper
* or Scissors sequence.
*/
class CRpsMoves : public CBase
    {
```

```
public:
        enum TRpsMessagePanic
            {
            ENoMovesArray,
            EIndexOutOfBounds,
            EChallengerMovesAlreadySet,
            EResponderMovesAlreadySet,
            ENoResponse,
            ENoChallenge,
            ESessionIdNotFound,
            EBadIndex1,
            EBadMoveIndex,
            ELastIsntUnansw,
            EBadType1
            };
        enum ERpsRespOrChallType
            {
            EResponse,
            EChallenge
            };
        enum ERpsMoveOwner
            {
            ESelf,
            EOpponent
            };
public: // construction & destruction
    IMPORT_C static CRpsMoves* NewLC();
    IMPORT_C static CRpsMoves* NewL();
    IMPORT_C ~CRpsMoves();
public:
    /**
    * Returns, whether this is a response or challenge move.
    * @return type of the move.
    */
    IMPORT_C ERpsRespOrChallType Type() const;
    IMPORT_C void SetType( ERpsRespOrChallType aType );

    IMPORT_C ERpsMoveOwner Owner() const;
    IMPORT_C void SetOwner( ERpsMoveOwner aOwner );

    /**
    * Adds a new move to the move array.
    */
    IMPORT_C void AddMoveL( TRpsMoveType aMove );
    IMPORT_C void DeleteMove( TInt aIndex );

    /**
    * @return the number of moves in the move array.
    */
    IMPORT_C TInt Count() const;
    /**
    * Returns the selected move (integer index)..
    * @return move type (paper, scissors, rock) of a selected move.
    */
    IMPORT_C TRpsMoveType Move( TInt aIndex ) const;
    /**
    * Resets the move array.
    */
    IMPORT_C void Reset();
```

```
    IMPORT_C void InternalizeL( RFileReadStream& aStream );
    IMPORT_C void ExternalizeL( RWriteStream& aStream )const;
private:
    CRpsMoves();
    void ConstructL();
private:
    CArrayFixFlat<TInt>* iMoves;
    CArrayFixFlat<TInt>* iResponseMoves;
    CArrayFixFlat<TInt>* iChallengeMoves;
    ERpsRespOrChallType iType;
    ERpsMoveOwner iOwner;
    };
```

`CRpsMessage` represents a data message, which is sent between players. In addition to the moves, it adds (writes) and reads (removes) the sender name field of the message. The message structure is defined in this class. If it is going to change, the `InternalizeL()` and `ExternalizeL()` functions have to be changed:

```
/**
 * Represents a data message that is sent between players.
 * The message can contain a response or/and a challenge.
 * If it contains both,
 * then the challenge is a new challenge.
 * The protocol allows only one outstanding challenge.
 */
class CRpsMessage : public CBase
    {
    public:
        IMPORT_C static CRpsMessage* NewLC();
        IMPORT_C static CRpsMessage* NewL();
        IMPORT_C ~CRpsMessage();
    public:
        IMPORT_C void SetSenderNameL( const TDesC& aName );
        IMPORT_C const TDesC& SenderName() const;

        /**
        * For accessing the challenge moves.
        */
        IMPORT_C const CRpsMoves& ChallengeMoves() const;

        /**
        * For accessing the response moves.
        */
        IMPORT_C const CRpsMoves& ResponseMoves() const;

        /**
        * Reads the message from the file.
        */
        IMPORT_C void InternalizeL( RFileReadStream& aStream );
        IMPORT_C void ExternalizeL( RWriteStream& aStream ) const;

    private:
        void Reset();
    private:
        CRpsMessage();
        void ConstructL();
```

```
private:
    HBufC* iSenderName;
    CRpsMoves* iChallengeMoves;
    CRpsMoves* iResponseMoves;
};
```

If the library is released, it is important to remember to **freeze** the interface to support **binary compatibility**. Freezing (using the command abld freeze) generates a .def file, from which the import library will be generated. The import library is used to map entry functions of the library to ordinals. The library contains an export table, mapping the ordinals to memory addresses (offsets from the beginning of the library) of the entry points. In Figure 5.1 three library types and the use of an import library and export table is shown.

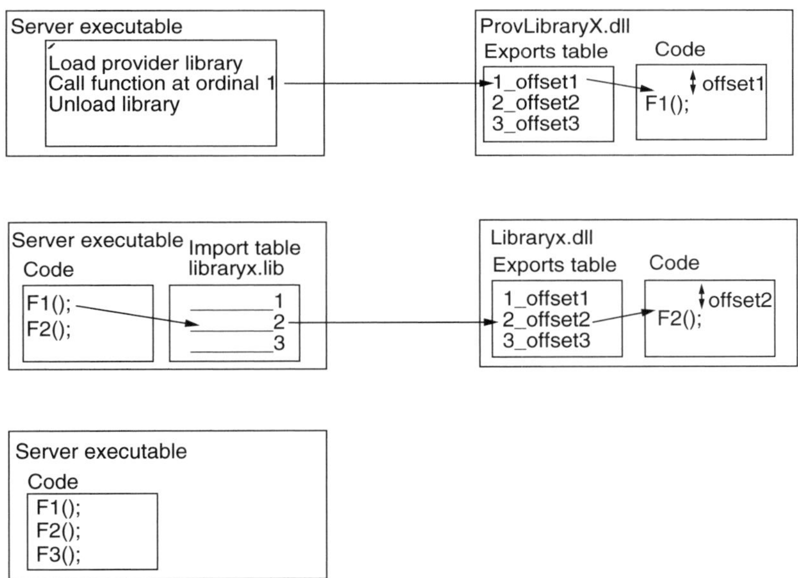

Figure 5.1 Shared and provider libraries and servers

There is one common thing shared by all libraries. They all have to provide an `E32Dll()` function. This function must also be implemented in provider libraries but it must not be the first entry point (ordinal 1) in the provider library. The difference from the shared library is that the interface has already been defined, and it typically contains only one entry function at ordinal 1. This function is called to create an object from a concrete class, which is derived from the abstract interface. There are several types of provider libraries [applications (.app); control panel applets (.ctl); MIME (multipurpose Internet mail extension) recognizers (.mdl); protocol modules (.prt); and device drivers (.pdd and .ldd)]. For application provider libraries, the entry point is the function `NewApplication()`, as shown in the example below; note also the existence of the `E32Dll()` function, although its implementation is empty:

```
EXPORT_C CApaApplication* NewApplication()
    {
    return new CRPSGameApp;
    }

// ---------------------------------------------------------
// E32Dll(TDllReason)
// Entry point function for EPOC Apps
// Returns: KErrNone: No error
// ---------------------------------------------------------
//
GLDEF_C TInt E32Dll( TDllReason )
// GLDEF_C used to prevent E32Dll to
// be at the first ordinal
    {
    return KErrNone;
    }
```

Server executables having an extension .exe do not typically have any entry points, because communication with the client is taken care of by using interthread communication (ITC). A server implements the `E32Main()` function, which provides an application programming interface (API) for the operating system to start the executable. The `E32Main()` function returns the address of the ThreadFunction to be called.

The RPS game uses several servers, for example, for accessing the communications services. One server, which practically all Symbian OS programs use, is the single-threaded kernel server. The kernel server provides, for example, access to device drivers. In addition to the kernel server, also the Window, Socket and File servers are used by the game. More information about using servers is provided later in this chapter, in Section 5.2.5.

5.1.1 Processes and Threads

In the WINS emulator it is not possible to create real processes. Each process definition is changed into a thread definition. **Processes** are units of memory protection, whereas **threads** are units of execution. So, one process cannot directly access the memory addresses of another process, unless a global memory chunk is created (discussed in more detail below). In the target, each application and server creates its own process, but also a user may create new processes. All processes can access the shared libraries and the system ROM.

It is possible to define four priority values for processes in the user mode (`EPriorityLow=150`, `EPriorityBackground=250`, `EPriorityForeground=350`, `EPriorityHigh=450`). This priority is used to calculate the overall priority for a thread, which is created in the process. The developer may also use an absolute priority value, in which case the process priority is not used. There are five priority levels available for threads executed in the user mode: `EPriorityMuchLess`, `EPriorityLess`, `EPriorityNormal`, `EPriorityMore`, and `EPriorityMuchMore`. The kernel has a privilege (i.e. higher priorities for processes and threads).

A thread is a unit of execution. Threads are scheduled based on the thread priorities or round-robin if the threads have the same priority. The scheduling is preemptive, which means that a higher-priority thread may interrupt a thread having a lower priority. In some cases, preemptive multi-tasking is not required; it is sufficient to implement cooperative multi-tasking. This is what we have used in our game, which is actually single-threaded. The switch between threads (especially when the threads are located in different processes, and a context switch is required) is 'heavier' than to schedule active objects, implementing the cooperative multi-tasking framework.

As mentioned above, there are two execution modes: user and kernel modes. The kernel always has a higher priority compared with the processes and threads in the user mode. Between these two modes, there are two libraries, EUser and kernel executive, both of which provide a large set of services to applications. The EUser is used to handle processes and threads, memory management, active objects, etc. The kernel executive is used to access, for example, device drivers. The EUser library is our next topic.

The RPS game is single-threaded. However, active objects are used in communications for cooperative multi-tasking. We will show the code examples in Part 3 of this book.

5.2 EUser: Services for Applications

EUser library provides services for user software. Frequently used services include the client–server framework, memory management, preemptive and cooperative multi-tasking, exception handling, and descriptors. Let us go through all these services with code examples to make Symbian OS familiar to all readers.

5.2.1 Memory Management

Each process must contain at least one thread. The bare process cannot be executed, but threads can. A new thread is created with a `Create()` function, whose signature is as follows:

```
TInt Create(const TDesC& aName, TThreadFunction aFunction,
            TInt aStackSize,TInt aHeapMinSize,
            TInt aHeapMaxSize,
            TAny *aPtr,TOwnerType aType=EOwnerProcess);
```

In the function parameters, a name for the thread, thread entry function, **stack** size (8 kB by default; see E32STD.h), minimum (256B) and maximum (no default value) sizes for the **heap**, and a pointer to the parameter of the thread entry function must be specified. When a thread is created, a new memory **chunk** is allocated for the thread. The chunk is a memory area having contiguous virtual memory addresses. Physically, the chunk contains the required number of allocated memory pages, the size of which depends on the processor architecture (4 kB in ARM).

The lower part of the chunk is reserved for the stack, and the upper part for the heap, as shown in Figure 5.2. The stack always grows downwards, so it is never possible for the stack to grow into the heap area. Heaps may be shared between threads by using a different version of the `Create()` function. Instead of determining the size limits for

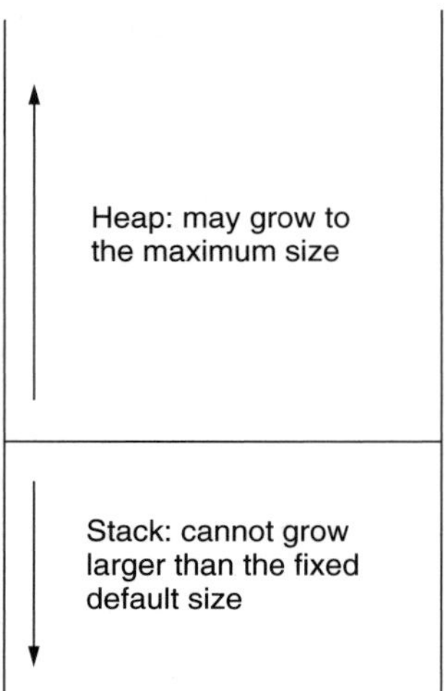

Figure 5.2 A memory chunk allocated for a thread

the heap, a pointer to an existing heap is supplied. This provides an easy way for communication between threads.

Stacks and Heaps

Stacks and heaps play an important role in storing created objects and other data. One should always be careful with the stack, because it may easily overflow. For example, in Unicode one cannot allocate more than 16 `TFileName` objects as the stack will overflow. In case of large memory allocations (`TFileName` is at the limit), it is preferable to use the heap.

All allocations in the stack are automatically deleted when not needed any more. However, for the objects allocated in the heap we must have a pointer to an allocated memory cell to be able to delete it. The `RHeap` class provides several functions for calculating the number of cells allocated in the heap or for determining the total space allocated to the heap. These functions are very useful in testing how the heap is used. Similar functions can be found in the User class as well.

CBase

All objects allocated in the heap should be of the type-C class, meaning that they are derived from a common base class `CBase`. There are two properties shared by all C classes. The `CBase` class provides a virtual destructor that allows instances of derived classes to be destroyed and cleaned up through a `CBase*` pointer. All objects derived from `CBase` can be pushed by `CBase*` pointers into the cleanup stack and destroyed by using the `PopAndDestory()` function of the `CleanupStack`.

There is also a new operator for the `CBase` base class. This operator calls `User::Alloc()` to allocate a memory cell in the heap. It also initializes the `CBase` derived object to binary zeros, i.e. members, whose initial value should be zero do not have to be initialized in the constructor. This, in turn, makes it safe to destruct partially constructed objects.

Most engine classes in the RPS game are C classes.

Cleanup Support

Since the heap is not automatically cleaned up there is a risk that the memory cells are not deleted, which means the application leaks memory. All resources, and obviously also allocated memory cells (and actually the whole memory chunk), are freed when the thread that has created that chunk dies. Of course, we must not use memory management in this way, because it is possible that some applications are run for years before the phone is rebooted. That is why developers should make a leak-free code.

A situation in which a memory leak may take place occurs if we use automatic variables to store pointers to allocated memory cells. If we leave the function in which the variables were defined, they are lost and we have no way to delete the cells anymore. In Symbian OS, the way to store pointers in situations such as this is to use the cleanup stack. The cleanup stack is a special stack for storing automatic variables that are handles (so that they can be closed after a possible leave) or pointers to allocated cells. If a leave occurs, each pointer stored in the stack is used to delete the allocated memory, and no memory leaks occur.

Below, we provide a typical example of two-phase construction and the use of a cleanup stack to create a C class:

```
EXPORT_C CRpsModel* CRpsModel::NewLC()
    {
    // Allocation of the objects in the heap
    CRpsModel* self = new (ELeave) CRpsModel;
    CleanupStack::PushL(self);
    // Allocation of the objects created by this object
    self->ConstructL();
    return self;
    }
```

Developer should use overloaded new operator, new `(ELeave)` instead of just new. The `ELeave` parameter tells us that this operation (allocation for object) may leave. By using `ELeave` the developer does not have to check whether the return value of the constructor is `NULL` or some other valid value. This problem is bigger in compound objects because although the allocation is successful for the object itself we cannot be sure, whether all other objects owned by this object have been properly created.

If the object has been successfully constructed, a pointer to that object is pushed into the cleanup stack. We then execute lines of code, which may cause a leave. If the leave has not occurred, the pointer is popped from the cleanup stack after we are sure no leaving can cause the loss of that pointer.

The second-phase constructor, shown below, creates new objects owned by the engine:

```
void CRpsModel::ConstructL()
    {
```

```
iChallengeMoves = CRpsMoves::NewL();
iResponseMoves  = CRpsMoves::NewL();
iResponse = EFalse;
iChallenge = EFalse;
iMessage = CRpsMessage::NewL();
}
```

Note that it would have been safe to assign two T classes already in the default constructor, because space has been allocated for them before assignment. However, challenge and response moves as well as an instance of the message are created in the constructor. All C classes are instantiated using `NewL()` or `NewLC()` factory functions instead of the default constructor. This ensures that memory allocations are made in a safe way.

Allocations in the stack are automatically freed, but the developer has to take care to free allocated cells in the heap memory. Destructor functions take care of deleting allocated heap objects:

```
EXPORT_C CRpsModel::~CRpsModel()
  {
  delete iMove;
  delete iResponseMoves;
  delete iChallengeMoves;
  delete iMessage;
  }
```

Note that the constructor should never trust that the allocation of an object has been complete. Trying to delete a non-existing object is a programming error that is why in `CBase`-based classes members are initialised to 0.

5.2.2 Exceptions

Exceptions are used in Symbian OS in a different way than they are used in standard C++, because the gcc version 2.7.2 used earlier did not support C++ exceptions. Symbian's TRAP macro and `User::Leave()` calls are analogous to try/catch and throw in C++. Any piece of code that might leave (i.e. call `User::Leave()` or any other leaving function) is executed inside the TRAP harness. When an error occurs, (e.g. memory allocation fails), `Leave()` is called, which obviously leaves the executed function, unwinds the call stack, and returns the most recent TRAP harness, which catches the exception. The software may try to recover or may then just panic (i.e. kill the current thread).

There is a clear difference between exceptions and panic situations. Exceptions are used in case, where it could be expected that something will go wrong. The most typical example occurs when we cannot allocate memory because of the `Out-Of-Memory` (OOM) situation. Another situation could be that the requested resource is temporarily unavailable. What should you do? You should not panic, because a temporal denial of service is not an error. Thus, you should throw an exception and handle the exception in the trap.

In cases where there is nothing to do to recover the software, panics are used. Panics typically reveal the existence of a programming error. A common error is that the developer exceeds the limits of a descriptor, for example.

The example code below shows, how a programming error may be revealed:

```
EXPORT_C TInt CRpsMoves::Count() const
    {
    __ASSERT_ALWAYS( iMoves, Panic(ENoMovesArray) );
    // has ConstructL been called?
    return iMoves->Count();
    }

void Panic( TInt aPanic )
    {
    User::Panic( _L("RpsModel.dll"), aPanic );
    }
```

The `Count()` function returns the number of moves if an instance of `CArrayFixFlat` has been created. It is a programming error, if the array does not exist, when the function is called. Asserts are very useful for checking the soundness of the external variables of the class. Note, also, the use of our own `Panic()` function, described earlier. By using it there is no need to specify the same category (RpsModel.dll) in each function call.

The second phase constructor for the `CRpsMoves` class is as follows:

```
void CRpsMoves::ConstructL()
    {
    iMoves = new (ELeave) CArrayFixFlat<TInt>(KNumberOfMoves);
    }
```

So, a panic occurs in `Count()` if the array to store the moves has not been created. However, if the creation fails in `ConstructL()`, a leave occurs already there. The constant `KNumberOfMoves` specifies how many moves must be selected before a challenge can be created or responded.

5.2.3 Descriptors

Descriptors are frequently used in Symbian OS programming instead of normal C++ strings. There are five descriptor classes that can be instantiated: `TBuf`, `TBufC`, `TPtr`, `TPtrC`, and `HBufC`. `TBuf` and

TBufC are called buffer descriptors and their differ from TPtr and TPtrC pointer descriptors with respect to the owned data. Buffer descriptors own the data, but pointer descriptors have only a pointer to the data located somewhere else (in a buffer descriptor or memory address). Descriptor classes ending with a letter C have a constant API, which means that the API does not contain functions to change the contents of the descriptor. All descriptors except HBufC can be allocated in the stack. HBufC is allocated in the heap, although it is not C class. Note that there is no HBuf descriptor. The contents of the heap descriptor may be changed through an instance of TPtr.

An example of using the heap descriptor is as follows:

```
EXPORT_C void CRpsMessage::InternalizeL( RFileReadStream& aStream )
    {
    Reset();
    aStream.ReadUint32L();
    // sender name
    TUint8 nameLength = aStream.ReadUint8L();
    if( nameLength )
        {
        HBufC* nameBuf = HBufC::NewLC( nameLength );
        TPtr namePtr( nameBuf->Des() );
        aStream.ReadL( namePtr, nameLength );
        CleanupStack::Pop( nameBuf );
        iSenderName = nameBuf;
        }
    // Challenge moves
    iChallengeMoves->InternalizeL( aStream );
    iResponseMoves->Internalize( aStream );
    }
```

We will use stack descriptors frequently in communication engines in Part 3 and will not repeat the same code example here. The InternalizeL() function reads the sender of the message, after which challenge and possible response (in case this is a response message) moves are read from the stream by using the corresponding functions of the CRpsMoves class.

The Reset() function empties previous moves from the move array. The first part of the message is an unsigned byte indicating the length of the sender's name. The name is read from a file to a heap descriptor. First, we create a descriptor object and allocate enough space for it to store the name. The Des() function returns a modifiable pointer descriptor to the data in the heap descriptor. This allows a change of contents of the buffer, although the programming API of HBufC does not allow it. ReadL() reads the name to the descriptor, which is then assigned to iSenderName member. Note that NewLC() leaves a pointer to the created heap descriptor in the cleanup stack, and it must be popped by the developer. Finally, the moves are read to move arrays by using the corresponding functions of the CRpsMoves class.

Package Descriptors for Interthread Communication

Descriptors have an important role in interthread communication (ITC). Although applications use ITC frequently, for example, when using operating system server, software developers may not have to use ITC mechanisms often, because it is embedded in the client-side API libraries of the servers. The basic function for ITC is SendReceive(), which is provided by the RSessionBase class. RSessionBase is a session handle that enables clients to communicate with servers in other processes. Details of the client–server framework will be described later in this chapter (Section 5.2.5).

The SendReceive() function takes two parameters, first of which is an integer identifying the requested service in the server. The second is a pointer to a TAny array of four elements. The elements are used to send and receive data between a client and a server. The size of each element is 32 bits. Thus, the maximum amount of data to be transferred is four times 32 bits. It is possible to send larger amounts of data by using descriptor addresses. Data of other classes can be sent as well, but the data must be first packaged into a descriptor. And this is exactly, what the package descriptors do.

There is one restriction in packaging data to descriptors. It is possible to package only flat data structures, (i.e. T classes and anything that does not have pointers to other classes or data). We have not used package descriptors in the RPS game, but a general example of their usage is as follows:

```
class TUser
    {
    public:
        TUser( void );
        TUser( const TDesC& aNickName );
        void SetUserName( const TDesC& aNickName );
        /* Other functions to get the user name */
    private:
        TBuf <KMaxNameLength> iNickName;
    };

void CClassDerivedFromRSessionBase::SendToServer( void )
    {
    // typedef TPckg<TUser> TUserPckg; Defined in header file
    TUserPckg* packageDescriptor;

    packageDescriptor->Set( iUser, sizeof(TUserPackg) );
    iMsgArgs[0] = (TAny*) packageDescriptor;
```

```
SendReceive(3, imsgArgs);       // 3 identifies the requested
// function
// Not a complete example
}
```

`TUser` is a class storing the nickname of a user in the chat application. We create a new type using a templated `TPckg`. Then we create a new object of this new type and store the content of an object `iUser` (of `TUser` type) to the created object. The address of the created package descriptor can be used in the `SendReceive()` function.

5.2.4 Multi-tasking with Active Objects

Threads may be used in Symbian OS for preemptive multi-tasking, but there is another mechanism – active objects for cooperative multi-tasking. Active objects are instantiated from a class derived from `CActive` or directly from `CActive`, which provides virtual functions (`RunL()`, `DoCancel()`) and member data [`iStatus (TRequestStatus)` and `TBool (iActive)`] used by the active objects framework. The active objects are executed inside one thread, which means that their scheduling does not require a change of thread or process context (which may, of course, happen in addition to active object scheduling). This makes active objects a 'lighter' technique compared with threads. Active objects are not pre-emptive; that is, the `RunL()` is executed to the end once the framework calls it. Thus, `RunL()` should be executable quickly in order not to block other active objects. `DoCancel()` is called by `Cancel()`, and developers must provide their own implementation of this function to cancel an outstanding request.

The active objects have their own scheduler – active scheduler. Before using this, one must to install the scheduler into the thread, issue a request from an active object, and call `Add()` function to add the active object into the scheduler. The last thing one should remember to do is to start the scheduler.

Typically, an active object is using an asynchronous service provided by some server. The active object has a handle to that server. Through the handle, the object has access to the server's interface and the object may issue asynchronous requests. There is always a parameter of type `TRequestStatus` (`iStatus`) in the asynchronous requests. When the request has been issued, the service provider changes the value of `iStatus` to `KRequestPending`. The active object must also call `SetActive()` to set `iActive` to `ETrue`. This notifies the active scheduler that this active object is waiting for a request to be completed.

The event loop has been implemented in the active scheduler. When the scheduler is started (and an active object has issued a request), the scheduler calls `User::WaitForAnyRequest()` function. This call decreases the value of the semaphore of the thread by one. If the value becomes negative, the thread starts waiting for any completed requests from the server.

After completing the service, the server signals the thread that requested the service. If the semaphore value becomes zero or positive, the thread may start execution. In addition to the `Signal()` call, the server also changes the value of the `iStatus` variable. If everything has succeeded, the value `KErrNone` is assigned to the variable. The active scheduler selects the active object, that has issued a request and has the highest priority. It calls the `RunL()` function of this object and the object may then handle the completion of the service. `RunL()` should be short, because it cannot be preemptied by other active objects in the same thread.

If the active object does not want to wait for the completion of the request, it should call `Cancel()`. This function calls `DoCancel()` implemented by the user. Typically, in the `DoCancel()` function, resource-specific cancels are called, and any handles that are open are closed. Figure 5.3 shows the whole lifecycle of the active objects.

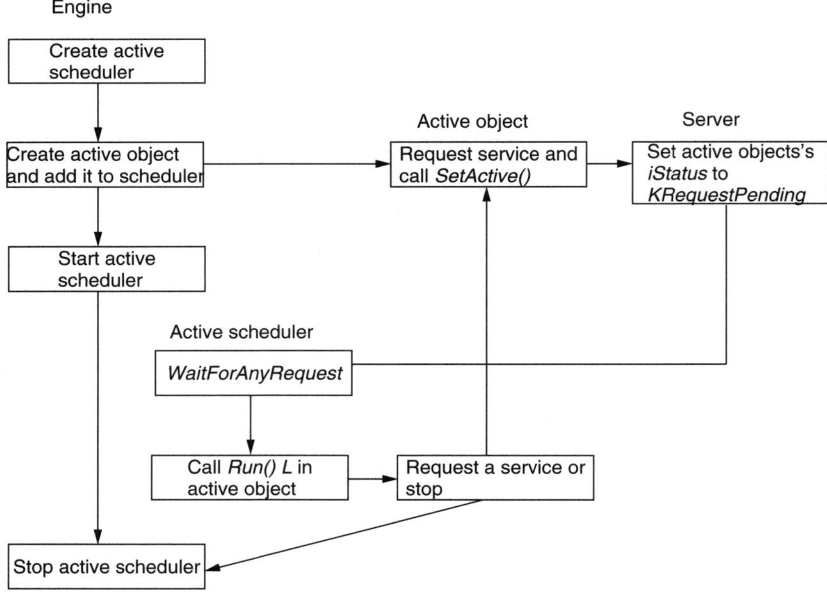

Figure 5.3 Lifecycle of active objects

5.2.5 Client–Server Framework

The use of servers is typical of microkernels. With microkernels, they help to keep the system small. Servers also allow the isolation of resources and a safe way to share resources compared with the situation in which each client uses the resource itself.

Servers are implemented in two parts. The server itself is implemented as an executable, for which a new process is created (in target builds). The other part is the client-side API, providing access functions to communicate with the server. The client-side API is implemented in a shared library, which any software component may use. The most important class is derived from the `RSessionBase` class, which provides a `CreateSession()` function. This function is used to create a session (i.e. a communication channel between the client and the server).

There is one drawback in the use of servers. Because servers are executed in another process, there is a context switch between the client and the server processes. The context switch in Symbian OS means storing the data from the data section to the home section when a new process is restored from the home section to the data section. The home section occupies the lower memory addresses of the address space.

Filing

Filing is also implemented according to the client–server framework in Symbian OS. In the RSP game we use files to read received messages. This is implemented in the `TryLoadDataWrittenByMsgViewerL()` function. The message application stores a message into a specific file, defined by the `KMsgFileName` constant. To access this file, we first need a session to the file server. The session is created with a `Connect()` function, which calls `RSessionBase::CreateConnection()`. After that we create a subsession. The creation is embedded in the `Open()` function of the `RFileReadStream` class. If the file cannot be found, there is no challenge, which means that we are starting the game. In the other case, the engine reads the contents of the message.

Note that the reading function in the engine may leave. That is why we should push the file session and file read stream handles to the cleanup stack. If the leave occurs and we lose the handles there is no way to close them but kill the thread. Because the engine is single-threaded this would mean killing the whole engine. By using the `CleanupPushL()` function, the `PopAndDestroy()` function takes care of closing the handles automatically:

```
void CRPSGameDocument::TryLoadDataWrittenByMsgViewerL()
    {
    TFileName name(KMsgFileName);

    RFs fsSession;
    fsSession.Connect();

    CleanupClosePushL(fsSession);
    RFileReadStream readStream;
    if ( readStream.Open( fsSession, name, EFileRead )
    != KErrNone )
        {
        // Couldn't find the file, so start a new game.
        iModel->SetChallenge(EFalse);
        CleanupStack::PopAndDestroy();
        return;
        }
    iModel->SetChallenge(ETrue);
    CleanupClosePushL(readStream);
    iModel->AddDataFromReceivedMsgL( readStream );

    CleanupStack::PopAndDestroy(2);
    // readStream and
    // file session
    }
```

The object of type `RFileReadStream` is given to the engine from the document. Engine calls the `Internalize()` function of the `CRpsMessage()`, which reads the sender, as shown earlier in the section "Descriptors", and calls `InternalizeL()` of the `CRpsMoves` class shown below:

```
EXPORT_C void CRpsMoves::InternalizeL
  ( RFileReadStream& aStream )
    {
    TUint8 count = aStream.ReadUint8L();
    for (TInt n = 0; n < count; n++)
        {
        AddMoveL( (TRpsMoveType)aStream.ReadUint8L() );
        }
    }
```

The implementation is rather straightforward. The first byte indicates the number of moves and each move occupies one byte.

5.3 Rock–Paper–Scissors Engine Functionality

We have shown the most important code samples from the RPS engine and will show a few more to clarify how the game works. Everything starts from the document class, which creates the engine and first checks, if there is a message sent from the other player. If there is a message, the player's role is responder. Otherwise, the player is challenger. The roles are set and get with trivial functions, as follows:

```
EXPORT_C TBool CRpsModel::HasChallenge() const
    {
    return iChallenge;
    }

EXPORT_C TBool CRpsModel::HasResponse() const
    {
    return iResponse;
    }

EXPORT_C void CRpsModel::SetChallenge( TBool aChallenge )
    {
    iChallenge = aChallenge;
    }

EXPORT_C void CRpsModel::SetResponse( TBool aResponse )
    {
    iResponse = aResponse;
    }
```

If the player is challenged, the challenger moves are read using the following function. First the moves are read into the member of the `CRpsMessage` class and copied from the message class to the engine one move at a time:

```
EXPORT_C TBool CRpsModel::AddDataFromReceivedMsgL
( RFileReadStream& aStream )
    {
    iMessage->InternalizeL( aStream );
    TInt nof_moves = (iMessage->ChallengeMoves()).Count();
    for( TInt i = 0; i < nof_moves; i++ )
        {
        iChallengeMoves->AddMoveL(
          (iMessage->ChallengeMoves()).Move( i ) );
        }
    return ETrue;
    }
```

The `AddMoveL()` function adds `TRpsMoveType` objects into a flat dynamic array capable of storing fixed-size elements (`CArray-FixFlat`):

```
EXPORT_C void CRpsMoves::AddMoveL( TRpsMoveType aMove )
    {
    iMoves->AppendL( (TInt)aMove );
    }
```

The storing is handled by the AppendL() function. If the array is full, AppendL() tries to expand the array buffer. It there is not enough memory, it leaves. In the other case, the array is increased by its granularity, for example, by five elements. That is why the CArrayFixFlat array is called a **dynamic array**. There are other classes for storing elements whose size changes and to store elements in a segmented rather than in a flat structure in the memory.

The functionality of the CRpsMessage class is almost nothing else but 'read and write the message contents to the file'. SetSenderNameL() is used to set the sender's name:

```
EXPORT_C void CRpsMessage::SetSenderNameL( const TDesC& aName )
    {
    HBufC* name = aName.AllocL();
    delete iSenderName;
    iSenderName = name;
    }
```

Other functions return the challenge-and-response moves and are implemented in the same way as the corresponding functions in the engine. The UI part of the application will be described in Chapter 8.

5.4 Summary

Compared with other operating systems, Symbian OS is rich with tricks and techniques which help the development of more robust software and make programming easier. These techniques originate from microkernel architecture and strict requirements set to software components with respect to memory management, for example. Symbian OS programming involves use of system servers, management of memory to prevent memory leaks, an own exception handling mechanism, descriptors, active objects, and so on. As we have shown in this chapter, even the implementation of a simple application engine requires that software developers be familiar with these techniques, not to mention GUI programming and communications programming, discussed later in this book. The next chapter is the first step towards GUI design, after which we discuss software testing before moving on to GUI programming.

6

User-centered Design for Series 60 Applications

User experience encompasses all aspects of the end-user's interaction with the company, its services, and its products. The first requirement for an exemplary user experience is to meet the exact needs of the customer, without fuss or bother. Next comes simplicity and elegance that produce products that are a joy to own, a joy to use (Nielsen and Norman, 2002)

In recent years, usability or user experience of the product has emerged as one of the key elements of market success in the software industry. Even though everybody can nowadays admit that a product must be easy to use, good user experience cannot be taken for granted. The product and its user experience have to be designed.

The discipline of designing usable products is called user-centered design (UCD). There has been much research in this area, and in addition to being termed UCD (Bevan, 2002), it is called usability engineering (Mayhew, 1999; Nielsen, 1994), contextual design (Beyer and Holtzblatt, 1998), human-centered development (Norman, 1999) or interaction design as termed by Alan Cooper (Nelson, 2002). The fundamental goal of these design processes is the same, although the methods may differ.

In the following sections we discuss how UCD can be applied in the application development for the Series 60 Platform. The first section introduces the overall process for UCD. There are then sections describing studies on end-user needs (Section 6.2) and concept design workshops (Section 6.3). Section 6.4 focuses on the challenges of interaction design for smartphones. Section 6.5 elaborates the process of writing user interface (UI) specifications. Finally, the setting of usability requirements and verifying usability are discussed (Section 6.6). An image editor application (Digia ImagePlus) developed by Digia is used as a case-study example of applying UCD, and an example of a UI specification is given in Appendix at the end of this book.

6.1 User-centered Design Process

Some important aspects of applying UCD successfully must be underlined: one is the iterative nature of product design; another is the multi-disciplinary teamwork. Also, the organization should build relevant user-experience expertise and skills.

6.1.1 Iterative Development

The nature of UCD involves iterative design and evaluation cycles, required to get the user experience of the product right. The process includes not only the design of the final user interface but also finding the fundamental needs for the product, setting the requirements for the development, and designing the product iteratively from concept to final user interface and user manual. Iterative evaluation means that the end-users verify the product prototypes as early as possible. In the early iteration cycles, the interaction and structure of the product is the focus. After the product meets the requirements of targeted work and user needs, designers can focus on detailed UI issues and usability verifications, and engineers can start the implementation project. The design requirements often need to be clarified during the design cycles from early design drivers to detailed usability requirements of the product. The UCD process is depicted in Figure 6.1 (ISO 13407, 1999).

How Many Design Iterations are Needed?

Since the iterating concept design requires time and resources, the design project should reserve time for the design iterations. The minimum requirement should be that the end-user evaluations are conducted once in the concept design phase to fix the product structure and once in the usability verification phase to fix UI details.

6.1.2 Multi-disciplinary Teamwork

To get the best results and fast iteration cycles, UCD requires multi-disciplinary teamwork. A good team includes representatives from marketing, engineering, and user experience functions. In the early concept design phase, the entire team gathers user-needs data and designs the concept in workshops iteratively until the structure of the product starts to become clear. After that, the team members may focus on detailed specification work in each area of expertise and meet regularly for clarifying the details and reviewing the specifications.

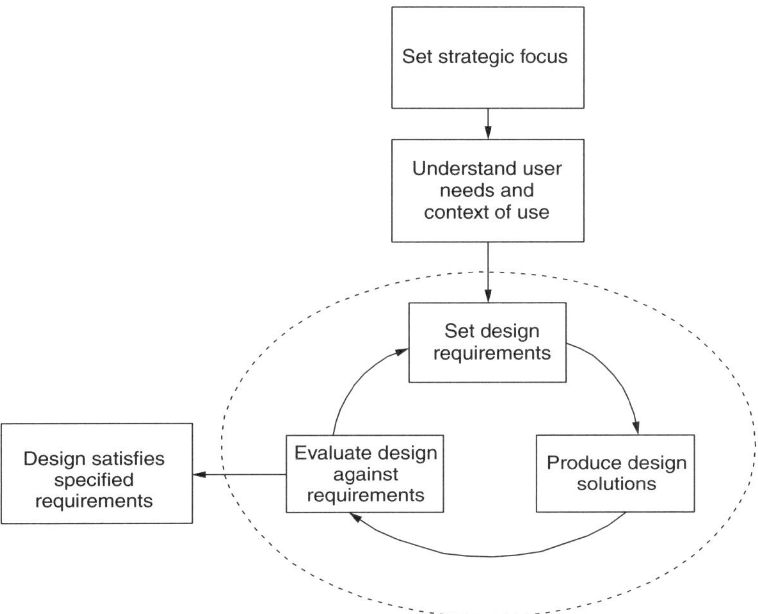

Figure 6.1 The user-centered design process

During the development phase, while the engineering project is developing the product based on the specifications, and marketing is focusing on product management, the user experience team designs UI details and graphics and verifies usability with developed prototypes or early product versions. Commonly, the display texts are localized and the user manual is written in the later phases of development. A solid change management process is needed to take care of those inevitable changes that occur when usability is verified.

6.1.3 User Experience Skills in Organization

Achieving a good user experience for the product requires expertise in product design and human factors. The following role names describe user-experience skills that should belong to a product development organization:

- User-needs researchers lead the multi-disciplinary team in user-needs studies and task analysis. Their background can be from sociological or psychological sciences.
- Interaction designers produce design solutions that meet user needs. They create design solutions in concept workshops and define detailed UI specifications.

- Graphic designers make the look of the product elegant and pleasing. Their skills are needed especially in products that contain new UI elements or graphical content.
- UI prototypers create rapid simulations of the developed products. With prototypes, usability verifications and design corrections can be made sufficiently early, before the actual product is ready.
- Usability specialists organize usability requirement setting and verification for the product. These may be the same persons as the user-needs researchers.
- User manual writers have skills for writing customer documentation for the product. They may also have responsibility for finishing and localizing the display text language.

Often the same person may take several roles and responsibilities in a design and development project. To get the best results, there should be at least two user-experience persons in the project working as a pair and helping each other in creating and verifying the user experience. A good pair consists of a person with usability evaluation skills and a person with interaction design skills.

UCD does not necessarily need to be a heavy and resource-intensive exercise, but, even with small efforts (e.g. the creation of paper prototypes) the design team can achieve better results than doing nothing in relation to end-users needs.

6.2 Understanding End-user Needs

Series 60 technologies enable many new product ideas, so developers may be tempted to rush in creating new products. However, each developer should wait a second and ask a few questions before starting the product development. Why is this product being created? Who is it meant for? This section discusses the importance of understanding end users' needs before creating any design.

Even if the product idea is something totally new and innovative, the developers of successful products should understand the needs of people's current life and create solutions that meet those needs. The key point is that, in the future, people will be doing the same things as nowadays with the developed product, or at least their intentions and goals will be the same. The way of working may not be exactly the same, but it will, it is hoped, be more effective, fun, or pleasing by using the developed application. The development team needs to find those intents and goals and base its design decisions on those and on enabling technologies. The team's responsibility is to develop a product that people can conveniently adapt to their lives and duties.

In the next section, the methods for observing end users and analyzing user-needs data are briefly introduced.

6.2.1 Observing People in a Mobile Environment

User-experience researchers have developed many useful methods for gathering good user-needs data for product development (Beyer and Holtzblatt, 1998; Dray, 1998). These methods apply the fundamental wisdom that finding user needs requires the observing and interviewing of end users in the places where they live and work. People cannot remember detailed needs in focus-group meetings or laboratory interviews. Seeing users in real-life situations helps designers to design products to fit into that context.

It is important that the observed users represent well the targeted end-user group of the product. First, the development team (i.e. the multi-disciplinary design team) clarifies 'the target users'. Who are the target users that will use the product? In which work or life situations will they need the product? Then, the team defines a focus for a user-needs study: the information they need to know of the user's life and tasks, related to the developed product. Having a focus helps in seeing the right things; the things may or may not be those expected. After that, the team can organize site visits to real-life situations and find out what people actually are doing and need.

Field-observation methods have been used successfully in office and home environments (Dray and Mrazek, 1996; Ruuska and Väänänen, 2000). They can be also used in a mobile and outdoor environment (Hynninen *et al.*, 1999; Wikberg and Keinonen, 2000). All that observers need are two focused eyes, a pen, and a notebook, as Figure 6.2 shows.

When observing a mobile person (e.g. a cyclist) a dictaphone and small digital camera are handy tools to capture information 'on the go'. Other ways to observe mobile or busy workers is to split the visit into an observation session and an interview session; the user is observed first while working and is then interviewed afterwards during a break. The work situation can also be videorecorded, and the recording be watched together with the end user as a basis for interview.

Even if the user-needs researchers are experts in field observation, everybody in the design team should have the opportunity to see end users. It should not be the privilege of the user-experience or marketing persons to meet the customers. Each team member may make one or two visits, or the team can work in pairs. The observers usually see different things, depending on their background (Dray, 1998). This helps to create a well-covered and shared understanding of user needs.

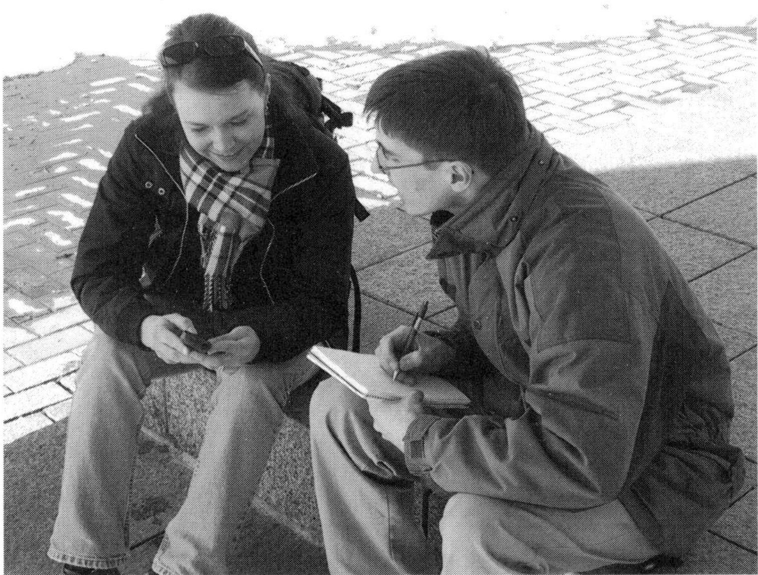

Figure 6.2 A mobile phone user (left) being observed 'in the field'

6.2.2 Analyzing User-needs Data

The information that the design team obtains from field observation is qualitative data. The data include stories about users doing tasks step by step, insights into information and communication flows in organizations or communities, pictures or drawings about users' environments sketched by the observer, copies or printouts about users' current tools, screen shots, and other relevant findings related to the mood or lifestyle of the users.

The data requires qualitative data analysis methods to bring the most relevant things to light to guide design work. The contextual design process provides some practical methods that Digia has found useful for designing applications and user interfaces for smartphones (Beyer and Holtzblatt, 1998).

Qualitative data analysis is most efficient when done by the entire team using paper, pen and post-it notes. Thus, the analysis requires resources and committed teamworking, but it is useful for 'priming the brain' for new ideas. The design team cannot prevent good design ideas popping up while creating shared understanding of user needs. Putting the data on the design room walls helps in consolidating the data. Figure 6.3 shows that the amount of data usually covers a lot of wall space.

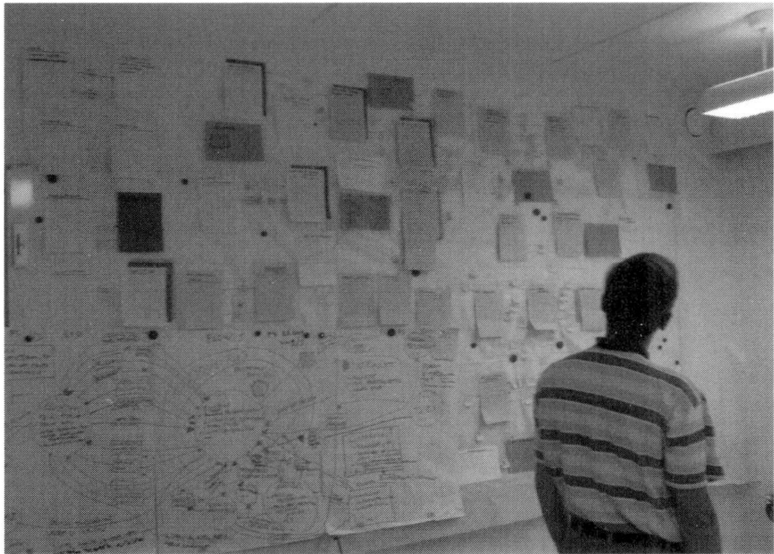

Figure 6.3 Data display on a design room wall: observing end users produces a lot of important data

The analyzed user-needs data is then a starting point for concept design, which is discussed next. The data can be transformed into electronic format for later use and sharing. Standard word-processing and chart-drawing PC tools can be used.

6.3 Concept Design Workshops

Requirements for a new successful product should not just be 'pulled out of the hat'. Organizing concept-design workshops after user-needs studies is a good way of developing product ideas into a meaningful concept definition and collecting requirements for the product. The workshops should be organized in iterative cycles, including design and end-user evaluation. Multi-disciplinary teamwork ensures that all the important viewpoints are taken into consideration. With focused workshops, the company may develop the proof of concept within a few calendar weeks.

It is worth reserving a few work days for one workshop so that the team can produce a design proposal and evaluate it with prototypes in the same workshop, or even iterate the design several times. We have observed that a team can work efficiently for a maximum of 3 to 4 consecutive days in one workshop, and then it is good to have, for instance, a week's break before continuing.

Methods for concept design workshops will be discussed in the following sections. The content refers a lot to the contextual design process that is used and applied in Digia's product design process. The workshops can be divided into the following phases: brainstorming, storyboarding, draft functional specification, and paper prototyping with end-user evaluation.

6.3.1 Brainstorming Product Ideas

In a hectic product development lifecycle it is quite easy to skip the user-needs studies and start a new product development by brainstorming the product ideas. If the earlier phase is skipped, the company may lose the opportunity of finding something special that makes the developed product move far ahead of competition, or the development team may develop a product that nobody wants to use. The product may lose the essential ingredient that can be seen only after the big picture is known. The user-needs studies also give relevance to the product ideas, which the management board requires to approve those ideas to move on to the development phase.

Concept design starts by brainstorming new ideas. The atmosphere of brainstorming should be open and free, not evaluative, so that everybody in the team can bring smaller and larger ideas to the design table. Everybody should have a positive attitude for developing and constructing new ideas on top of the ideas of others. To lead the brainstorming in the right direction, the analyzed user-needs data should be placed on the design room wall so that the brainstorming can be focused on the data. If the design team does not have a dedicated design room, a company meeting room can be reserved for a few days.

Next, a few possibilities for brainstorming topics are listed: design ideas, scenarios, and design drivers.

Design Ideas

Reading the user-needs data on the wall, each team member can develop dozens of concept, technology or UI ideas that solve the user needs or problems. It is good to start from individual idea creation with design ideas and to continue with the team vision, as described in the next section.

Vision

A useful target for brainstorming is to create a team vision; that is, scenarios about the developed product as used in the future. The team creates stories about end users using the developed product or system. The best design ideas can be gathered as starting points for the scenarios. One team member starts the story, describing the end user and the task, and the others continue adding new ideas, while the vision is drawn out on a flipchart. When the brainstorming is based on the user-needs data it usually stays on the right track. If the design team is developing a product that is part of a larger system, it creates a story that covers the entire system in a way that is coherent within a user's life.

Design Drivers

One brainstorming possibility is to pick up the design drivers from the user-needs data (Wikberg and Keinonen, 2000). The design team can identify a number of design drivers that emerge from the user-needs data. One example of a design driver for an image editor is, for example, the importance of emotional and humorous messaging among teenage users. These drivers are the design guidelines that are kept in mind throughout the development. They can be the starting point for usability and design requirements that are formalized with more detail after a few concept design iterations. Setting usability requirements is discussed in the section on usability verification (Section 6.6).

6.3.2 Storyboarding

After brainstorming, the team continues to design the detailed task sequences of the new product. They create storyboards or high-level use cases from each intended task for the planned product. In these step-by-step storyboards, each piece of paper describes one step of the user's task or the response of the product to the user's action. When the storyboarding is based on the task sequences from the user-needs study, the new design can be adapted to natural task flows and context of use.

One does not need to be an artist to create simple storyboards, as seen in Figure 6.4; everybody in the team can participate in the drawing. A larger team can be divided into subteams, and each subteam can be assigned a different task to be drawn. The created storyboards are then evaluated by the entire team, issues are collected, and stories fixed. It is important that the evaluation be focused on functional and interaction problems; detailed UI issues should be handled later, after the structure of the product is clear.

After the evaluation, the team can continue to analyze the product structure and create a draft functional specification, as described in the next section. Storyboards can also be finalized by graphic designers and can be used for communicating product ideas to management.

6.3.3 Functional Specification: User Environment Design

Storyboards describe the product through high-level use cases. Another useful item of output is a functional specification that describes the structure and functionality of the product, without the UI elements.

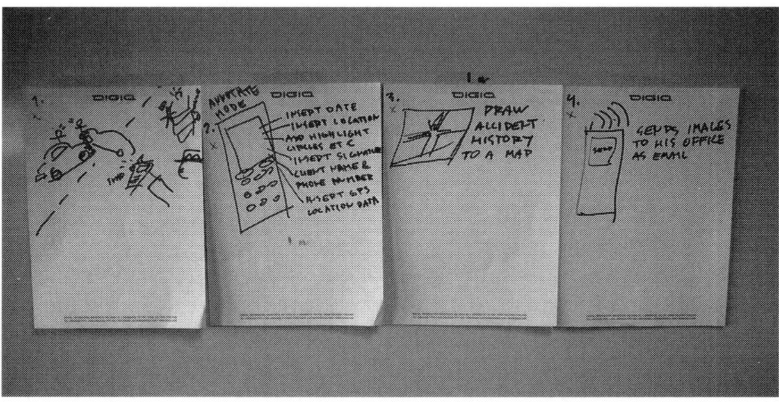

Figure 6.4 Storyboard for an image editor application

A good way to specify and evaluate the structure of the application is to create a user environment design (UED), with a functional specification notation used in contextual design. UED is a quick way to write down the functionality of the design. It allows the team to concentrate on designing, not documenting, and to move on to paper prototyping. The UED structures the states of the product into focus areas, based on the coherent work or task that the user is doing. The UED is created from storyboards by identifying the focus areas, assigning the user and system functions to each focus area, as well as highlighting the links between focus areas based on the natural task flow. It excludes detailed UI elements, thus turning the focus onto structural issues. The UED is maintained during the concept iterations and it can be the basis for the detailed UI specification.

A well-structured UED is a good starting point for designing UI variants to other platforms. Engineers like the UED, since it describes the essential requirements in a form that can be used for detailed use cases and architectural design.

A Quick Introduction to the Notation of User Environment Design

UED is a notation format that can be used to describe a user interface very quickly. Table 6.1 provides a brief description of UED notation. For more detail, see Beyer and Holtzblatt, (1998).

6.3.4 Paper Prototyping with End-user Evaluation

As soon as the first storyboards and the UED draft are on paper it is time to get the first end-user feedback on the product; this is before anybody gets fixed in the design and before a lot of technical design is done.

Table 6.1 Notation used in user environment design

Heading or symbol	Description
Focus area	The focus area is a collection of functions that are a coherent part of the user's work
Purpose	This describes briefly, in the user's words, what work is done in the focus area
○	An open circle indicates a system function; this refers to the case where the system does something automatically (e.g. displays a screen or plays a tone)
●	A closed circle indicates a user function; this is invoked to do the user's task (e.g. a command in the Options menu)
(global function)	A function is written in parentheses when it is needed in multiple focus areas [e.g. (copy/paste function)]
→	A single arrow indicates a link between two different focus areas (e.g. as when selection of a list item opens a new view with different functions)
→→	A double arrow indicates a double link, where two different focus areas exist in the same context. This is not very common in the Series 60 Platform because of the small screen size but some examples exist (e.g. an active call indication is displayed at the top right-hand side of the screen, while a calendar is open at the same time)

A quick way to evaluate a concept design is to create a paper prototype. It is a very simple tool and is a not very 'cool' technology but it suits well the concept evaluation purposes (Snyder, 2001). The design team sketches draft UI layouts on the basis of the storyboards and the UED. Post-it notes are good for representing Series 60 displays, and the phone mockup can be made of cardboard. Figure 6.5 illustrates a startup view of the image editor paper prototype. Each focus area usually means one application view on the display. User functions can be assigned to menu options or key commands. System functions are items on the display, information notes, or hidden functions occurring inside the software or network. Alerts and other audio components can be prototyped by human voice. The Series 60 UI style guide can be used as a reference, although the design team should not get stuck on UI details; these can be fixed later (Nokia, 2001g).

As soon as the first paper prototype is created, it is then given to a sample of end users for evaluation, preferably in a real context of use. Three or four end users are enough for one evaluation cycle. The team can work in pairs; one person is a 'computer', changing paper displays while a test user interacts with the prototype, and the other person takes notes, as shown in Figure 6.6. The cardboard prototype is given

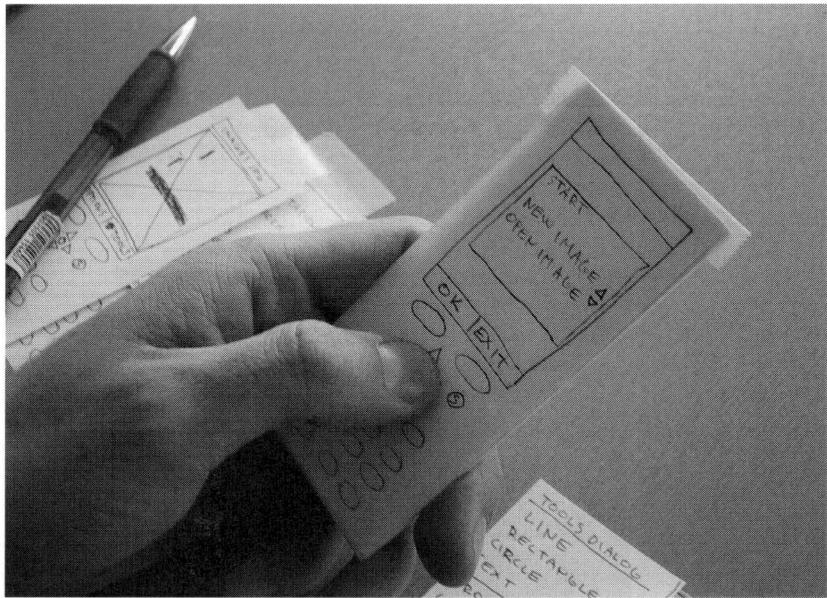

Figure 6.5 Image editor paper prototype

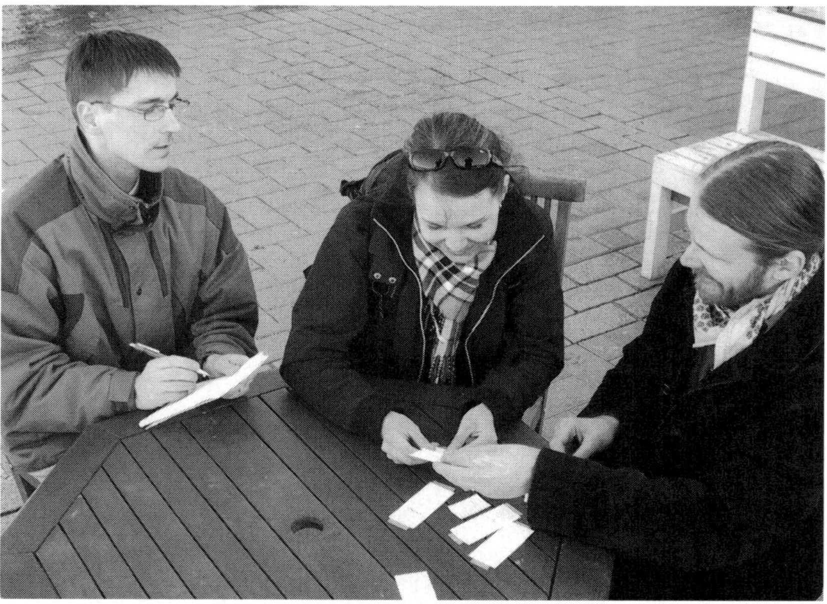

Figure 6.6 Paper prototype test made in a real context: note taker (left), user (middle) and 'computer' (right)

to test users, and they conduct their own tasks with the prototype; the designers can do some interviewing or observation before the test to find out the relevant tasks from the users. The users' own data from their current tools is filled into the prototype to make interacting with the prototype relevant.

When testing the paper prototype in a mobile environment, 'the computer' can use removable tape for attaching paper displays to the notebook pages and a cardboard device. Carrying the entire prototype is thus easier, and displays will not fly away with the wind.

The test users generally respond positively to the test situation and are eager to work with the prototype. Since the prototype is 'low-tech' they understand that it is just an idea of the final product. This invites them to give improvement ideas, to notify the test pair of structural problems, and to ignore irrelevant UI details in the early phase of the design. Often, a test session changes into a co-design mode, where the designers redesign the concept with the users.

After the test sessions the team analyses the results, redesigns the concept, and updates the UED based on the findings. If it seems that there are still structural issues in the concept, the paper prototype testing is iterated again. After a few test iterations, the structure of the application should remain stable and only UI issues should arise in tests. This means that the concept is ready, and product development with detailed interaction design and technical specifications can be started. A specific requirement specification can be elicited from the user-needs data and the UED, in addition to business and technical requirements.

6.4 Interaction Design for Smartphones

Devices with small displays, sometimes affectionately called 'baby faces', have been in the consumer market for some time now. These devices pose different design challenges for the UI designers compared with desktop devices. Taking the design of a desktop computer and cramming that into a smaller interface is not the solution. Good usability for the application is not guaranteed when following the guidelines and the UI style, since it is merely a tool that lets the designer focus on supporting the end user's tasks with the application.

6.4.1 Challenges in Smartphone Interaction Design

When designing an application for a smartphone, the designer's main problem is to find a usable solution that works for the small screen size and different environment of usage. Use of a portable device is totally different from that of a stationary desktop computer. This should always be kept in mind when designing applications for smartphones.

Some of the main characteristics affecting interaction design of the smartphones are as follows:

- They are usually designed for special purposes. Mobile phones are used mainly as telephones, but they do support some other tasks. The tasks supported by the device are simple information management tasks, such as taking notes, saving telephone numbers, managing meeting times, and to-do lists.
- An increasing number of devices can connect to various data sources with different protocols. Also, the peer-to-peer connectivity is an important function to allow the sharing of data with other users. Smartphones can usually connect to a desktop computer to synchronize or backup data. With data connectivity, devices are becoming the main communication device for some people, with short message service (SMS), e-mail, and a multimedia message service (MMS). With multimedia messaging, the user interface needs to be flexible to accommodate picture and text viewing and sound and video replaying at the same time.
- The text input is slower when compared with desktop computers. Two basic ways of inserting text can be identified: telephone keypad, and stylus with character-recognizing software or a software keyboard that is used by tapping the stylus. The UIQ UI style uses a stylus, and the Series 60 UI style uses a telephone keypad (See Figure 6.7). Whatever the method of text input, it is important for designers to recognize the need of the application. For example, an instant messaging application will not be usable if the text input is difficult and slow for the user. (For more information on UIQ, see Symbian, 2002a).
- Some devices are designed to be used with one hand; at least the most important tasks such as answering an incoming call can be done with one hand only. Sometimes, the stylus is used for the more complex tasks. Here also, the need of the application has to be recognized, for example, a phone with a stylus interface has to have some kind of hands-free function to enable note taking while a call is active.
- The mobile context of the use is a big issue when designing the interface, and sometimes it is a neglected aspect. The user might be walking, carrying things, and using the device at the same time. Changing light conditions and noise can also affect the usability of the interface by distracting the user. These are all usability factors that cannot be verified in normal laboratory tests. To be on the safe side, it is good to build an ergonomic model by collecting information on how the users use the products while on the go,

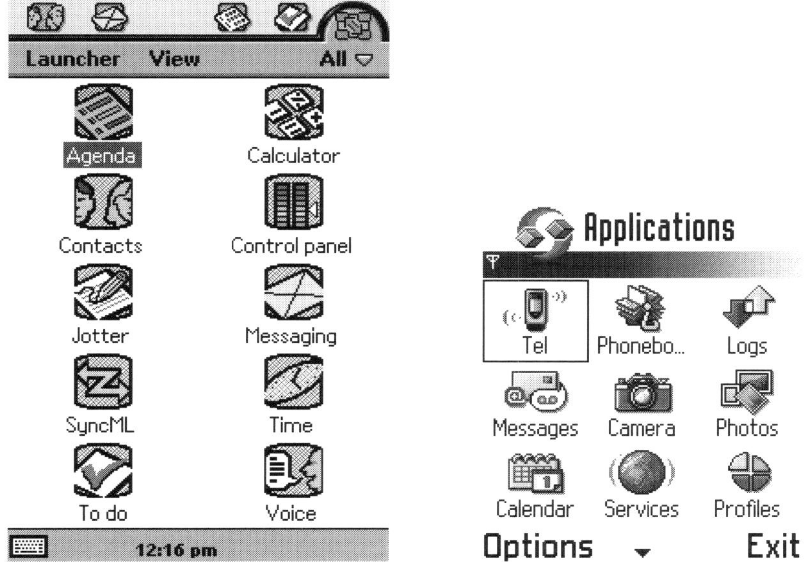

Figure 6.7 Two Symbian OS smartphone user interface styles: UIQ (left) and Nokia Series 60 (right)

and what the disturbing factors in the environment are that may affect the use of the product. The disturbing factors could be, for example, bumpy roads, changing lighting, or rush hour on public transport. If the application increases the stress of the user in an already stressful environment, it is a good reason to redesign.

- Designing a small interface often results in a modal interface. The application is said to be modal if, for instance, it does not let the user begin a new task when in the middle of an unfinished task. The application contains different 'modes' for different tasks. In the desktop computer world, designing a modal interface is usually considered a bad idea for usability (Raskin, 2000), but most graphical interfaces have some modality. With small screens it is almost impossible to avoid.
- The same visual language used in the desktop computer will not be effective on the smartphone. The pixels reserved to give a three-dimensional effect on the user interface usually take too much space of the small screen to make the use effective. For the same reason, the window paradigm is usually not used (Holmqvist, 1999).

As seen in Figure 6.7, Symbian UIQ has a slightly larger display, with 208 × 320 pixels compared with Nokia Series 60, with 176 × 208 pixels. UIQ is designed mainly for two-hand stylus interaction, whereas

Series 60 is based on single-handed usage with a joystick (for the navigation and selection keys), soft keys, and phone keypad. Most of the UI controls such as dialog boxes or menus of the UI libraries (Avkon for Series 60, and Qikon for UIQ) are also different.

Owing to the differences, an application designed for Series 60 cannot be compiled directly to UIQ, and vice versa. And as applications designed for them are not fully source compatible, binary compatibility also does not exist. Porting from one UI style to another is, however, straightforward and does not necessarily involve great effort.

Many parts of the user interface need to be redesigned for making a variant of a third-party application (e.g. to another UI style). Different screen size may require splitting application data to different views. Joystick interaction and pen interaction may require a totally different approach for applications such as the image editor. Also, menu functions can be arranged differently from Series 60 Options menus and the UIQ menubar. Basically, all the UI functions need to be checked and the usability of the changes verified when creating a version of the application to another UI style – even if the functional structure and UED of the application is intended to remain the same.

6.4.2 Using Series 60 User Interface Style

Nokia Series 60 UI style is an interaction style with ready made UI elements such as menus, different types of lists, notes, tabs, etc. Series 60 UI style is specially designed to be used with one-hand-operated mobile phones. Using the UI style, the resulting product is easier to design into a consistent and easy-to-learn user experience. In their work, the UI designers can focus on supporting the user's tasks.

When to Stick to the Style

The Series 60 user interface has gone through an exhaustive design process with task analyses and user testing. The UI style is devoid of usability errors, is consistent, and works for the tasks for which it is intended. The UI style is documented in the UI style guide. In case of Series 60, it is the Series 60 User Interface Style Guide (Nokia, 2001g). Following the UI style does not automatically grant good usability, but it helps tremendously by letting the designer focus on the right issues and preserve consistency with other applications. The general UI guidelines have to be implemented with sense, bearing in the mind the user's specialized tasks and the need of the application. The UI style is always a compromise between many different tasks that have to be served by the generalized UI style.

There are some cases where the Series 60 UI style does not provide sufficient interaction mechanics to allow for the building of different applications. Such new interaction mechanics are demanded in the direct manipulation of objects, such as image editing, moving a pointer in the screen, dragging and dropping objects, and so on (see Figure 6.8). These are not originally used in the Series 60 UI style.

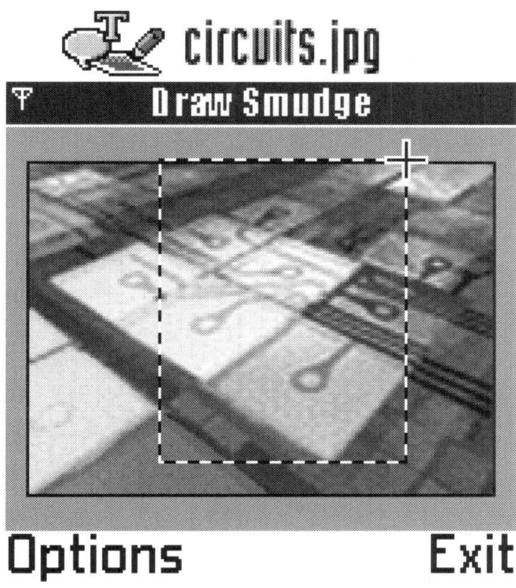

Figure 6.8 Example of a new interaction mechanic in Digia ImagePlus application

Applications with totally new interaction mechanics, such as applications with the above-mentioned direct manipulation, need more careful usability testing than do applications that use an existing UI style. However, once the interaction has been tested it can be used for other purpose too. All new interaction mechanics should be verified with a software prototype. Paper prototyping may be a reasonably effective way to verify the usability of applications using an existing UI style. A software prototype is the best tool for verifying complex and detailed interactions that cannot be simulated with a paper prototype.

6.4.3 Application Design for Series 60

The best way to get familiar with the UI style is to use it, but in order to begin serious design work it is necessary for every interaction designer to read the style guide. In addition to using devices extensively with the target UI style, every possible way of getting familiar with the UI style should be used.

The existing user interface of a Series 60 device can be used as a starting point when designing new applications. When arranging a design workshop, a sample device or an emulator can be taken into the workshop to inspire the design process. The emulator, a hardware device, or the Series 60 UI style guide can be used to track any conflicts between the new application and the UI style of the existing applications.

When new applications are designed, the following issues, at least, should be checked:

- If any of the existing applications (also on desktop computers) already have similar keypad shortcuts they should be used for consistency.
- When system functions such as folder views, image browsing, contact-book browsing, and calendar browsing are needed, the Series 60 system default views should be used to preserve consistency with other applications. In addition, the Notes, Queries, and the Options menus should be used as they are used in the original software. The style guide gives very good examples of these (Nokia, 2001g).
- Basic UI elements should not be used in a non-standard way unless there is a very good reason. For example, if the application needs a view that opens automatically after the application launch, with multiple input options, the Options menu should not be used; a query with appropriate prompt text is the correct display to use. For example, selecting 'New message' in the Messaging application opens a query with a prompt and three options.
- Consistent language throughout the whole application should be used. For example, the system command for closing the application is Exit, not Quit or End. All other commands should be as logical, such as Edit for modifying existing data. If the application uses a language other than English, the existing conventions for naming should be used. If the application is, for instance, a calculator for accountants, it is a good idea to use the terminology found in spreadsheet software.
- The logic of the menu structure should be checked with tests. Press of the select key can be used as a quick and natural selection key instead of picking a command from the Options menu. The select key can also open a collection of important functions in the OK options menu. This should be used sparingly; the OK options menu is even more task-specific than the Option menu.
- Back stepping with the Back softkey should be natural and bring the user to the expected view. This is a good case for paper prototyping to test.

In the application design, it should be remembered that the system focus is always moved to the incoming call, which can interrupt any task at hand. The user has to choose either to answer the call or to reject the call. Only after that can the work be continued. This kind of an interruption in a desktop computer might be annoying, but in smartphones it can be accepted; after all, they function mainly as telephones (see Figure 6.9).

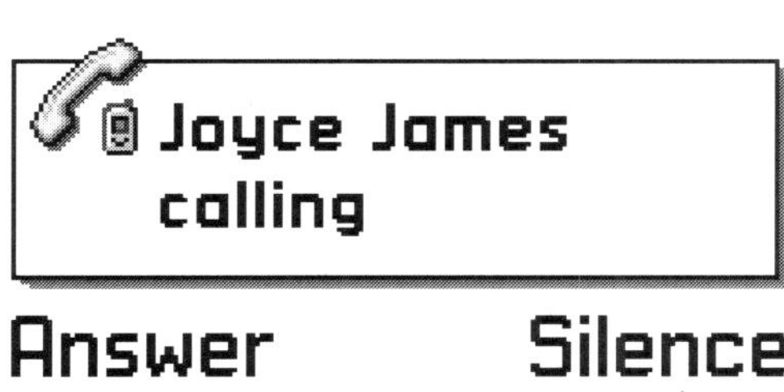

Figure 6.9 Incoming call on Nokia 7650

The Options menus in Series 60 are always context-sensitive. Only the most relevant functions for the view or task are displayed. In this way, excessively long option lists are avoided; the correct functions can be reached quickly. Designing the Options menu requires very careful planning and task analysis. A paper prototype should be used to verify the correct order of the functions. Pressing the select key is generally used for the most important function of the Options menu. Pressing the select key can also open an OK options menu that contains functions strictly related to the selected object. For the difference between the Options menu and the OK options menu in the Contacts application, see Figure 6.10.

It is advisable to avoid designing an application that forces the user to look for information in other applications to complete a task. Pressing and holding the applications key provides one way to overcome this restriction; the user can jump to another location in the device and

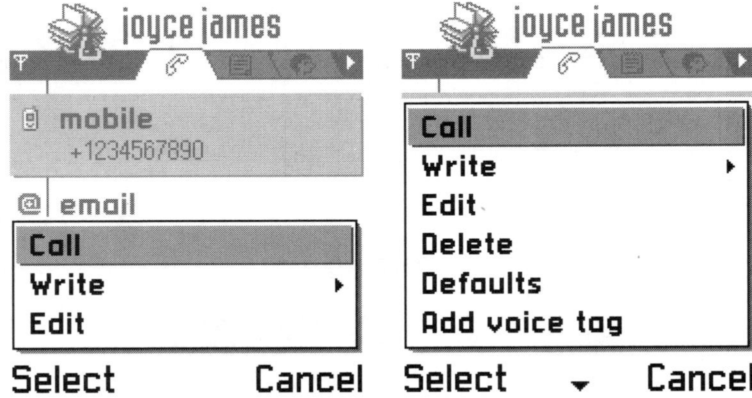

Figure 6.10 The OK options menu and the Options menu in the contacts application: the OK options menu (left) contains only the most important functions from the Options menu (right)

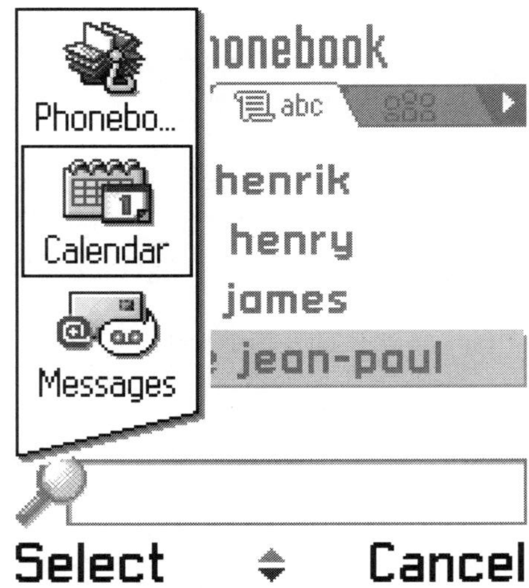

Figure 6.11 The applications key of Nokia 7650: this key can switch from one application to another

switch back to the task that was interrupted (see Figure 6.11). The copy/paste function also helps to avoid tedious writing of text and therefore also minimizes spelling errors. Both the applications key and the copy/paste function are significant advances in the interaction of Series 60, since the early mobile phone user interfaces did not allow either.

If the application uses modes, they should be used in such a way that the user is informed when the mode is on or that the user holds a modifier key pressed down. Raskin (2000) states that pressing and holding a modifier key is easier for the user to remember than pressing once to switch on and pressing another time to switch the mode off. One example of the mode indication is the text input modifier indication in the navi pane (Figure 6.12). The edit key (ABC) can also be used to switch between modes in new applications other than the text input modes. The edit key is the only key that can be pressed down simultaneously with another key. Other mode indicators can be: changing the application icon, changing the application title, changing the text in the navi pane, and so on.

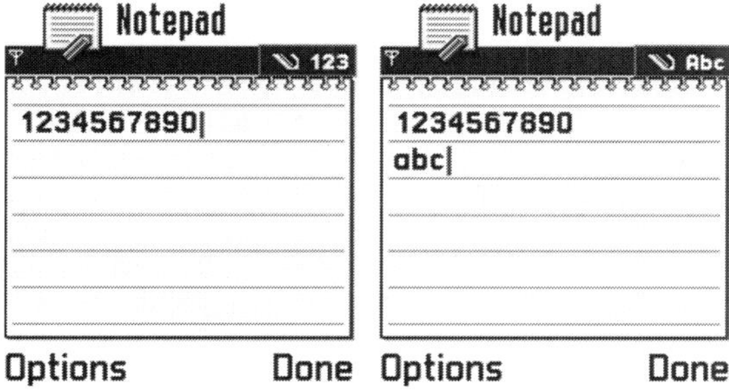

Figure 6.12 The text input modifier indication in different states: number input mode (left) and text input mode (right) (highlight by author)

6.5 Writing the User Interface Specification

The User Interface specification is a document that defines the user interface of a software product. The software engineers use the specification as a tool for programming. The UI specification also documents the design of the product as a result of a UCD process. We first discuss UI specification as a tool for communicating design decisions to a multi-disciplinary software production team. We then introduce how to use Series 60 UI style in writing a specification.

6.5.1 Who Needs the User Interface Specification?

The UI specification is a tool for the production team to identify, document, and implement the properties that compose the user experience of the product. The properties of the user interface are not confined to the graphical user interface (GUI). Other properties can be sound, data input and output, how the application connects with the operating system in the device, and how the application connects with the cellular network and network servers. From the properties described in the UI specification we know how the application behaves in the real world. A typical UI specification is a documentation of how the end user is going to use the product.

Consumer software products are often produced by a team of experts, each with their own area of specialized knowledge. The UI specification document can serve as a common ground for discussion and communication on the product. The UI specification is representative of the product when the product is under development. Each member of the team has his or her own interest in the product, and therefore different expectations of the UI specification exist:

- The customers want to see that their requirements are all covered and that a solution for their problem is found. The customer then accepts the funding and timetable of the project.
- The person responsible for the usability plans the usability test according to the UI specification and use cases. A heuristic evaluation can be made, too, with the UI specification, using the usability aspects list, described in the section on usability verification (Section 6.6).
- The testing engineer makes the functional test plans with the help of the UI specification.
- The project manager assigns the relevant resources to work on different parts of the product.
- The marketing and sales managers get an idea of what the application does so that they can sell it to the right people.
- The manual and help writer writes the manual with the help of the UI specification.
- The software architect defines the technical architecture of the product, and the software engineers implement the features of the product according to the architecture and UI specifications.
- The translator needs the UI specification to translate the user interface to different languages.
- The interaction designer needs the UI specification to maintain and document the design decisions between iterations.

6.5.2 Who Can Write the User Interface Specification?

The task of writing the UI specifications falls naturally to an interaction designer. It can be written by a person who has some experience in

UCD or by a person who has been evaluating the particular user needs for the product or who is otherwise familiar with a typical user. It is also possible that a software engineer writes the UI specification. However, it is not recommended that the UI specification be written by someone who does technical implementation simultaneously with the same application (Norman, 1999). The difference between programming and UCD make it very difficult to keep good user experience in mind when at the same time thinking of the internal programming logic of the software.

Although one person writes the UI specification, the result cannot be optimal if there is no communication between other team members. Good results can be achieved only with team effort and user testing with the intended users. The reviewing of the UI specification should be done mainly by product-management and engineering staff. Even when bringing in input and comments from outside the team it is important to keep the focus on a good user experience of the product. The writer of the UI specification can also arrange walkthrough sessions with other experts to evaluate design ideas at an early stage.

6.5.3 Characteristics of a Good User Interface Specification

The UI specification should be unambiguous to the reader. It is important that there be only one interpretation of the described features and therefore only one possibility for the software engineer in the implementation of them. In conditional features, the exact conditions for each feature should be given so that they can be implemented without ambiguity.

The specification should be consistent in terminology and should not use multiple terms for the same feature; for example, the terms select key and select button should not be used interchangeably.

The UI specification should be easy to update, even if the person doing this is not the original author. The same features should not be specified repeatedly in the document, as changes opens the possibility for errors, and the same feature can be specified differently elsewhere in the document.

The specification is ready when a sufficient level of detail is reached and when the addition of more details to the specification does not give any additional necessary information to the reader.

6.5.4 Lifecycle of a Specification: From Concept to User Interface Specification

In the early stage of the product concept the UI specification can be only a sketch of the product. Usually, it is sufficient to identify the raw structure of the application and a couple of commands that link from one view to another. This is usually sufficient for making a paper prototype and testing the concept with end users or for discussing the concept with customers. Avoid making too detailed a specification of the user interface too early. Sketchy designs command a more creative approach from the designers and test users. If the design looks too 'fine' it is mentally difficult to make the necessary changes to features. Also, from an efficiency point of view, it is a waste of work to finalize a design that is going to be changed after a usability test. Only after a number of usability test cycles can it be ascertained that the design works for the users. It is more cost-efficient that the changes to the design be made to the UI specification than to the program code after several weeks of programming.

6.5.5 Early Draft of the User Interface Specification

The draft version of the UI specification can use the UED notation and rough line drawings as the screen layouts (Beyer and Holtzblatt, 1998). The UED notation is easy to update and quick to read within a short learning time. Later, when the details find their place, the UI specification can be written in standard language (a common language to all readers), and the line drawings can be changed to layouts designed by a graphic designer.

After the usability problems are tackled with new designs that have passed the test of users, the details can be added to the structure. Main views of the application can be documented to a specification draft. Images of the views can be drawn with a drawing application and embedded into the UI specification; a drawing application widely used in the organization should be used so that others can reuse the drawings. Screen layouts should not be used without descriptive text. Functions should be added to the views with a description of what they do.

6.5.6 From Structure to Details

There are different combinations of specification styles used in the industry. We think that it is best to use standard language to make the specification understandable to most readers; use of descriptions in charts and layout images increase the readability of the specification. The charts can be used to give an overview of the applications structure; state charts can also be used to describe complex interactions. Layout images can describe the application views in detail. The UI specification can also contain use cases created from the storyboards created in the concept workshop. Use cases are a great help not only when planning the functional testing of the application but also when demonstrating how the application works in real-world situations. Thus, for example, it can be verified whether any changes are really needed. An example of a use case is given in Appendix (page 493).

In the early stage of product development it is important to get the structure of the application reviewed and approved by the team. When this is done, the fine adjustment of the details can begin. This is typically the most time-consuming stage of the work. Usually, it helps if the engineers are closely engaged in the project from the start so that they can start to plan the architecture while the UI specification is being written. The engineers can also help to clarify the details. The specification should be reviewed by the team, or walkthrough sessions should be arranged if discussion of the details is needed. If someone in the organization has previous experience of an application of the same type, they can act as outside reviewers. Another interaction designer can also help by reviewing the UI specification. When the specification is considered complete it can be tested with a prototype. Refer to the section on usability verification for more information (Section 6.6).

6.5.7 Maintaining the User Interface Specification

When the specification is approved by all the readers there should not be anything to add to it or to remove from it. However, things that need changing always occur. If the change is considered important for the product it can be made into a 'change request' and evaluated by the team (or change management board, steering group, or similar). If the change request is approved, the change is applied to the UI specification. Changes to the UI specification affect many later documents, so the change management process needs to be highly visible. It is important that the usability of the product not be compromised by changes that are not verified by usability testing. Changes should not be done if they work against the previously tested good usability.

When the programming of the product is beginning, the interaction designer is responsible for the communication of the UI specification to the programming team. The UI specification is also the documentation of the product for the next versions of the software, so it should be kept up to date during the programming.

Reviewing the User Interface Specification

The reviewing should always be done in written form. A good practice for reviewing the documents is that every reviewer writes his or her comments on the circulating copy of the document. The reviewers should be informed what are the most important issues that need reviewing in the UI specification, and, additionally, what issues can be ignored. The reviewers' time should not be wasted in irrelevant issues.

6.5.8 User Interface Elements in Series 60 Platform

The Nokia Series 60 UI style guide defines how the basic elements of the Series 60 style should be used (Nokia, 2001g). The interaction designer should first read this document before starting the design work. The following is a non-exhaustive list of how the UI elements should be specified in a UI specification. Appendix is part of a UI specification of Digia ImagePlus application.

Lists

Lists are used to display ordered data. Sometimes the list view can be displayed in grid form (e.g. the application grid in Nokia 7650). The lists are often customized for a particular purpose; compare for example, the week view with the day view in the calendar application (Figures 6.13 and 6.14). The specification should answer the following questions:

- What kind of a list type is used? Is the list item one-lined or multi-lined?
- What is displayed on the line or lines? Where do the data come from?
- Is there an icon column used with the list?
- Can the list be marked?

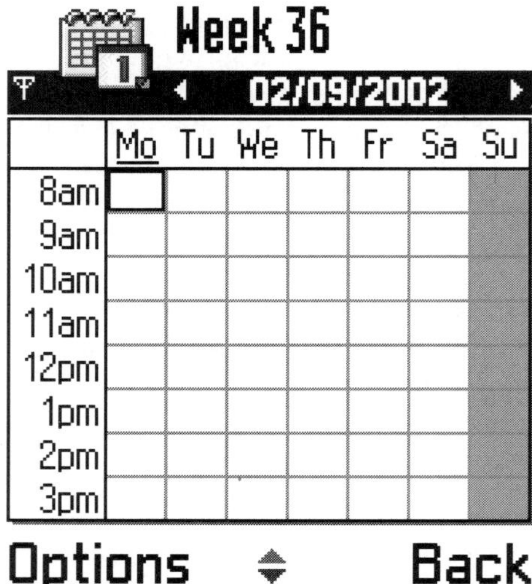

Figure 6.13 Week view in the calendar application

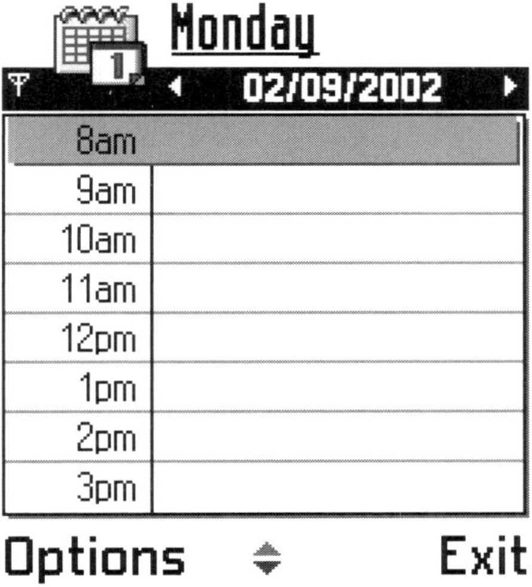

Figure 6.14 Day view in the calendar application

Icons

Icons are used to differentiate list items from one another. The list icons can also change to indicate a changed state of the list item (e.g. the Inbox icon in Messaging application changes appearance when a message is received). An application icon is used for branding of the application or to display indication of the temporary application modes. The specification should describe the icon name and the logical name of the icon, and how and where it is used.

Menus

Menus are lists of functions that can be displayed by user action. Menus can be invoked by the press of the Left softkey (Options menu) or by the Select key (OK options menu). The submenus can be described in the UI specification with the intended text style (see Appendix). The specification should describe:

- the function names and logical names in the menu, and the order of the functions in the menu;
- the function names and logical names of the submenus;
- when the functions are available, if they are context sensitive, and what they do.

Softkeys

Softkeys are located at the bottom of the screen. Their functions can change depending on the application state. A left softkey and right softkey are available [see also the section the control pane for a description of the softkeys (Section 9.2)]. The specification should describe:

- labeling and logical names of the softkeys;
- when the softkeys are available and what they do.

Notes

Notes are messages that communicate temporary states of an application. Notes contain text, animation (optional), graphics (optional), and softkeys (optional); a sound can also be associated with a note. For a detailed description of the different types of notes see the section in notifications (Section 11.4). The specification should describe:

- the note name and when the note is displayed;
- the type of the note (e.g. confirmation note, information note, warning note, error note, permanent note, wait and progress note);
- what string of text is displayed in the note and the logical name of the string;
- when the note disappears;
- what softkeys are displayed, their labels, and their logical names;
- details of whether there is a specific image, animation or sound associated with the note.

Queries

A query is a view where the user can submit information to the application. The information can be inserted into a text input box (data query), selected from a list (list query, multi-selection list query), or selected with the softkey (confirmation query). [For a detailed description of different types of queries, see Section 11.5; for different editor case modes (upper-case text/lower-case text/automatic/user selectable), and input modes (text/number), see Section 11.3.] The specification should describe:

- prompt text string of the query and the logical name of the string;
- whether the input line is automatically filled with text or is highlighted;
- the number of input lines used (in the data query);
- the number of characters that can be inserted (in the data query);
- the input mode of the query, text mode, or number mode;
- what softkeys are displayed, their labels, and their logical names.

The Hardware Keys

The hardware keys can be different, depending of the device model. The Nokia 7650 keypad is described in Appendix. The table of the hardware keys should be used as a checklist of hardware key functions in every view.

Logical Names

The logical name is a label or name of an element in the user interface, such as an icon or a text string. Logical names are used to help the translation of the application. The logical name strings can be composed of words that help to identify what the name signifies (e.g. for the right softkey: <right_softkey_cancel>; for a note: <note_waiting_connection>).

6.6 Usability Verification

Almost every software product has been designed with 'ease of use' as one of the design requirements. However, not all products are successful in terms of usability. Getting the usability of the final product on the right level requires setting usability requirements and verifying them with usability verification tests.

In the following sections we discuss the possibilities for setting and verifying the usability requirements of a software application. The setting of measurable requirements and the use of usability checklists are introduced. There is also a section on the possibilities to test usability at a sufficiently early stage. Finally, a lead adapter test for the final product is presented.

6.6.1 Setting Usability Requirements

It is useful to set usability requirements in a measurable form so that they can be verified in a usability test or evaluation. By verifying usability requirements, the product development team knows when the iterative design process can be ended. Also, the product management and the customers get meaningful information on how well the product meets the defined usability targets.

Attaining unambiguous usability requirements for a new product is not very easy at the beginning of a project. Sometimes, the design team starts with rough design drivers, and more detailed usability requirements are stated after the first concept iterations. For the next product version, the usability requirements can be based on the previous version. Also, benchmarking competing products and alternative designs may help in clarifying the requirements. A method for comparing usability of competing products is introduced in the Section 6.6.2.

After the user-needs study is completed, the target users and the main use cases for the product are defined. These data can be used for identifying user tasks related to the application and analyzing their criticality or frequency (Ruuska and Väänänen, 2000). This information helps to set criteria on how efficient each task should be. Tasks under pressure are critical and should be able to be conducted with simple actions and without careful looking at the display. Tasks seldomly occurring may be placed deep in the Options menu.

A Method for Analyzing the Criticality of a Task

Tasks my be scored as follows:

- Score 1: critical task. This is a task that is conducted under pressure, or two tasks that are conducted simultaneously (e.g. making a note while talking on the phone).
- Score 2: routine task. This is a task that is conducted frequently or several times when the application is used.
- Score 3: often task. This is a task that is conducted often or at least once in every application usage session.
- Score 4: occasional task. This is a task that is conducted only from time to time (e.g. changing a setting item).

Often, finding suitable measurements and setting the targets is not an easy job. Another way to assure usability is to use usability checklists while 'walking through' the product in heuristic evaluations (Nielsen, 1992) or when verifying the usability in the usability test. Each problem that is found is categorized based on a scale of criticality, which reports the expert's assessment of the criticality of the usability problem.

A Method for Categorizing Usability Issues

Issues may be scored as follows:

- Score 1: catastrophic; hinders the user from using the product in a feasible manner and has to be repaired before the product is launched;
- Score 2: major; complicates the use significantly and has to be repaired;
- Score 3: minor; complicates matters for the user of the product and should be repaired;
- Score 4: cosmetic; can be repaired in the future.

Table 6.2 describes different usability aspects, their definitions, checklists, and ways to measure the usability aspects. All usability requirements can be collected into one table that contains the definition of tasks, criticality, measurement, target criteria, verification plan, and verification results. The same table can serve as a requirement definition and verification plan and can provide information on verification results. The verification results can include how well the set targets have been reached and what kind of usability issues have been found. Table 6.3 gives an example of a usability requirements table for an image editor product.

6.6.2 Verifying Usability

The usability requirements are verified through usability evaluations. The usability verification plan should be part of the development project plan, so that resources and product increments can be allocated to the right time. The usability verification tests often reveal usability problems that should be corrected in the product. That is why it is recommended to start usability verification as early as possible, before it is too late to correct the errors without missing the planned launch date.

The best way to verify usability is to organize usability tests with a sample of end users and predefined test tasks, based on the verification plan. A group of 6–8 test users is usually sufficient to reveal all relevant usability issues and is significant enough for verifying the usability, even if the number is not statistically significant. Laboratory testing with videorecording equipment also allows accurate time measurements for analyzing the efficiency of the tasks. In the absence of a usability lab, a normal meeting room can be used. Alternatively, the evaluation can also be conducted by external usability experts, who 'walk through' the product with a usability checklist and identify possible usability problems.

The person evaluating the usability or organizing the usability tests should preferably not be the designer of the product. The designers sometimes tend to be too subjective and love their design; which may affect the test situation. However, the designers and developers should participate in the test session by observing it, since it is an efficient way to 'wake them up' if the product has usability problems.

If the design team is small, the observers may sit in the same room and talk with the user about the design issues after the test is over (see Figure 6.13). The observation room can also be a separate room with a video monitor covering the test situation. The developers can also watch videorecordings of the session afterwards.

The test results are updated in the usability requirements table for communicating the project management about the state of usability.

Table 6.2 Usability aspects

Aspect	Definition	Checklist	Method of measurement
Familiarity	Match between system and the real world	The system should speak the user's language, with words, phrases, and concepts familiar to the user, rather than system-oriented terms. Follow real-world conventions, making information appear in a natural and logical order.	Percentage of users who can conduct the task for the first time without help.
Efficiency	Efficiency of use, and flexibility	Accelerators (unseen by the novice user) may often speed up the interaction for the expert user in a way that the system can cater to both inexperienced and experienced users. They allow users to tailor frequent actions. They are important in often and routine tasks.	Number of key presses required to conduct the task Speed, related to competing products Number of provided shortcuts in routine tasks
Mobility	Ability for users to use the product in mobile context of use.	Important in often and routine tasks. Design product for single-hand usage, use auditive and tangible feedback for interaction.	Percentage of tasks that can be conducted with a single hand Percentage of tasks that can be conducted without looking
Error handling	Error prevention and error messages	There help users recognize, diagnose, and recover from errors. Error messages should be expressed in plain language (no codes), indicate precisely the problem, and constructively suggest a solution. Even better than a good error messages is a careful design that prevents a problem from occurring in the first place	Number of errors Percentage of users that recover from errors without help

Learnability	Recognition rather than recall	Make objects, actions, and options visible. The user should not have to remember information from one part of the dialogue to another. Instructions for use of the system should be visible or easily retrievable whenever appropriate.	Percentage of users remembering how to do the task independently when they use it the second time after used or introduction.
Memorability	Users remember how to use it once they have been away from it for some time	The product guides the user to conduct the task. This is important in tasks used with occasional frequency	Percentage of users to remember how to do the task independently after a period of time (week) since they have used the product
Help	Help and documentation	Even though it is better if the system can be used without documentation, it may be necessary to provide help and documentation. Any such information should be easy to search, focused on the user's task, list concrete steps to be carried out, and not be too large.	Percentage of users to find help from user documentation
Consistency	Consistency and standards	Users should not have to wonder whether words, situations, or actions in the user interface mean the same or different thing. Follow platform conventions.	

(*continued overleaf*)

Table 6.2 *(continued)*

Aspect	Definition	Checklist	Method of measurement
Freedom of choice	User control and freedom	Users often choose system functions by mistake, and will need a clearly marked 'emergency exit' to leave the unwanted state without having to go through an extended dialogue. The device should support undo and redo	
Visibility	Visibility of system status	The system should always keep users informed about what is going on, through appropriate feedback within reasonable time	
Aesthetic	Aesthetic and minimal design	Dialogues should not contain information that is irrelevant or rarely needed. Every extra unit of information in a dialogue competes with the relevant units of information and diminishes their relative visibility	
Subjective user satisfaction	Users' overall feelings about the product		Scale questionnaire after the evaluation (1–5): unpleasant–pleasant boring–entertaining useless–useful (practically, socially) inefficient–efficient ugly–beautiful

Table 6.3 Usability requirement table

ID	Task	Frequency[a]	Measurement	Criteria	Verified	Result	Usability issue[b]
1	Opening image for editing	3	Learnability	100% of users can do it the second time	Test 1	Learnability 100% Familiarity	
2	Annotating image with text	2	Familiarity Learnability	75% of users can enter and edit text to the image without help 100% of users can do it the second time	Test 1	Familiarity: 4 of 6 users can do it Learnability: 100%	2
3	Add icon to image	3	Familiarity Satisfaction	Familiarity: 75% of users can add an icon to their image Satisfaction: All users can find 20% of the provided icons useful			
4	Moving and resizing	2	Learnability Error handling	Learnability: 100% of users can learn to do it Error handling: 100% of users understand what the cursor highlight or pointer indicates	Test 1	Learnability: 100% Error handling: 5/6	

[a]1, critical; 2, routine; 3, often; 4, occasional.
[b]1, catastrophic; 2, major; 3, minor; 4, cosmetic.

Figure 6.13 Usability test setting: the observer follows the test via a video projector

The usability issues found can be categorized according to a checklist and rated based on a scale of criticality.

If the set usability criteria are not met, the project management may require more design iterations, or they may decide to ease the criteria intentionally. Some of the issues found may cause a change request to the requirement specification or UI specification; these need to be handled in the change management of the project. It is also useful to write a test report with detailed findings, correction proposals, and new design ideas. It serves as a future knowledge base for other projects.

Comparing the Usability of Competing Products

It is sometimes difficult to measure the usability in numeric terms. It is easier to judge whether one product is easier to use than another when two products are compared. If competing products exist on the market, or if the design team has two competing design alternatives, the end users can be asked to judge the usability in a comparative usability test.

In the comparative usability test, the test users do a set of tasks with the first product and then repeat the same tasks with another product. The users can rate the product after each task and set preferences regarding the products at the end of the session. The most relevant results are obtained when two products are compared; if three or more products are used, users tend to forget how each one worked.

The test order should be changed between users, and the product brand should be hidden so as to avoid bias in the test results. The competing products must also be on a similar maturity level; if one product is a fully working end product and another is an unfinished prototype, the end users will vote for the end product, no matter how good are the interaction and design of the prototype. If the comparison can be conducted before the product sales, possible usability problems of the product can be corrected if it falls behind the competitor in the test.

6.6.3 Paper Prototyping or User Interface Simulation Tests

The challenge in a usability verification is to conduct it sufficiently early in order to make it useful for the developed product. One way to avoid structural usability issues is paper prototyping. Paper prototyping is a useful tool for concept design (see Section 6.3), but it can also be used in later phases.

Paper prototyping also helps and speeds up the specification phase, since the design team does not need to think and argue about whether the design works or not. They can create a paper prototype and test it in a day or two. Paper prototypes can also be used in usability verifications at the development phase if a UI simulation or a final product cannot be used sufficiently early. The UI graphics can be tested by printing the UI layouts with a color printer and creating a more detailed paper prototype from them.

However, the paper prototype cannot reveal the detailed interaction issues that come from system feedback or keyboard interaction. For detailed interaction, a computer simulation may be a way to solve design issues before the final product is 'born'. Creating a UI simulation often takes significant time and additional resources; it may be essential to give the design team enough time for design iterations, especially if the team is creating new ways for interaction. Sometimes, the project may deliver a demo product for sales purposes before the final product. The demo can be used for usability verification if it works in a similar way to the actual product.

6.6.4 Lead Adapter Test

Before the product is launched, a beta version or final product can be given to a sample of lead adapters for a few weeks. A lead adapter can be described as an early adopter of the product within the target user segment or an otherwise easily accessed target user who is willing to try out the product. A lead adapter test can give realistic information on how the actual product works in a real life situation before the product is sent to a larger audience. To guarantee a good user experience, it is better to have end-user contact with the product far earlier than this, but the company should not miss the opportunity to get good feedback and information in order to set usability requirements for the next product.

Since the product is given to end users to test in real situations over a long period of time, users cannot be observed all the time and simultaneously. A good way to collect detailed, qualitative information is to ask the end users to keep a diary of their usage. With a well-planned diary, the company gets information about how, when, and where the users use the product, the frequency of used functions, errors and problems, as well as improvement ideas. After the test period, the test users can be interviewed based on the diary notes. The diary helps the end users to remember even small details of events that occurred in the past. Often, those details are very relevant to the interests of the product team.

6.7 Summary

In this chapter User-centered design for Series 60 applications a practical process creating applications of high usability has been detailed. The process started from understanding end-user needs and continued with iterative design and evaluation cycles. The first iterations are conducted in concept design workshops; later iterations are focused more on interaction design and UI details.

We addressed the challenges of interaction design for smartphone user interfaces and described some useful guidelines for writing UI specifications. We have also concentrated on various usability verification methods for assuring high-quality user experiences. Once application developers have addressed these issues of user experience design, they may focus on UI implementation issues, discussed, in the following chapters.

7

Testing Software

Testing means using and operating a system or a component in a specific environment, observing the results, and then evaluating the results of some aspect of the system or component (Binder, 1999, page 1113). The evaluation causes the test to be passed or failed, which is typically due to a flaw in the tested item. Test **plans**, **suites**, and **cases** are designed with the aid of requirements, standards, and other development products, such as specifications. Deviations or variations found in the development products during the testing of the software and investigation of the failures and failed test cases are called **errors** and **defects**.

Software testing scope is usually categorized as follows: **white box**, the source code and internal details of the implementation are known and used; **black box**, only specifications are used, but no internal details, in the testing. **Unit** and **module** testing are used as white box testing. **Functional** testing and **system** testing belong to black box testing. The scope of unit testing is a unit in a module; the scope of module testing is the whole module; the scope of functional testing is an implementation; and the scope of system testing is the whole system. The scope of integration testing is a little less clear, but usually its scope is testing the interoperations of an implementation in the whole system including hardware. In small systems, integration testing is seen as testing the interfaces of classes and modules in an application or implementation.

Traditionally, these scopes are implemented and executed, one following the other, in time order, to form the test phases of a project. Every test phase has its own advantages compared with the others. Test phases are quite inefficient if they are executed in the incorrect phase of a project. Testing culture defines the exact to which terms are for white box and black box techniques and scopes. Whereas unit testing is highly code-dependent and **verification** is made at a very basic level (the unit is a class), functional testing is focused on features and implementation in an almost complete environment. Integration testing tests the interoperability of the application or modules after integrating them into the system. System testing tests the system requirements and functionalities overall and also the volume, performance, and overload situations of the whole environment. The system testing task can also include building and integrating software into a complete environment.

In testing, it is essential that everything is done systematically. The testing process proceeds in steps from test case design and planning, test execution, result validation and verification, and failure reporting to bug correction verification. These steps should be made in a systematic manner, so that every step can be reproduced later for a specific software build. Testing for a certain software build means that the results and the reports are valid only for that build. Especially in iterative software development, the need for **regression** testing is great and it is often done for every new build.

7.1 Validation and Verification in Testing

All testing phases use verification and validation to check the correct results against other testing environments, specifications, standards, and requirements. **Verification** means that the execution output of the software is compared with a specified, expected result. The expected result is derived from the specification. **Validation** is the process of checking whether the software or system is doing 'the right thing' (Fewster and Graham, 1999, page 562). These methods – validation and verification – form the testing work. Usually, the **V-model** is used to show the relation between development, test design and testing of the software (Fewster, Graham, 1999 pp. 6–7). In Figure 7.1, these relations are represented in five development and testing levels. Verification and validation can be also related to the V-testing concept. (Perry William E. 2000, pp. 64–69, p. 166 and 170).

Validation and verification are not mutually exclusive. Since testing needs specifications as inputs, their quality substantially affects testing. In the worst case, there are no specifications and the testing is then done with very subjective conclusions. Usually, when making a test plan and test cases it is possible to find defects without execution of the software; simply verifying the specification is enough, as it is done through inspection. For example, missing error text and error recovery functions are easy to find. The implementation can be defective or incomplete with respect to the requirements. The boundary values or performance can also be out of date.

When a test fails, the reason for the deviation is checked from the specification and then the deviation is stated in the failure report. It is more effective for the designer to see the reason of the failed test

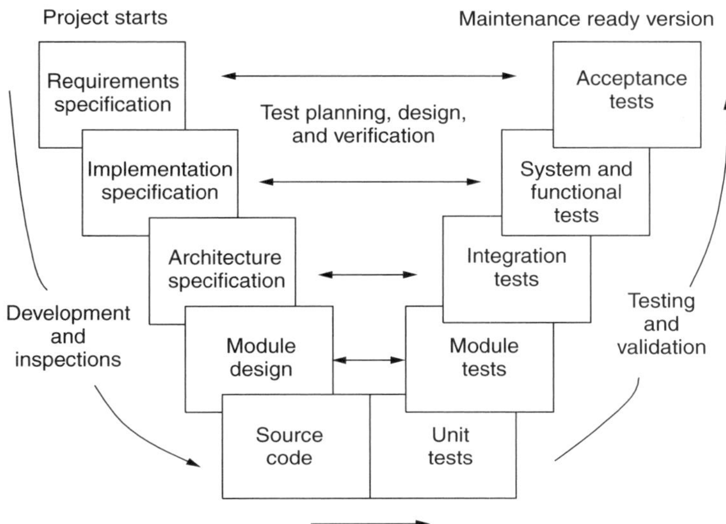

Figure 7.1 Relations between development, test design, and testing

verified against the specification than simply to read the test result, which is validated as incorrect.

Even if the testing itself works fine, we need validation of the process, too. If testing is done, for example, in the environment, and has wrong, old, or faulty components, or components that have not been tested or updated, we easily get many failures. Testing can also be much harder to perform in such an environment. Because of a high number of failures, the need for regression testing is greater and thus takes more time. To be effective, testing should only be done on software and in an environment leads to that is valid for the test phase. This validation is implemented in the process by entry end exit criteria and can be defined in the test plans too. More about entry and exit criteria is found in Section 7.1.2.

7.1.1 Planning and Designing Tests

In the Series 60 Platform environment there are certain aspects that can be mapped onto different testing scopes and methods. Despite the fact that in this chapter the focus is on testing, it should be noted that the overall quality of software processes (assessment, inspection, development, and testing) significantly affects testing in terms of how it is done and its quality. There is a difference in testing when there are dozens or hundreds of open failure reports and unsolved failures.

As with iterative development, testing is done in phases. In testing, it should be noticed that when parts are missing in the software the need for **stubs** and other test **drivers** is bigger. **Stub** is a piece of code that has the same interface as the lower layers and possibly returns pre-determined values, but it does not contain code to performany real computations or data manipulations. Since the development and coding is done in slices, parts, and versions, the test should take account of changes made through the process to the final tests. Also, it may be necessary to focus the testing goals on the ready parts of software only. In the process, it should be noticed that test execution can be finished and accepted only partly if some features are missing. Functional or system testing can rarely be done if integration is not ready.

Test plans, test suites, and test cases are designed with the aid of software development specifications, design files, requirements, and standards. Usually, different specifications and design files are used as input for different test plans. Standards are available on test documentation (e.g. IEEE standard 829; See Binder, 1999, pages 340–44). The test plan contains information about environment, responsibilities, schedule, and scope of testing. The test suite and cases contain detailed instructions for test execution and references to documentation used. The test case defines also the state of the implementation when starting the test, the test set up, inputs, conditions, procedures and the expected results. When the test suite and cases are executed, each test is evaluated as 'pass' or 'no-pass'. The test report contains the result of the test. The used environment depends a lot on the testing phase and the maturity of the system.

Testing scopes, used documentation, and environment are, for example:

- unit and module testing: source code and design files, integrated development environment (IDE);
- integration: emulator environment (IDE), real environment, prototypes, interaction, and architecture design;
- functional testing: emulator environment (IDE), real environment, prototypes, requirements, functional and implementation specifications, and standards;
- system testing: final environment;
- customer acceptance testing: final product.

Used tools for testing are, for example:

- Debug asserts, invariants, and debugger;
- test drivers, stubs, and debug viewers (e.g. unit testing framework);
- test automation tools, test applications, console programs, emulator, and hardware.

Assertions should be used as built-in tests to indicate the defect directly. Assertion is a boolean expression, which is evaluated as 'true' or 'false'. Assertions can be used in a debug build, but they can be left to code for testing in real use and to prevent the execution of code in failure situations. For test personnel, assertion panics are easy to identify and report. See, for example, Nokia (2001b, 2001f). Section assertions provide more information on assertions. The debugger is used in integration, functional, and system testing for bug solving after the failure can be repeated securely. However, debugging itself does not belong to the testing activities.

7.1.2 Entry and Exit Criteria

In a project, the entry and exit criteria are defined for testing phases. Every testing phase should be checked for validity; the phase cannot be started if the quality or exit criteria of the previous state are not acceptable.

The entry criterion defines the conditions for starting the test phase. It can be very subjective, but, for example, open error reports are usually quite easy to check from the failure database. One entry criterion can be, for example, that no critical open failure reports exist, but here it should be remembered that defects always exist; the zero tolerance of failure count tells that testing has not been done rather than the software process is of a good quality.

The entry criteria can validate the process too. Are the documents needed in the test phase reviewed and accepted? Is the source code frozen and versioned for using in this testing phase, and is the tested implementation already in the build? If there is unfinished work in the development left, the testing cannot be finished completely.

The exit criteria are used to check the testing goals and results. The exit criteria can be stated for coverage and results, and, of course, a testing report should be filled when exiting the test phase.

7.1.3 Failure Corrections

How can one make a correction then? Usually, testing finds many deviations, which, of course, should be corrected. The failure rate depends greatly on the software and project in question. A new code usually has more defects than a recycled code. Failure management is essential in every software project and its value increases when the project scale gets bigger.

For bug corrections, the test phases are iterative. The testing is done, reports are filled, failures are reported, and development gets back to the V-model for the current correction to remove the defect. Then, the testing phases are executed again, but now also the corrections are verified. The tests for corrections can be added as new regression test cases. If the testing phases have already been accepted and the exit criteria are met, only a subset of the tests may be needed. Usually, the scope of regression testing is subjective and its definition is based on experience. It would be more secure to redo all tests in the regression testing.

7.2 Sources of Failures

A **failure** – an error in operation – is typically caused by a defect or a fault in the software or specification. The same defect can appear as many different kinds of failures and thus errors in operation. Defects are usually classified as wrong (incorrect implementation), missing (implementation lacks something), or extra (implementation contains something that is not specified) (Perry, 2000 page 6).

Which defect is the most common? In functional testing, one of the most probable errors lies in the boundary values. In object-oriented design it is surprising that when the development environment should be efficient and clean the faults turn out to be of the simplest class of minimum and maximum values. Memory leaks and not so well optimized user interfaces or databases are also a problem in integration. So, in development, all boundary values should be checked at a very early phase; the correct technique for this is white box testing among code reviews.

7.2.1 Architectural Problems

Surprisingly, architectural problems are one of the biggest sources of errors. Even if the design looks fine, interoperability can easily cause severe problems in a multi-tasking environment. Also, code inefficiency and performance problems can be nasty surprises in the integration phase. Integration testing is the right phase to detect these problems easily, but it does not mean that it is the only possible means of detection. Using test stubs and a simulated environment it is possible to estimate the performance of single routines and thus to validate the performance before moving on to other test phases.

7.2.2 Boundary Values

Boundary values are easiest to test in unit testing, although boundary values are used in all testing scopes. All error codes, null pointers, values of zero, and boundaries near to minimum and maximum values should be executed and tested at least once. In practice, a code that is not executed especially at its boundaries often fails in integration and functional testing. It is possible to use integration or functional testing only for these values, which are defined by equivalence partitioning. An equivalence class is a set of data or inputs, which are treated in the same way. Equivalence classes are usually stated in the specifications (Perry, 2000 page 125).

Most basic defects of this kind should preferably be caught at unit testing. Some values can, in fact, be difficult and troublesome to initialize in the working environment, and regression testing should be used in preference.

The strategy for testing near or at equivalence classes and their boundary values is called **partition testing** (Binder, 1999, page 47). It is applicable for all testing scopes and integration.

7.2.3 Combinations

We get a little deeper into the internal design of the program when we use combinational logic. Again, we use boundary values in different sets. These sets are formed by, for instance, equivalence classes, and every set has its own features, which are implemented in different parts of a program's source code. An easy way to access these combinations is to list the values in a table and then choose the right values to test any set in the table. A common error is to write the source code without any limitations or conditions. These kinds of errors are usually found with one value of the erroneous set, so there is no need to use a full range of inputs in the testing.

A decision table links inputs and their conditions to the correct actions or functions (Binder, 1999, page 121–73). Conditions are evaluated as true or false. An action for a combination is selected, when all its conditions are true. With a decision table it is quite easy to find errors in the combinational logic; for example, some inputs (or their combinations, if conditions exists) execute a wrong function, or some functions are unimplemented. A truth table is used to show all values for a Boolean function. It is like a decision table, but in a truth table, all values are represented as a truth value (e.g. true or false).

In Table 7.1, a simple decision table for different speeds and bearer types is shown as an example. The action on the right-hand side of the table is to initialize the speed in implementation. In the implementation, the input range is partitioned to sets, which are processed by different actions.

The initialized speed is not the same as the input value, because the implementation allows only a maximum speed of 28 800 bps

Table 7.1 Decision table for right data the call speed initialization when the call type is analogue

Call speed (bps)[a]	Bearer type	Initialized speed (bps)[b]
Automatic	Automatic	Automatic
9 600	CSD	9 600
14 400	CSD	14 000
19 200	HSCSD	19 200
28 800–56 700	HSCSD	28 800

[a]Call speed as an input (call type is analogue).
[b]Speed in implementaion.
Note: CSD, circuit-switched data; HSCSD, high-speed circuit-switched data.

(bits per second). The input range also determines the bearer type, which is 9600 bps or 14 400 bps for circuit-switched data (CSD), and over 14 400 bps for high-speed circuit-switched data (HSCSD). In the table, it can be seen that inputs are partitioned by the call speeds to different bearer types (CSD and HSCSD), and the HSCSD speed rate is defined as 28 800 bps when the input speed is 28 800 bps or more for the analogue call type. The error case in the situation presented in Table 7.1 might be, for example, initializing a speed rate of over 28 800 bps in implementation. When **equivalence partition** testing and boundary values are used, at least six separate tests are needed to test the sets in Table 7.1. Depending on the implementation of the automatic speed initialization, even more tests are needed.

In Table 7.1, the conditions for the input are done only for the input speed. The call type is analogue. If we want more conditions for the inputs, for example, by adding a condition for the call type (analogue or ISDN), the table size will increase (see Table 7.2) or we can make a separate table for the ISDN call type.

The decision tables are applicable over the whole testing scope and development. In particular, the boundary values are easier to find in the decision tables. In functional testing, decision tables can be valuable for designing test cases for special situations and checking the coverage of those test cases.

7.2.4 Memory Management

In memory management, and often in using a file system, pointers and handles are used exclusively. Null pointers and zero values and, of course, faulty cleaning are common sources of errors. Memory management and file system usage also have a physical limit. A defect in memory management can be hard to find. Memory management failures usually do not give many clues to the situation, just that the data area is somehow corrupted and it may happen occasionally.

Table 7.2 Decision table for right data call-speed initialization; call type is analogue or ISDN

Call speed (bps)[a]	Call type[b]	Initialized bearer and call type	Initialized speed (bps)[c]
Automatic	Analogue	Automatic	Automatic
9 600	Analogue	CSD, analogue	9 600
14 400	Analogue	CSD, analoque	14 400
19 200	Analogue	HSCSD, analogue	19 200
28 800–56 700	Analogue	HSCSD, analogue	28 800
Automatic	ISDN	CSD, ISDN v1.1	9 600
9 600	ISDN	CSD, ISDN v1.1	9 600
14 400	ISDN	CSD, ISDN v1.1	14 400
19 200	ISDN	HSCSD, ISDN v1.1	19 200
28 800	ISDN	HSCSD, ISDN v1.1	28 800
38 400	ISDN	HSCSD, ISDN v1.1	38 400
43 200–56 700	ISDN	HSCSD, ISDN v1.2	43 200

[a]Call speed as an input.
[b]Call type as an input.
[c]Initialized speed in implementation.
Note: CSD, circuit-switched data; HSCSD, high-speed circuit-switched data.

In memory management, the correct testing phase for error situations is first in white box testing (unit testing). The tests should at least provide an answer to the question regarding what to do when the memory allocation fails. Also, some receiving values are interesting. Maybe a pointer has not been updated, or there is a pointer to the memory allocation that is not in use anymore (especially when there is some failure or special situation in the function called). The Null value of a pointer should always be checked in case the call to the used interface fails. Trap harness, cleanup stack, and cleaning are used for better memory management and to indicate and prevent the execution of the faulty code.

For a smartphone the memory resources are often insufficient. 'Out-of-memory' (OOM) situations should be tested for error recovery (see the unit testing example in Section 7.4). Also, the 'error recovery' architecture for OOM errors may be complicated and so it is important to design it well.

7.2.5 Recovery Functions and Fault Tolerance

The importance of fault tolerance depends on the application used and the environment. It is not enough that the program handles error situations and is capable of terminating; it may also need additional recovery and failure preventative functions. Usually, these depend on the system and its resources, but in, for instance, the smartphone environment there are 'low memory', 'battery low', and 'missing signal' indications. Also, weak performance can cause design problems.

When the used environment is a smartphone its most wanted feature is the ability to make and receive calls. At the system design level, all error situations should be designed in a way that no part of system can disturb the major feature (i.e. calling and receiving a call).

Databases have features that should be considered at the system level. At the system level, the database design should answer to the following questions: What if the database is corrupted? What if the database is used simultaneously? Is the database efficient enough? How big can the database be? In system testing, performance and overload situations are usually tested, and, in a complete environment, multi-tasking is used.

The Symbian OS environment offers the possibility of using traps, cleanup stacks, and cleaning in order to handle error situations. The 'trap harness' mechanism offers the user at least some information about the error situation and prevents the faulty code being executed. Also, assertions can terminate the program when an error situation is detected. Termination should be used only when the code detects a programming error ('should never happen' situation), never run-time, or, especially, resource allocation failures. The failure of an assertion usually triggers program termination; the coding standards and inspections should enforce this.

Memory and battery low messages, and other system messages to the applications are needed to prevent error situations, such as running out of memory. If these signals or messages are processed in advance, at least the user will receives a notification from the system, and he or she can react to the situation.

7.3 Testing and Debugging Tools

Using of testing tools can very easily turn out to be a development project of its own. If the testing process does not handle systematic test planning, test case design, verification of specifications, and validation of the whole system, tools can hardly make the project more effective. Too often, error databases are missing, and so the whole error correction and maintenance process can be too subjective. It is very easy to make wrong decisions if knowledge of and information on software quality are already defective.

Test tools can be categorized into those that capture and replay, those that analyze, and, of course, automation tools. All these tools and software support testing in someway. Close to the software building phase we can use different kinds of switches or methods, such as debug printouts and built-in tests with assertions. In the unit testing phase, when the code is developed, the test stubs and drivers and other helper applications are used extensively. For the source code, there is a possibility of using analyzers (such as LeaveScan). A system of test tools, drivers, and software supporting test execution is sometimes called a 'test harness'.

Whereas integration and functional testing use tools not present in the software development environment, white box testing (unit and module testing) uses exactly the same IDE and software development kit (SDK) to build the unit testing environment, and the unit testing environment uses heavily the same source code to make an instrumented environment for testing.

The test automation and scripting usually include their own development environment, with their own scripting or coding language and a possible executable environment (emulators and possible hardware). Also, differentiation tools should exist for test result analysis. When test automation and scripting is implemented correctly, great efficiency is achieved in regression testing, because there is less reworking. Even if the test result verification is to be made manually, the automated test execution can save time; of course, the same skills are needed for testing as if it were done completely manually. Something that is easily forgotten is that test scripting and maintenance takes time; make sure that all benefits and advantages are gained by using automation before starting test automation, even with the simplest goals.

7.3.1 Test Automation

Test automation covers all testing work. Test execution and result verification can be automated, and in some cases it is possible to generate inputs automatically. Usually, testing is only partly automated. For example, scripts are made manually, but tests are executed and the result is compared automatically with expected results verified in advance.

In test automation, good results are achieved easily in regression and **stress** testing. In stress testing, the automated execution is done for long periods (hours and days), or a great amount of data is input, and it cannot usually be done manually.

Unit and module testing are perhaps easier to automate than other testing phases. If unit testing uses executable test drivers and cases (code which implements the tests) it is possible to automate the execution and result validation. See the unit testing example in Section 7.4 in which the test cases are implemented as an executable code and results can automatically be validated during the execution of the test suite and a test report be generated.

The goals for automating testing should be defined in advance. Automating the whole process can be very time-consuming and even impossible, so the first goal could be, for example, automating the test execution and checking the results manually. This is a much easier job and gives quicker results.

7.3.2 Capture and Playback

Capture and playback tools are used to record test input for later use. By capturing the test execution it is possible to achieve efficiency in regression and stress testing. If result verification is done manually, no great advance to manual testing is achieved. Capture and replay tools are handy for making scripts for relatively stable software. For example, stress testing is a good phase for using automated execution for test input, if the final result can be verified easily manually.

With just capturing and playback, the result can contain information on memory consumption. This kind of monitoring along the time axis can give important information on the efficiency of smartphone software in real use.

7.3.3 Comparison

For the **comparison** with expected results, a wide range of tools is available. If test execution results and output can be compared with expected results, the test status can be defined. A good comparison tool also offers masks for comparison, where the comparison is focused on certain conditions. For instance, the time and date stamps can be left out, the comparison criteria can be focused on strict lines or columns, and so on. Whenever the comparison fails, the comparison tool should be synchronized to continue the comparison. If the test environment and tool support writing logs that include information of the test status a simple search function is also applicable.

7.3.4 Debugger

Usually, the development environment includes a **debugger** for the execution of software. Debuggers are used for solving errors and finding defects. Debuggers usually have capabilities of showing the register and variable values and making breakpoints in the wanted code address or place in code. If the values of the variable can be changed, it can be used as a test-support tool by making test situations that otherwise would be hard to prepare.

A debugger is a powerful tool for bug solving and also suits unit testing. The call stack usually points at the exact source of error, because the call stack contains the stack of currently called code. The debug prints are used to generate the test results. A debugger can usually be used in all test driver environments; only some real time environments may be hard to debug.

7.3.5 Debug Output

When software is built with a special debug switch, the software built-in debug outputs can be read from the output viewer. Such a viewer can be part of the built-in environment or a special program. For example, the unit testing environment can use the debug output for outputting the test result. Here, debug output for `CArrayTests` unit tests in a unit testing environment is defined:

```
// define DEBUG_PRINT
#ifdef _DEBUG
     /// Prints to debugger output
     #define DEBUG_PRINT RDebug::Print
#else
    inline void EmptyDebugPrint(TRefByValue<const TDesC>, ...)
        {
        }

     #define DEBUG_PRINT EmptyDebugPrint
#endif
```

The debug output is a clean way to print information on the internal state of the programs. Never use 'hard-coded' test printings to the real user interface, because they might be forgotten to the source code and to the final product.

Also, when solving failures, debug prints can be used. Debug prints can be added, for example, to decision statements and, in this way, solve the path and values of variables, when a program has been executed. In a real time environment, this may be the only way to investigate the state of the program because use of a debugger may disturb and slow down the execution of the software.

7.3.6 Test Applications

Test applications are needed to create test input and possibly to validate the result. They are made by designers for testing the interfaces; there is no need to wait for all software to be integrated before testing. Test applications can also be used for performance evaluation in emulator and hardware environments. Symbian OS and Series 60 environments are provided with graphical user interfaces to be used in test environments before the final user interface is ready.

For a test application it is possible to use the debug output printouts for displaying information on execution, and so the results can be verified more accurately. Test applications are not programs to be released, so the debug printouts can be used. The nature of the test application is between white box and black box testing; there are no limitations for using the code and program internal values, although the test application is used as a replacement for the real application or interface.

When the whole system is not ready but the interfaces need to be tested in integration or functional testing, should we just wait? No, it is possible to use test applications to complement the unfinished environment and make the integration 'bottom-up' or 'top-down'. Thus, testing, for example, is not dependent on the schedules of the rest of the project for integration, and we have the possibility of making a performance evaluation and bug corrections right away. When the real integration to hardware starts, we have a mature code ready, which answers questions of performance, usability, and for which the functions have been executed at least once.

In the emulator environment it is possible to use the debugger and other source-code-dependent information, such as call stacks, in error situations. When preparation for integration or functional testing is done in the emulator environment with the aid of test applications, use of, for instance, a debugger in error situations is a great advantage.

7.3.7 Console Programs

Console programs can be used as test support tools when integration to the environment is complete. They can be used, for example, for database initializations, filling file systems, or, possibly, for generating test data in the emulator environment. In addition to these, they are used when the environment lacks a component that normally provides the needed output.

Console programs are test applications with the advantage that it is possible to use interfaces before a graphical user interface (GUI) is available. A console is usually implemented in the environment before a GUI, and can be the first usable user interface in the environment. If console programs are used in the hardware, a command shell is needed. From the command line the program and its arguments can be typed and executed.

7.4 Unit Testing

White box testing, where the code is used as a direct test material, is mostly the developer's work. In white box testing (module or unit testing and partly application testing) the tests themselves may be written in the source code. Almost all of the documentation of unit testing can be done at the source code level, and a failed test points the place of error directly. In the unit testing environment, for example, assertions can be used for testing the conditions. If a unit testing framework is used it usually has the ability to make result printouts, for example, to the console or to a debug window.

As the unit tests are used to test the smallest parts of the software, testing by only the module interface (module testing) or user interface can be difficult. So, a test driver is required to execute and test the parts of a module (e.g. classes). Usually, a class itself is documented by, at the least, comments in the source code. The source code's internal logic (e.g. decision statements, classes) and its features are then tested.

The focus in unit testing is:

- class construction and destruction;
- leaving, especially in out-of-memory situations;
- selection, decision, and iterative statements (in C++ these are, for example, if, else, switch, while, do, for, break, continue, return, and goto);
- execution paths of selection and decision statements;
- function calls, all arguments, and returning values;
- boundary, error, and null values;
- empty lists, a list with one item, and the first and last items of the list.

Appending a new element in an array is tested by initializing the array, appending the first element, verifying that the element is really appended, and that the counter (if one exists) of the array is increased by one element. There is also a limit for the array, and the last element is also appended and, after that, appending a new element causes an error to happen. The error should be specified by a return value, for instance, and this value is then verified.

In unit testing, the test stubs are also involved in the testing environment. Usually, the test stubs are used for initializing the test before the code under test is executed. They can also replace the interface and return, for example, the test values. Stubs are also used to simulate missing environment and libraries. Here, we come to an important situation: whenever unit testing is done and possible test stubs have been written, it is possible immediately to test the boundary and error values of the code. If unit testing is left out, we soon notice it in functional testing and integration testing (bottom-up or top-down) as numerous errors in boundary areas. There are also problems in making test situations for error values and states. The required time for the integration or functional testing phase is increased along with the number of found failures, and the need for regression testing is thus greater. White box testing is very valuable when the boundary and error values, at least, are tested. The time required for correcting a wrong index, for instance, in coding phase is much shorter than in the integration testing phase.

7.4.1 Unit Testing Framework

The following unit testing examples are based on Digia's own unit testing framework developed for testing Symbian OS software. However, these examples apply to unit testing in general. Test suites and cases are executable, and verification of results is done by using assertion conditions. The test suites and cases are packed to dynamic link libraries (DLLs), as shown in Figure 7.2. The DLLs are then executed by a test driver.

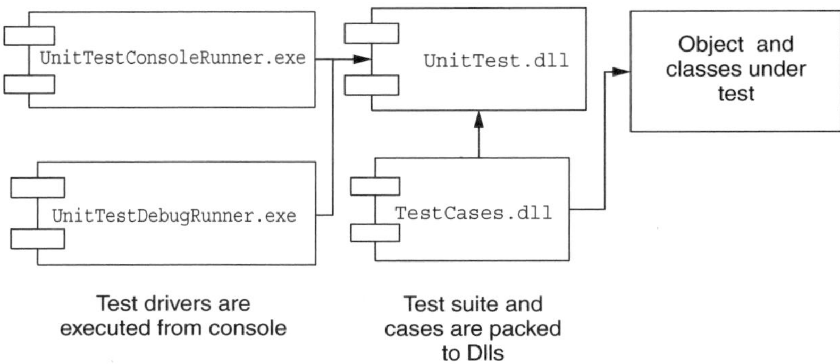

Figure 7.2 Example of a unit testing environment and structure; DLL, dynamic link library

The first step is to install the unit testing framework to the environment. After that, the framework is directly in use. The IDE debug window can be used for displaying the debug output and results. The next phase is to create a testing project, such as `CArrayTests`. In the example unit test, `CArrayTests` is used for basic unit testing of the Symbian and Series 60 template class `CArrayFixFlat`. It is surprising how easily index (boundary value) errors can appear when coding with the arrays.

Test DLLs can be executed using the installed test drivers. The drivers are executed from the console with a driver (e.g. `EUnitDebugRunner` or EUnitConsoleRunner). When building is done for the udeb environment, in error situations (by means of a EUNIT_ASSERT breakpoint), a debugger can be used. A demonstration for normal building procedure in the Series 60 environment is as follows:

```
bldmake bldfiles
abld build wins udeb CArrayTests
```

Then we can execute tests with the test driver:

```
\epoc32\release\wins\udeb\EUnitDebugRunner CArrayTests.dll
```

7.4.2 Macros

For test cases, macros are used for writing the test case table and the test cases. Here, the EUNIT_TESTCASE macro defines a test case. The test case table contains functions defined in the test suite:

```
// Test case table for CArrayTests
EUNIT_BEGIN_TESTCASE_TABLE(CArrayTests, "CArrayTests test suite")
    EUNIT_TESTCASE("Test empty", SetupL, TestEmptyL, Teardown)
    EUNIT_TESTCASE("Test append", SetupL, TestAppendL, Teardown)
    EUNIT_TESTCASE("Test append and OOM", SetupL,
      TestAppendOutOfMemoryL, Teardown)
    EUNIT_TESTCASE("Test at", SetupL, TestAt, Teardown)
    EUNIT_TESTCASE("Test delete", SetupL, TestDeleteL, Teardown)
    EUNIT_TESTCASE("Test append KMaxElements", SetupL,
      TestMaxElementsL, Teardown)
    EUNIT_TESTCASE("Test at KMaxElements-1", SetupL,
      TestAtMax, Teardown)
    //EUNIT_TESTCASE("Test overflow", SetupL, TestOverflowL, Teardown)
EUNIT_END_TESTCASE_TABLE
```

Every test case contains setup and teardown functions for test initialization. In the CArrayTests example, the tested class is instantiated in the setup and deleted in the teardown.

In the example above, test cases defined by the EUNIT_TESTCASE macro have output that gives the test name and other information to the user. In the macros, the functions used in test setup, executing the test, and finishing the test are defined.

7.4.3 Test Suite

A unit test suite for CArrayTests is derived from the unit testing framework. Now it is possible to add methods as test cases to class CArrayTests derived from the framework. Also, the iArray member variable is declared in the CArrayFixFlat template class. CArrayFixFlat is one of the used application interface classes in the Series 60 platfrom.

The CArrayTests unit tests do not implement all tests for CArrayFixFlat. Here, the unit tests are done for basic array manipulations and boundary values. For TestOverflow() causes the E32USER-CBase 21 panic to be raised, as Series 60 application programming interface (API) reference documentation specifies. In the examples, the basic idea of testing with boundary values and overflows by using class and its functions can be seen.

In the following test suite example, an iArray member variable is declared in the class CArrayFixFlat<TInt>. For simplicity, the type is TInt and the granularity is one. All test cases are declared as functions, which are then called when the whole test suite is executed. The test case implementation follows this example:

```
/*
 * CArrayTests test suite
 */
class CArrayTests :
        public CEUnitTestSuiteClass
    {
    public:
        static CArrayTests* NewLC();
    public:
        ~CArrayTests() {
           delete iArray;
           }
    private:     // Test case functions
        void SetupL();
        void TestEmptyL();
        void TestAppendL();
        void TestAppendOutOfMemoryL();
        void TestAt();
        void TestDeleteL();
        void TestMaxElementsL();
        void TestAtMax();
        void TestOverflowL();
        void Teardown();
    private:     // Implementation
        EUNIT_DECLARE_TESTCASE_TABLE;

    private:     // Data
        enum { KMaxElements=5,
               KGranularity=1,
               KTestValue=99
             };
        // Tested array is type of TInt
        CArrayFixFlat<TInt>* iArray;
    };
```

In the example above, the CArrayTests class is derived from the unit testing framework. The test suite contains the test cases as member functions in the class. The `iArray` is the pointer to the tested object. The `EUNIT_DECLARE_TESTCASE_TABLE` then contains the test case table, which defines the member functions to be executed.

7.4.4 Executable Test Cases

The test cases written in the source language do not necessarily need other test documentation; the used environment is the same as for the implementation under test. Assert clauses implement the 'post' conditions for the tests cases.

In the more complicated test functions, the debug print (`RDebug::Print`) function is used for a better test result display. Tests are used so that the execution in boundary values is checked. In the overflow test case, `CArrayTests::TestOverflowL()` raises the E32USER-CBase 21 panic by the member function `InsertL()`. This is specified in the Series 60 documentation, and the overflow should be tested at least once in this way; usually, assertion could be commented for continuous execution of the test program, because the panic stops the test execution.

Test cases are presented in the next example. The test result is checked with the `EUNIT_ASSERT` macro. If the clause in `EUNIT_ASSERT` evaluates to false (= 0), a breakpoint is raised and so the test is failed. If the `EUNIT_ASSERT` evaluates to true (! = 0), the test case execution is successful and the test is passed. The test case is as follows:

```
// Test cases for CArrayFixFlat<Tint>

void CArrayTests::SetupL()
    {
    iArray = new(ELeave) CArrayFixFlat<TInt>(KGranularity);
    }

void CArrayTests::TestEmptyL()
    {
    EUNIT_ASSERT(iArray->Count()==0);
    }

void CArrayTests::TestAppendL()
    {
    // Append value
    const TInt value = KTestValue;
    const TInt countBefore = iArray->Count();
    iArray->AppendL(value);
    EUNIT_ASSERT(iArray->Count() == countBefore+1);
    EUNIT_ASSERT(iArray->At(iArray->Count()-1) == value);
    }

void CArrayTests::TestAppendOutOfMemoryL()
    {
    // Append value
    const TInt value = KTestValue;
```

```
    const TInt countBefore = iArray->Count();
    __UHEAP_FAILNEXT(1);
    TRAPD(err, iArray->AppendL(value));
    EUNIT_ASSERT(err == KErrNoMemory);
    EUNIT_ASSERT(iArray->Count() == countBefore);
    }

void CArrayTests::TestAt()
    {
    // Test value At()
    const TInt value = KTestValue;
    iArray->AppendL(value);
    EUNIT_ASSERT(value == iArray->At(iArray->Count()-1));
    }

void CArrayTests::TestDeleteL()
    // Index is starting from 0
    {
    const TInt value = KTestValue;
    iArray->AppendL(value);
    iArray->Delete(0);
    EUNIT_ASSERT(iArray->Count()==0);
    }

void CArrayTests::TestMaxElementsL()
    {
    // Append max count of elements
    // Index starts from 1

    for (TInt i=1; i<=KMaxElements; i++)
    {
       iArray->AppendL(i);
       EUNIT_ASSERT( i == iArray->Count());
       DEBUG_PRINT(_L("  CArrayTests: Testing KMaxElements, count
         is %d"),i);
    }
    }

void CArrayTests::TestAtMax()
    {
    // Test value At(KMaxElements)
    for (TInt i=1; i<=KMaxElements; i++)
    {
       iArray->AppendL(i);
    }

    const TInt value = iArray->At(KMaxElements-1);
    EUNIT_ASSERT(value == KMaxElements);
    DEBUG_PRINT(_L("  CArrayTests: Value At(KMaxElements-1) = %d"),
      value);
    }

void CArrayTests::TestOverflowL()
    {
    // Fill array to KMaxElements
    for (TInt i=1; i<=KMaxElements; i++)
    {
       iArray->AppendL(i);
    }
```

```
    // Append to max elements+1, use Insert
    // E32USER-CBase 21 panic is raised
    const TInt value = KTestValue;
    iArray->InsertL(KMaxElements+1, value);
    }

void CArrayTests::Teardown()
    {
    delete iArray;
    iArray = 0;
    }
```

For automated execution of test cases, the test cases are implemented as functions in the source code. When the cases are executed, the unit testing framework is used to print a summary of the test results to the debug output. If assertion in the test case fails, a breakpoint is raised by the framework. Then, with the aid of the debugger, the reason for the failed execution can be solved.

In the previous example, the test cases are implemented independently. So, every test case can be executed without disturbing the other test cases. The `CArrayTests::SetupL()` is used to initialize the class under test and then `CArrayTests::Teardown()` to delete it.

7.4.5 Test Execution Results

The unit testing framework prints the results and a summary of the test execution to the debug output window. Also, other test execution output can be seen in the printout. The results should show directly whether the test case is passed or not.

With `EUnitDebugRunner` the execution output of `CArrayTests` looks like this:

```
EUnitDebugRunner: Loading test dll carraytests.dll
AddLibrary carraytests[1000af59]
EUnitDebugRunner: Running 7 tests
EUnitDebugRunner: "CArrayTests test suite" Started
  EUnitDebugRunner: "Test empty" Started
  EUnitDebugRunner: "Test empty" executed.
  EUnitDebugRunner: "Test append" Started
  EUnitDebugRunner: "Test append" executed.
  EUnitDebugRunner: "Test append and OOM" Started
  EUnitDebugRunner: "Test append and OOM" executed.
  EUnitDebugRunner: "Test at" Started
  EUnitDebugRunner: "Test at" executed.
  EUnitDebugRunner: "Test delete" Started
  EUnitDebugRunner: "Test delete" executed.
  EUnitDebugRunner: "Test append KMaxElements" Started
  CArrayTests: Testing KMaxElements, count is 1
  CArrayTests: Testing KMaxElements, count is 2
  CArrayTests: Testing KMaxElements, count is 3
  CArrayTests: Testing KMaxElements, count is 4
  CArrayTests: Testing KMaxElements, count is 5
```

```
  EUnitDebugRunner: "Test append KMaxElements" executed.
  EUnitDebugRunner: "Test at KMaxElements-1" Started
  CArrayTests: Value At(KMaxElements-1) = 5
  EUnitDebugRunner: "Test at KMaxElements-1" executed.
  EUnitDebugRunner: "CArrayTests test suite" executed.
EUnitDebugRunner: 7/7, 100% of tests succeeded.
RemoveLibrary carraytests[1000af59]
```

In error cases, a breakpoint is raised to show the failed `EUNIT_ASSERT`. So, the error can be investigated directly. Of course, some test cases cause a panic to be raised; in such a case, the call stack should be checked. The call stack contains the stack of the current function calls.

An example of the call stack when a user breakpoint (failed assertion) is raised is given below; the assertion was modified to fail in the example to show the error situation:

```
TEUnitDebugAssertionFailureHandler::HandleAssertionFailureL(const
TEUnitAssertionInfo & {...}) line 147 + 9 bytes
EUnit::AssertionFailedL(const TEUnitAssertionInfo & {...}) line 19
CArrayTests::TestAppendL() line 114 + 75 bytes
CEUnitTestSuiteClass::TTestCase::RunTestL() line 241
MEUnitTestCase::Run(MEUnitTestResult & {...}) line 56 + 47 bytes
CEUnitTestSuiteClass::Run(MEUnitTestResult & {...}) line 148
CEUnitTestDll::Run(MEUnitTestResult & {...}) line 65
MainL() line 286
E32Main() line 310 + 29 bytes
startupThread() line 211 + 5 bytes
RunMainThread(void * 0x00402949 startupThread(void)) line 341 + 5 bytes
User::StartThread(int (void *)* 0x5b001604 RunMainThread(void *),
  void *
0x00402949 startupThread(void)) line 438 + 7 bytes
runThread(void * 0x008623e8) line 173 + 23 bytes
KERNEL32! 77e8758a()
```

In the first line of the call stack, it can be seen that the panic is caused by an assertion failure. In line 4, the call stack shows the function `CArrayTests::TestAppendL()`, where the panic occurred. Now, the developer can jump directly to the error situation in the debugger and, for example, investigate the value of the variables.

After continuing execution of the test cases in the debugger from the breakpoint, the result in the debug output also contains the failed test case. In the next example, the whole execution result is presented:

```
AddLibrary carraytests[1000af59]
EUnitDebugRunner: Running 7 tests
EUnitDebugRunner: "CArrayTests test suite" Started
  EUnitDebugRunner: "Test empty" Started
  EUnitDebugRunner: "Test empty" executed.
  EUnitDebugRunner: "Test append" Started
  EUnitDebugRunner: "Test append": Assertion failure: (iArray->
    At(iArray->Count()-1) != value), \SAMPLE\CARRAYTESTS\
    Carraytests.cpp, line 114 ""
```

```
  EUnitDebugRunner: "Test append" executed.
  EUnitDebugRunner: "Test append and OOM" Started
  EUnitDebugRunner: "Test append and OOM" executed.
  EUnitDebugRunner: "Test at" Started
  EUnitDebugRunner: "Test at" executed.
  EUnitDebugRunner: "Test delete" Started
  EUnitDebugRunner: "Test delete" executed.
  EUnitDebugRunner: "Test append KMaxElements" Started
  CArrayTests: Testing KMaxElements, count is 1
  CArrayTests: Testing KMaxElements, count is 2
  CArrayTests: Testing KMaxElements, count is 3
  CArrayTests: Testing KMaxElements, count is 4
  CArrayTests: Testing KMaxElements, count is 5
  EUnitDebugRunner: "Test append KMaxElements" executed.
  EUnitDebugRunner: "Test at KMaxElements-1" Started
  CArrayTests: Value At(KMaxElements-1) = 5
  EUnitDebugRunner: "Test at KMaxElements-1" executed.
  EUnitDebugRunner: "CArrayTests test suite" executed.
EUnitDebugRunner: 6/7, 85% of tests succeeded.
RemoveLibrary carraytests[1000af59]
```

In the previous example, the failure of the test case 'test append' was caused intentionally by a wrong assertion in line 7. The overflow test is commented in the source test code as a panic is raised when it is executed. The failed test case results are also displayed. The result can be used directly in the test report. Lines 17–21 and line 24 contain debug outputs which support and give more information on the test execution.

7.4.6 Summary of the Unit Testing

Unit testing is probably the most efficient way to test an executable code. Since unit tests are made in the implementation phase, the errors found in the code can be corrected right away. When the unit tests themselves are made executable they can also be used as automated regression test cases. In the Series 60 environment, assertions can be used to check the results and conditions in the test cases.

The unit tests can be documented as source code. Since the assertion clause is used to check the condition in the test case, other result documentation is not needed. Also, the error situations can be solved directly, because the test cases focus the testing on small parts of the program.

Why is unit testing so important? Good reasons are, for example:

- the code is executed and tested at least once before other testing phases;
- the boundary value errors are found and corrected directly;
- experimentation with the code is possible.

The other testing phases cannot test the internal design as well as unit testing can. For example, functional testing is focused on the whole implementation and its features. The correction of an error is also more time-consuming in the other testing phases and may not require any debugging.

7.5 Application Tester

Capture, scripting, and test automation tools are thought to solve resource problems in regression testing. When making test cases, it should be noticed that if it is possible to automate the test case execution, it can later be a very time-saving and economic test case in regression testing. So, the test cases should be written as if they were automated, and then the other test cases will be made manually. Even if the test is not automated (so there is no automated test input, execution, or result verification) it can be a great advantage to have the capture or the other scripts to do the uninteresting initialization job. For regression testing, the exact initialization of test cases is a great help, even when test result verification is done manually. Also, the capture and scripting tools are handy for making a large amount of input, and that is usually enough for stress and smoke testing. Smoke tests are used in testing the minimum executability (Binder, 1999, pages 325 and 628).

The testing scripts are made with the aid of test cases. For example, functional test cases can be used. The first thing is to initialize the environment for a test. This is usually done through a subscript. Then, the environment is used by an application in a way that the desired test situation occurs. If a capture tool is used the user interface events are recorded so that they may later be played back by the subscripts. After that, the actual test is executed. With GUIs, result verification may be a problem. The user can be informed by a note about the need for result verification, and so result verification is done manually. Another possibility is to take a screenshot of the situation and then compare it with a picture taken beforehand.

Functional test case automation can be troublesome. In functional testing, capture and replay may be used for initialization and setup only. In stress and smoke testing, the situation is different. There is a need to execute the application for a long time or to input a large amount of data. In the smoke tests, the applications may only be started and then closed to verify the minimum execution.

An example of an application tester, called Digia AppTest, is shown in Figure 7.3. It is developed in Digia to support functional, stress, and smoke testing. It contains recording utilities and script language and it has many inbuilt events, for example, for creating an OOM error.

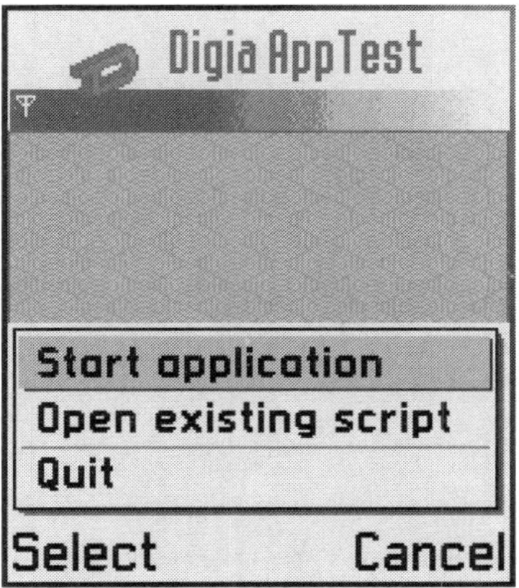

Figure 7.3 Application tester in the emulator environment

It also writes log files with memory information and timestamp, if a memory tracker DLL is used. These features make it quite effective for easy capture and script writing for stress tests. Test suites can be made and call to other test scripts is provided for modularity. AppTest can be used in hardware as well. The playback and, for instance, taking of screenshots are used as in the emulator environment.

7.5.1 DLL Command

Digia AppTest has verification support provided by a custom DLL interface. It makes it possible to extend the language for the exact verification purpose, and has complete custom DLLs for screenshots and memory-tracking commands. In the Series 60 GUI, it is quite hard to perform direct result verification automatically, but with a custom DLL this is possible. Of course, use of DLLs needs programming skills, but, in general, test drivers, stubs, and test applications that require programming skills have already been used before the functional testing phase.

Use of DLLs in result verification makes the testing a little closer to unit testing (the term 'gray box' may be used) and does not eliminate the need for test verification by the GUI. In practice, automated GUI content verification should be done by comparison, optical character recognition (OCR), text recognition, or by a GUI content support output (e.g. in plain text):

```
class MDLLInterface
    {
    public:
        virtual TInt Initialize( CScript* aScript, CLog* aLog ) = 0;
        virtual void Teardown() = 0;
```

```
        virtual TBool StartL( CVariables* aArray,
          CCoeEnv* aCoeEnv ) = 0;
        virtual TBool EndL( CCoeEnv* aCoeEnv ) = 0;
};
```

DLL command execution returns information on the execution of the DLL command. For return values, blocks are defined, for example, for failed situations. The DLL command has interfaces for starting, initializing, tearing down, and ending the command. The DLL is then called from the script with a DLL tag.

7.5.2 Scripting Language

Digia AppTest has a language for making **scripts**. It uses tags for script commands, special key codes for scanned events, blocks, and subscripts for modularity. The language also supports variables and loops. An exception mechanism is used to catch, for example, leaves in OOM situations. The reason (value) of the leave can be displayed by a variable.

With blocks, it is possible to make the script more modular. Also, by including subscripts, script maintenance and script programming is easier. Of course, blocks are also used as procedures for common tasks in the scripts.

In the following example, a simple test suite for a smoke test is presented. The used application is Phonebook, which is defined with the APP tag. This suite is a collection of subscripts PbkTest01, PbkTest02, and PbkTest03, which are executed in a loop defined by the LOOP tag. The first subscript makes a new entry in the phonebook, the second opens and views the entry, and the third one deletes it. If the application leaves during the execution, the leave is caught and displayed in the screen. The test suite is as follows:

```
<BODY>
     <NAME>Pbksmoke</NAME>
     <AUTHOR>It's me</AUTHOR>
     <DESCRIPTION>Smoke tests and loops</DESCRIPTION>
     <DATE>15.04.2002 10:12</DATE>
     <MAIN>
           <APP name="Phonebook.app">
               <LOOP start="0" stop="5" interval="1" var="i">
                      <LOG>Loop count ${i}</LOG>
                      <INCLUDE name="c:\documents\PbkTest01.txt"/>
                      <INCLUDE name="c:\documents\PbkTest02.txt"/>
                      <INCLUDE name="c:\documents\PbkTest03.txt"/>
               </LOOP>
           </APP>
     </MAIN>
     <EXCEPTION type="Leave">
            <NOTE>Leave.. ${value}</NOTE>
     </EXCEPTION>
</BODY>
```

A test suite is used to collect many individual scripts together. With smaller subscripts, the script maintenance is much easier. The subscripts are used to make a library for test suites and so to make modular test suites. In the example above, the included subscripts contain captured input for replaying the execution. The subscripts are designed in a way that all views of the application are checked, and then the subscripts return to the main view. The execution can be looped. With this script, the application is smoke tested and the execution can be made for a long period (e.g. one day), or the amount of used data can be large.

The advantages of automated smoke and stress tests are that they are easy to do and they can be executed often, even when there is no time to test manually. Manual test may not even find those errors that can be detected by executing stress tests.

The script files contain the captured and possibly modified information and commands in application tester scripting language. A part of the captured events script file is as follows:

```
<BODY>
<NAME>PbkTest01</NAME>
<AUTHOR>It's me</AUTHOR>
<DESCRIPTION>Phonebook views</DESCRIPTION>
<DATE>04.06.2002 01:02</DATE>
     <MAIN>
          <APP name="PHONEBOOK.app">
               <KEYDOWN interval="1.58">EStdKeyLeftFunc</KEYDOWN>
               <KEYDOWN interval="0.48">EStdKeyDevice0</KEYDOWN>
               <KEY interval="0.00">EKeyDevice0</KEY>
               <KEYUP interval="0.05">EStdKeyDevice0</KEYUP>
               <KEYUP interval="0.04">EStdKeyLeftFunc</KEYUP>
               <KEYDOWN interval="0.86">EStdKeyLeftFunc</KEYDOWN>
               <KEYDOWN interval="0.17">EStdKeyDevice0</KEYDOWN>
               <KEY interval="0.00">EKeyDevice0</KEY>
               <KEYUP interval="0.08">EStdKeyDevice0</KEYUP>
               <KEYUP interval="0.02">EStdKeyLeftFunc</KEYUP>
               <KEYDOWN interval="1.23">EStdKeyRightArrow</KEYDOWN>
               <KEY interval="0.00">EKeyRightArrow</KEY>
               <KEYUP interval="0.13">EStdKeyRightArrow</KEYUP>
               <KEYDOWN interval="1.28">EStdKeyLeftArrow</KEYDOWN>
               <KEY interval="0.00">EKeyLeftArrow</KEY>
               <KEYUP interval="0.00">EStdKeyLeftArrow</KEYUP>
               <KEYDOWN interval="1.01">EStdKeyLeftFunc</KEYDOWN>
               <KEYDOWN interval="0.25">EStdKeyDevice1</KEYDOWN>
               <KEY interval="0.00">EKeyDevice1</KEY>
               <KEYUP interval="0.06">EStdKeyDevice1</KEYUP>
               <KEYUP interval="0.03">EStdKeyLeftFunc</KEYUP>

          </APP>
     </MAIN>
</BODY>
```

The timestamp is seen in this sample. There is a possibility of using timing when doing the playback. On left of the script there are timestamps (intervals) that are used for timing the playback. The captured events are used to open and view an entry in the Phonebook.

For example, in the emulator environment, pressing key 'c' causes the following events to be captured:

```
<KEYDOWN interval="0.47">C</KEYDOWN>
<KEY interval="0.00">c</KEY>
<KEYUP interval="0.09">C</KEYUP>
```

Events can be read and written in plain text, and the captured input can be used directly for playback. For clarity, the capture and playback script files should be used in small subscripts for modular design of the tests.

App Test is capable of catching leaves:

```
<BODY>
<NAME>Phonebook</NAME>
<AUTHOR>It's me</AUTHOR>
<DESCRIPTION></DESCRIPTION>
<DATE>21.05.2002 12:43</DATE>
     <MAIN>
           <APP name="PHONEBOOK.app">
           </APP>
                <APP name="PHONEBOOK.app">
                <OOM type="EDeterministic" rate="50">
                <KEYDOWN interval="2.02">EStdKeyLeftFunc</KEYDOWN>
                <KEYDOWN interval="0.40">EStdKeyDevice0</KEYDOWN>

                (...Here are the captured events)

                <KEYDOWN interval="0.55">EStdKeyDevice1</KEYDOWN>
                <KEY interval="0.00">EKeyDevice1</KEY>
                <KEYUP interval="0.04">EStdKeyDevice1</KEYUP>
                <KEYUP interval="0.04">EStdKeyLeftFunc</KEYUP>
                </OOM>
           </APP>
     </MAIN>

     <EXCEPTION type="Leave">
            <NOTE>Leave.. ${value}</NOTE>
     </EXCEPTION>
</BODY>
```

With an exception tag, it is possible to define that, for example, a leave is caught. This is very convenient in OOM error situations, for instance, because execution of the script can be continued after OOM. Usually, the OOM error note is displayed by the system to the user, as in Figure 7.4. The OOM rate value can be initialized in the loop, and so the OOM test coverage can be created. The OOM errors are used to test the capability of the application to handle memory allocation failures.

Figure 7.4 Out-of-memory error note displayed by the system

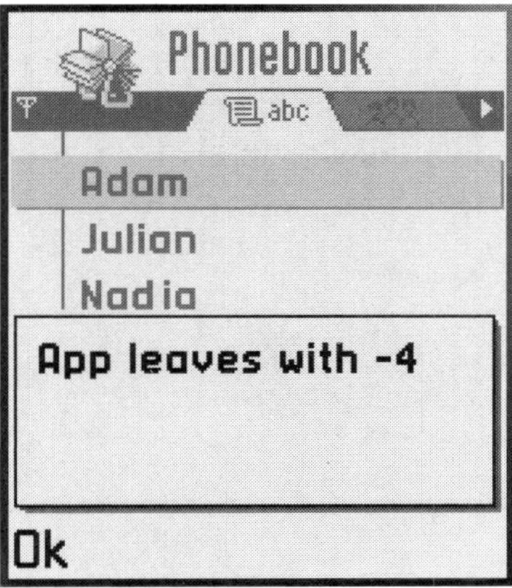

Figure 7.5 Example of the out-of-memory execution output to the screen

When an OOM error happens, the exception can be displayed to the user by a note. In this example (Figure 7.5), the application does not show the OOM error note when the AppTest is active, since instead of the system's OOM error note, the AppTest's exception block is called.

A note 'Leave.. −4' is displayed to the user. It is important that the OOM error can be caught, and the memory tests can continue. The global error value (−4) means `KErrNoMemory` in the Series 60 and Symbian environment.

In the following example, the memory consumption of the application is investigated during execution. In line 9, the memory tracker DLL is started. The memory tracking is done at one-second intervals, and the log is written to a file `MemScan.log`. The used subscript in this example is PbkTest02, which opens and views one entry in the Phonebook:

```
<BODY>
<NAME>Phonebook</NAME>
<AUTHOR>It's me</AUTHOR>
<DESCRIPTION></DESCRIPTION>
<DATE>21.05.2002 12:43</DATE>
     <MAIN>
           <APP name="PHONEBOOK.app">
           </APP>
           <DLL interval="1.0" dll="MemTrackerDLL.Dll"
              app="phonebook.app" log="c:\MemScan.log">
                <APP name="PHONEBOOK.app">
                     <INCLUDE name="c:\documents\Pbk\PbkTest02.txt"/>
                </APP>
           </DLL>
     </MAIN>
</BODY>
```

Memory tracker is one of the extension DDLs provided with Digia AppTest. With the memory tracker it is possible to track the memory consumption of the application in some time interval. The memory tracker makes a log file, where the memory consumption and heap size are displayed:

```
Tracking application "phonebook.app"
04th Jun 2002 12:58:10.98 AM      Bytes allocated = 65216,
  Heap size = 85892, Available = 4236, Cells allocated = 1337
04th Jun 2002 12:58:12.14 AM      Bytes allocated = 93584,
  Heap size = 122756, Available = 6648, Cells allocated = 1839
04th Jun 2002 12:58:13.24 AM      Bytes allocated = 93584,
  Heap size = 122756, Available = 6648, Cells allocated = 1839
04th Jun 2002 12:58:14.35 AM      Bytes allocated = 93584,
  Heap size = 122756, Available = 6648, Cells allocated = 1839
04th Jun 2002 12:58:15.45 AM      Bytes allocated = 93584,
  Heap size = 122756, Available = 6648, Cells allocated = 1839
04th Jun 2002 12:58:16.55 AM      Bytes allocated = 97828,
  Heap size = 122756, Available = 1612, Cells allocated = 1907
04th Jun 2002 12:58:17.65 AM      Bytes allocated = 97828,
  Heap size = 122756, Available = 1612, Cells allocated = 1907
```

The log file contains information on the execution. The timestamp and used memory are displayed. The memory consumption can be monitored for long periods and it can be useful in optimization of the application, for example. In the memory tracker log, for example, increasing memory consumption can be detected.

7.5.3 Summary of Digia AppTest

Digia AppTest is used in functional, stress, and smoke testing. The testing is made on a complete application and its user interface.

Usually, complete automation is impossible, and this should be notified when starting to make scripts for the testing. It is wise to start with the simplest possible goals. In smoke and stress testing, capture and replay capability can be sufficient to make the scripts if the execution can be looped.

AppTest is suitable for the following situations:

- smoke tests,
- stress tests,
- functional test support in initialization and setup,
- out-of-memory tests of an application,
- functional testing, when the result verification is done by DLLs or manually.

Scripting all the tests is difficult, because the maintenance work can increase during the project. Scripts should be modular, and subscripts should be used. When the application or program changes, for example, by the user interface it is necessary only to modify that subscript.

Digia AppTest provides DLL extension support. If the result verification can be made by a DLL it is possible to automatize the functional tests too. Sometimes, the term 'grey box' is used in this testing situation.

7.6 Summary

Although testing should be systematic work, it needs not be uninteresting. When the schedule is tight and there are many instructions in the process there is also room for creativity. As in software development, testing needs skill and creativity for good results to be obtained. Experimentation is more than allowed, especially in unit testing.

Although testing finds nasty-sounding things, such as errors, during the project, and shows that something does not work, it should be remembered that, finally, all tests should be passed and the testing results should be accepted. This means that the program, too, has reached an acceptable quality and it is confirmed that it does what is required. To find and correct an error is not enough, since we know that every error correction should be verified, and every change in the software can be a place for a new defect.

Testing as a process should be modified for the current software project, and it should be developed during and afterwards. If some process fits the application development, it may not fit a low-level device driver implementation. Also, the scale of the project can be a challenge.

When starting a software project one should also define how the testing should be done. This includes, for example:

- the testing techniques and methods;
- the testing process and the way the testing is implemented in the project;
- responsibilities;
- error database and change management;
- documentation, test plans, suites, and cases;
- the version control system;
- testing tools and environment.

Testing skills may also include development capability. In iterative software development there is a need for a testing environment where the environment is capable of executing software from almost the starting phase of the project. This increases the need for test support software (e.g. test drivers, stubs, and applications) and, of course, the need to develop them.

Part 2

Graphics, Audio, and User Interfaces

8

Application Framework

The benefits of separating a graphical user interface (GUI) application into an engine part and user interface (UI) part were discussed in several chapters in Part 1 of this book. The application **engine**, also known as an application **model**, contains all data of the application and the structures and algorithms dealing with it. The engine is responsible for interacting with the system devices, such as the file system, communications, etc. The application **user interface**, usually called simply the **application** in Symbian OS, has two main tasks: it deals with the visual representation of the application data on the screen (picked up from the engine), and it sets the overall behavior of the application, defining how the users interact with the application.

This chapter provides a description of the Uikon **application framework**, in addition to its Avkon **extension**. Uikon provides the application launching service, a set of standard UI components and user input handlers for the applications. Avkon provides the concrete look-and-feel for UI components and new UI components for smartphones based on the Series 60 Platform. The other chapters in Part 2 describe the use of Avkon UI components in the applications.

8.1 User Interface Architecture

A high-level description of the UI architecture of any GUI application is shown in Figure 8.1. The system level contains the base (EUser library, device drivers, etc.) and system servers (communications servers, Window server, etc.). The graphics layer takes care of drawing all the graphics on the screen. The core contains the application framework, called Uikon. The Uikon framework consists of several libraries providing services for application launching and **state persistence**, which allows the restoration of the same state in which the application was previously closed. Other libraries provide classes for event handling

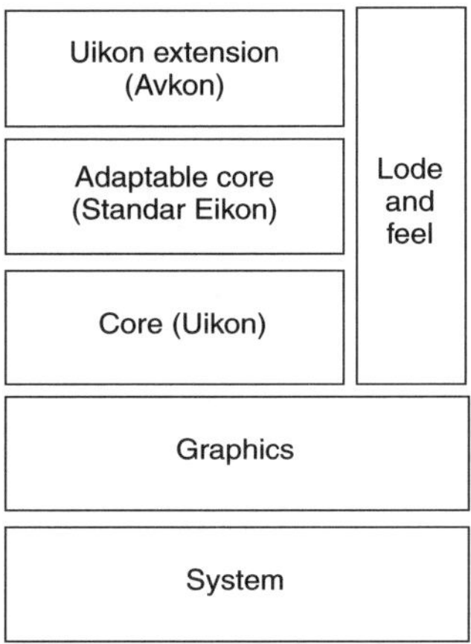

Figure 8.1 User interface architecture

and standard GUI components (also known as controls) with basic look and feel for all Uikon applications. An adaptable core (standard Eikon) is implemented by the provider of the UI style. In the case of the Series 60 Platform, standard Eikon has been implemented by Nokia. Together with a style-dependent framework extension (Avkon), standard Eikon provides the style-dependent set of UI components with their own look and feel.

In Figure 8.2, the contents of the Uikon framework is depicted in more detail. **Application architecture** (App Arc) and **control environment** (CONE) provide a great deal of UI functionality in a general and an abstract form. Uikon makes the UI functionality concrete. In addition, Uikon supplies the look and feel for the standard UI components. The Uikon and CONE frameworks have virtual functions through which the rest of Symbian OS can be reused. Application developers derive classes implementing these virtual functions with any specific functionality required. Writing GUI applications involves a great deal of derivation from the base classes of Uikon and CONE. The role of the CONE in event handling and graphics drawing will be described later in this chapter (Section 8.3.2).

The basic classes of any GUI application are shown on the left-hand side of Figure 8.3. Sometimes, applications may require a separate **controller** object, if the engine provides asynchronous services, taking a long time to execute, for example. The controller is added to the set of basic classes on the right-hand side of Figure 8.3. If a separate controller is not required, the **AppUI** (`CMyAppUi`) also contains the controller's functionality. The **application** class (`CMyApplication`) is used in the launching of the application. The **document** class creates the application user interface. In addition, it provides a layer between

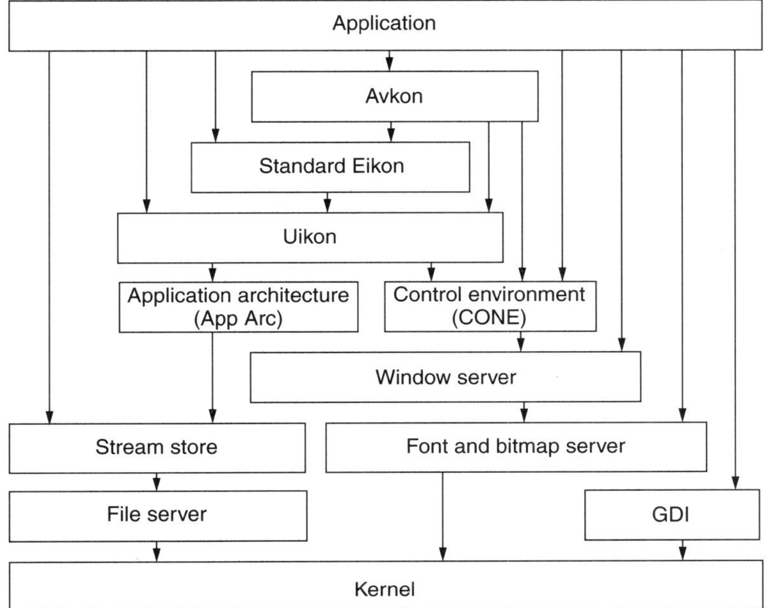

Figure 8.2 Uikon application framework; GDI, graphics device interface

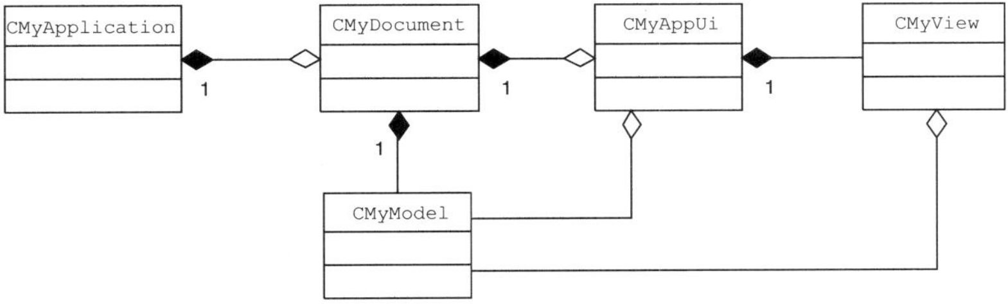

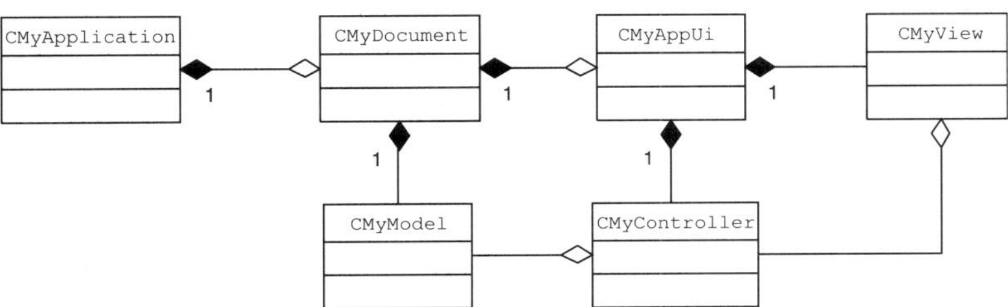

Figure 8.3 Basic application classes

the user interface, the model, and the file in file-based applications. The commands invoked by the user are handled in the AppUI class. The **view** class (`CMyView`) displays the user's data on the screen.

Note that there is not a base class for engine (`CMyModel`) in the application framework. The engine can be derived, for example, directly from `CBase`, which is the base class of any compound class, or `CActive`, which is the base class of active objects.

8.1.1 Uikon Application Classes

The Uikon framework is used by deriving own application classes from its base classes and specializing their behavior if required. The base classes of applications and their relationship to the Ukon application architecture and control environment are shown in Figure 8.4.

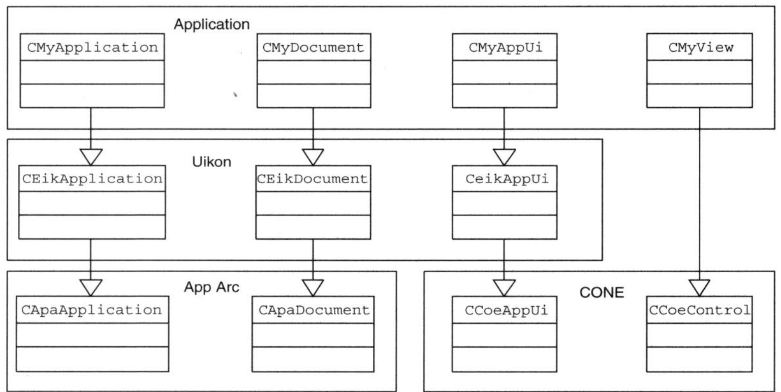

Figure 8.4 Uikon application classes; App Arc, application architecture; CONE, control environment

Application

The application class is derived from `CEikApplication`, which is derived from a class declared by the application architecture. There are two functions to be implemented in this class, `CreateDocumentL()`, which creates the application's document, and `AppDllUid()`, which identifies the application by returning its unique identifier.

The declaration of the application class of the rock–paper–scissors (RPS) game is as follows:

```
// INCLUDES
#include <eikapp.h>

// CONSTANTS
// UID of the application
const TUid KUidRPSGame = { 0x101F5BBD };
```

```
// CLASS DECLARATION

/**
* CRPSGameApp application class.
* Provides factory to create concrete document object.
*
*/
class CRPSGameApp : public CEikApplication
    {
    private:
        /**
        * From CApaApplication, creates CRPSGameDocument document
        * object.
        * @return A pointer to the created document object.
        */
        CApaDocument* CreateDocumentL();
        /**
        * From CApaApplication, returns application's
        * UID (KUidRPSGame).
        * @return The value of KUidRPSGame.
        */
        TUid AppDllUid() const;
    };
```

The implementation of two functions is quite straightforward:

```
// INCLUDE FILES
#include    "RPSGameApp.h"
#include    "RPSGameDocument.h"
// ================= MEMBER FUNCTIONS =======================
// ---------------------------------------------------------
// CRPSGameApp::AppDllUid()
// Returns application UID
// ---------------------------------------------------------
//
TUid CRPSGameApp::AppDllUid() const
    {
    return KUidRPSGame;
    }
// ---------------------------------------------------------
// CRPSGameApp::CreateDocumentL()
// Creates CRPSGameDocument object
// ---------------------------------------------------------
//
CApaDocument* CRPSGameApp::CreateDocumentL()
    {
    return CRPSGameDocument::NewL( *this );
    }
```

`AppDllUid()` returns the unique identifier (UID) number (0x101F5BBD) and `CreateDocumentL()` uses document's `NewL()` factory function to create a document. Note that a pointer to the application class is passed to the document. The document needs the UID value from the application class when handling application files.

Document

The document class is derived from `CEikDocument`, which is derived from a class in the application architecture. The document class instantiates the application engine (model) and creates the AppUI object using the `CreateAppUiL()` function. The `TryLoadDataWrittenByMsgViewerL()` function is used to read any existing message data (i.e. moves from a file) if another player has challenged this player earlier.

```
// INCLUDES
#include <eikdoc.h>

// CONSTANTS

// FORWARD DECLARATIONS
class CEikAppUi;
class CRPSGameAppUi;
class CRpsMessage;
class CRpsModel;

// CLASS DECLARATION
/**
*  CRPSGameDocument application class.
*/
class CRPSGameDocument : public CEikDocument
    {
    public:     // Constructors and destructor
        /**
        * Two-phased constructor.
        */
        static CRPSGameDocument* NewL(CEikApplication& aApp);
        /**
        * Destructor.
        */
        virtual ~CRPSGameDocument();

    public:     // New functions
        CRpsModel& Model();

    protected:      // New functions
        /**
        * Checks, if there is a received message. Reads it, if it
        * exists. Otherwise starts a new game.
        */
        void TryLoadDataWrittenByMsgViewerL();

    private:     // from CEikDocument
        /**
         * From CEikDocument, create CRPSGameAppUi "App UI" object.
         */
        CEikAppUi* CreateAppUiL();
    private:
        CRPSGameDocument( CEikApplication& aApp );
        void ConstructL();
    private:
        CRpsModel* iModel;
    };
```

In addition to the model and application UI creation, the document stores (and restores) the application's data, including the user data and application settings, such as the selected language of the application. All filing operations are also performed from the document.

Data is stored using a **stream store**, which is built on top of the file server. Stream store defines services for adding and deleting streams to a store. A **stream** is a sequence of any binary data, representing, for example, the state of the object, and a **store** is a collection of streams. The base class for the document (`CEikDocument`) has two functions for storing and restoring the state of the application: `StoreL()` and `RestoreL()`. The functions have an empty implementation and must be implemented by the developer who wishes to persist application data.

To avoid unnecessary storing in cases where document data have not been changed, `SaveL()` can be used. The document has a flag indicating whether its content has changed or not. This flag may be set by using a `SetChanged()` function and get using a `HasChanged()` function, respectively. `SaveL()` calls `StoreL()` only if `HasChanged()` returns `ETrue`.

A complicated application may require several streams to store its state. First, document data are stored. Document's `StoreL()` calls `ExternalizeL()` in the model, which may further call the `Externalize()` function of the components. We saw the opposite in Chapter 5. The document asked the model to read message data by calling its `AddDataFromReceivedMsgL()` function, which called `Internalize()` first from the `CRpsMessage` class and then from the `CRpsMoves` class.

Each stored (or restored) stream is identified by an identification number (ID). In a store, the first stream is a dictionary stream, containing the IDs of all the streams it contains. The individual streams are accessed in the store, using the dictionary stream.

The implementation of the document class of the RPS game is shown below. We have omitted constructor and destructor functions because of their simplicity and `TryLoadDataWrittenByMsgViewerL()`, because it has already been shown in Chapter 5 (page 91). After the creation of the model, the document checks first if there is a new message received. Application user interface is created in a standard way using two-phase construction:

```
// INCLUDE FILES
#include "RPSGameDocument.h"
#include "RPSGameAppUi.h"
#include "RpsModel.h"

// ================= MEMBER FUNCTIONS =======================
void CRPSGameDocument::ConstructL()
```

```
    {
    iModel = CRpsModel::NewL();
    TryLoadDataWrittenByMsgViewerL();
    }

CEikAppUi* CRPSGameDocument::CreateAppUiL()
    {
    return new (ELeave) CRPSGameAppUi;
    }
```

AppUI

The AppUI class is derived from `CEikAppUI`, which is derived from the base class in the CONE (`CCoeAppUi`). The main task of this class is to handle application-wide UI details. Responses to various kinds of events can be handled at the AppUI level by implementing virtual functions of the base class. The functions are not purely virtual, but their implementation is empty. Virtual functions to be implemented are:

- `HandleKeyEvent()`, which is called if a key press event occurs (and all the controls on the control stack return `EKeyWasNotConsumed` from the `OfferKeyEventL()` function); we will come back to event handling in Section 8.3.2;
- `HandleForegroundEventL()`, which handles changes in keyboard focus when an application is switched to the foreground or from the foreground;
- `HandleSwitchOnEventL()`, which handles an event, when the smartphone is switched on;
- `HandleSystemEventL()`, which handles events generated from the Window server;
- `HandleMessageReadyL()`, which handles the message ready message generated by the Window server;
- `HandeApplicationSpecificEventL()`, which handles any application-specific events; this function is called when the framework has an event ID beyond the range of Window server events;
- `HandleCommandL()`, which is called when the user issues commands; this may be the only function that is implemented from the event handling functions.

The AppUI class also takes care of creating application views. The views are created in the `ConstructL()` function. In the RPS game, the application UI class is derived from Avkon extension, and we will show the code in later examples in this chapter (see Section 8.1.2).

View

The application view is derived from the CONE control class (`CCoeControl`) or any class for which the base class is CCoeControl. All controls need to draw themselves (unless they are simply containers having controls). Thus, they have to implement a virtual `Draw()` function, defined in `CCoeControl`. The `Draw()` function receives a parameter, which is a reference to `TRect` object generated by the windows server. The object contains the smallest bounding box to need redrawing. `TRect` classes represent rectangles, the sides of which are parallel to the axes of the coordinate system.

8.1.2 Avkon Application Classes

The Avkon extension provides one layer of classes more between application and Uikon classes, as shown in Figure 8.5. The Series 60 applications are expected not to support file persistence, or .ini files. `CAknApplication` removes .ini file opening functionality from the applications by over writing the `OpenIniFile()` function so that it leaves if called. `CAknDocument` on the other hand removes the automatic file opening functionality during launching of the application by overwriting the `OpenFileL()` with an empty implementation of that function. The Rps game application class derives directly from the Uikon application framework classes even if it doesn't use .ini files to

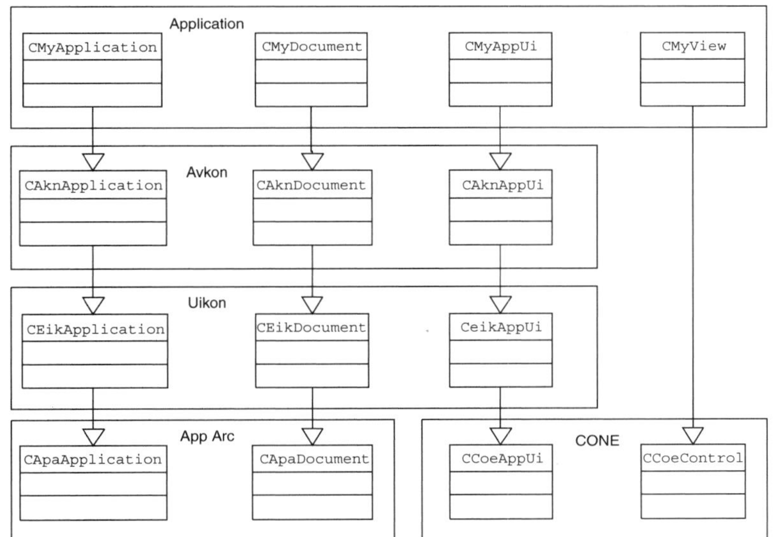

Figure 8.5 Avkon application classes; App Arc, application architecture; CONE, control environment

reduce the derivation chain, but it could have been made to be sub class of the `CAknApplication` as well.

Avkon AppUI

The `CAknAppUi` class provides color scheme support and look and feel for all generic controls. It also provides a few new accessories specific to the Series 60 Platform:

- status pane: located on the top part of the screen, this displays status information on the current application and state as well as general information on the device status (the signal strength and battery charging); the status pane may be invisible;
- command button array (CBA): located at the bottom of the screen, this shows the text associated with softkeys.

In addition, there is a functionality for handling system events in general and, for example, `HandleForegroundEventL()` implementation. There is another AppUI base class (in addition to `CAknAppUi`) in the Series 60 Platform: `CAknViewAppUi`. This class is used, if view classes are derived from the `CAknView` base class rather than from `CCoeControl`. The Series 60 view architecture uses these classes as well, as we will describe in Chapter 13.

The header file of the application UI class is shown below. The `KMsgBioUidRpsMessage` constant defines the RPS game smart message type. The other constant, `KNumberOfMoves`, defines how many moves must be selected before a challenge can be created or a challenge can be responded to `TRpsGameState` defines the states of a simple state machine. We have used just one variable (`iState`) to indicate the state-dependent behavior instead of applying a state design pattern. The initial state is `EStateNull`, which is changed to `EStateChallenge` or `EStateNewGame`, depending on whether the player has received a new challenge or not. When a response message has been received, the state is changed to `EStateResponse`.

`DoChallengeL()` checks if the player has made enough selections, after which it calls `SendChallengeMsgL()` to send the message. The `SendChallengeMsgL()` function uses the engine's communications services to send a message, or it sends it directly using the `CSendAppUi` class, which creates a message and uses the message framework to send the message to the other player.

`DoPlayL()` redraws the view and calls `CheckWinnerL()` to calculate the score for each player. The `InformationNoteL()` function shows the result of the game as a note (see Chapter 11). Finally, `DynInitMenuPaneL()` is called just before a menu is opened. This function determines, which **menu items** are available in the menu, based on the state of the game. For example, the challenge cannot be sent before a minimum number of moves has been selected.

Note that the view created by the application user interface is CRPS-GameContainer, indicating that there are several views in the container:

```
// INCLUDES
#include <eikapp.h>
#include <eikdoc.h>
#include <e32std.h>
#include <coeccntx.h>
#include <aknappui.h>

// FORWARD DECLARATIONS
class CRPSGameContainer;
class CRPSGameDocument;
class CRichText;
class CSendAppUi;

// CONSTANTS
const TUid KMsgBioUidRpsMessage = {0x0FFFFFFB};
const TInt KNumberOfMoves = 3;

_LIT(KLose, "You lose!");
_LIT(KWin, "You won!");
_LIT(KDraw, "It's a draw!");

// CLASS DECLARATION

/**
* Application UI class.
* Provides support for the following features:
* - Eikon control architecture
* - view architecture
* - status pane
*
*/
class CRPSGameAppUi : public CAknAppUi
    {
    public:
        enum TRpsGameState
            {
            EStateNull,
            EStateNewGame,
            EStateChallenge,
            EStateResponse
            };
    // // Constructors and destructor
    /**
    * EPOC default constructor.
    */
    void ConstructL();

    /**
    * Destructor.
    */
    CRPSGameAppUi();
```

```
public:     // New functions
    void SetState( TRpsGameState aState );
public:     // Functions from base classes

private:
    // From MEikMenuObserver
    void DynInitMenuPaneL( TInt aMenuId, CEikMenuPane* aMenuPane );

private:
    /**
    * From CEikAppUi, takes care of command handling.
    * @param aCommand command to be handled
    */
    void HandleCommandL( TInt aCommand );

    /**
    * From CEikAppUi, handles key events.
    * @param aKeyEvent Event to handled.
    * @param aType Type of the key event.
    * @return Response code (EKeyWasConsumed, EKeyWasNotConsumed).
    */
    TKeyResponse HandleKeyEventL( const TKeyEvent& aKeyEvent,
      TEventCode aType );

private:
    /**
    * Challenges another player (no challenge message received)
    */
    void DoChallengeL();

    /**
    * Sends the challenge message e.g. in the smart message
    */
    void SendChallengeMsgL();
    /**
    * Checks the result of the game
    * received and at least KNumberOfMoves made.
    */
    void DoPlayL();
    void CheckWinnerL();
    /**
    * Tells the winner in the UI.
    */
    void InformationNoteL( const TDesC& aText ) const;
    CRPSGameDocument& Document() const;

private:     //Data
    CRPSGameContainer* iAppContainer;
    CSendAppUi* iSendAppUi;
    TRpsGameState iState;
};
```

Avkon View

`CAppView` may be derived from `CAknView` or from `CCoeControl` or any of its subclasses. `CAknView` is not a control. It is not derived from `CCoeControl`, but directly from `CBase`. Although it is not a control, it is still a CONE view. When using objects instantiated from

CAknView-derived classes, we typically define a CCoeControl-derived container with no or a trivial Draw() method. The container then consists of all the visible controls.

In the RPS game, we have not derived our 'view' class from CAknView but rather from CCoeControl, as shown in the header file:

```
// INCLUDES
#include <aknview.h>

// CONSTANTS
const TInt KMaxNumOfItems = 3;

_LIT(KImagePath, "z:\\system\\apps\\rpsgame\\rpsimages.mbm");

// FORWARD DECLARATIONS
class CEikLabel;      // for example labels
class CRpsMoves;

// CLASS DECLARATION
enum TRpsContainerState
    {
    ESelecting,
    EPlaying
    };

/**
*  CRPSGameContainer  container control class.
*
*/
class CRPSGameContainer : public CCoeControl, MCoeControlObserver
    {
    public:     // Constructors and destructor

        /**
        * EPOC default constructor.
        * @param aRect Frame rectangle for container.
        */
        void ConstructL(const TRect& aRect);

        /**
        * Destructor.
        */
        ~CRPSGameContainer();
    public:     // New functions
        /**
        * State setting function.
        */
        void SetState( TRpsContainerState aState );

        /**
        * @return KNumberOfMoves
        */
        TInt MaxCount() const;
        /**
        * Adds a new move to selected set of moves.
        */
        void AddCurrentToSelectedItemsL();
        /**
```

```
        * @return the selected moves
        */
        const CRpsMoves& MyItems() const;
        void UndoLastItem();

        void SetMyItemsL( const CRpsMoves& aMoves );
        void SetOpponentItemsL( const CRpsMoves& aMoves );

    public:     // Functions from base classes
        TKeyResponse OfferKeyEventL(const TKeyEvent& aKeyEvent,
          TEventCode aType);

    private:     // from CCoeControl
        void SizeChanged();
        TInt CountComponentControls() const;
        CCoeControl* ComponentControl( TInt aIndex ) const;
        void Draw( const TRect& aRect ) const;
        void HandleControlEventL( CCoeControl* aControl,
          TCoeEvent aEventType );

    private:     // new functions
        void LoadImagesL( const TDesC& aImageFile );
        void ShowNextItem();
        void ShowPreviousItem();

    private:     //data
        CArrayPtrFlat<CFbsBitmap>* iBitmaps;
        TInt iCurrentItemId;
        CRpsMoves* iMyItems;
        CRpsMoves* iOpponentItems;
        TRpsContainerState iState;
        TInt iMyItemIndex;
        TInt iOpponentItemIndex;
    };
```

KMaxNumOfItems represents the number of moves that must be drawn on the screen. Also, the container class has states to restrict possible functionality. In the ESelecting state the player does not yet have enough moves to make a challenge (or to check the result). When the minimum number of required moves has been reached, the state is changed to EPlaying.

The AddCurrentToSelectedItemsL() function adds a new move to a set of moves and redraws the screen. MyItems() returns the move set, and UndoLastItem() removes the last move from the set of moves. SetMyItemsL() and SetOpponenItemsL() set the set of moves according to the data read from the message. The moves are drawn on the screen using the function PlayNextMyItem() and PlayNextOpponentItem().

The function of LoadImagesL() is simply to load the rock, paper, and scissors bitmaps from files. Functions overridden from CCoeControl will be described later in this Chapter 8.3.1.

8.2 Launching an Application

The application is started by selecting it from the shell. The shell gets the name of the application and puts it into a CApaCommandLine. After that, the shell creates a new process, passing apprun.exe (part of Eikon) as the name of the executable to run in the new process. Apprun.exe calls the EikDll::RunAppInsideThread() static function, which creates CEikonEnv environment. RunAppInsideThread() calls the CEikonEnv::ConstructAppFromCommandLineL() function, which loads the dynamic linh library (DLL; actually, it calls a function in the application architecture to do this).

ConstructAppFromCommandLineL() calls NewApplication(), which creates a new application process. The implementation of NewApplication() is as follows:

```
// ============================== OTHER EXPORTED FUNCTIONS
//
// -------------------------------------------------------
// NewApplication()
// Constructs CRPSGameApp
// Returns: CApaDocument*: created application object
// -------------------------------------------------------
//
EXPORT_C CApaApplication* NewApplication()
    {
    return new CRPSGameApp;
    }

// -------------------------------------------------------
// E32Dll(TDllReason)
// Entry point function for EPOC Apps
// Returns: KErrNone: No error
// -------------------------------------------------------
//
GLDEF_C TInt E32Dll( TDllReason )
    {
    return KErrNone;
    }
```

The application creates the document. The CreateAppUiL() function in the document class is called by ConstructAppFromCommandLineL(). The AppUI takes care of creating application views. Finally, RunAppInsideThread() calls the CCoeEnv::ExecuteD() function, which starts the active scheduler that runs the event loop waiting for the user input.

Note that new(ELeave) is not used in NewApplication() function and thus NewApplication() is not a leaving function. The framework checks whether the return value of the constructor is different from null to check the success of object creation. E32Dll() is required by all libraries. In our case, it does nothing.

8.3 Implementation of the User Interface of the Rock–Paper–Scissors Game

Let us describe the implementation of the rest of the functions of the RPS game. So far, we have not shown the implementation of two classes: CRpsGameAppUi and CRpsGameContainer. Nokia has written instructions for game development for the Series 60 Platform (Nokia, 2001e). Note that the example RPS game does not conform to these instructions and is not complete in that sense.

8.3.1 Creating the User Interface

In ConstructL() of CRPSGameAppUi, the BaseConstructL() function is called first to initialize the application user interface with standard values. The application resource file will be read as well. Next, the view instance is created (CRpsGameContainer). ClientRect() gets the area of the screen available for the application to draw views. By default, this is the main pane (i.e. main application view). This area is given to the view to indicate the area in which it can draw itself. AddToStackL() adds the created view into a **control stack**, which is used to determine to which control a key event should be given. An important thing to be done in the destructor is to remove the control from the control stack. We will explain the control stack in more detail in the event-handling section (Section 8.3.2). In case the messaging framework is used in the RPS message delivery, CSendAppUi can be used. It allows an easy way of creating and sending messages. The parameter specifies a menu item command. In our example, the value is 1; always 1 (i.e. information on the command is not used).

If the player is challenged, there already exists a challenge message, from which the opponent moves are copied. In the other case, both challenge and response messages are empty. After setting the moves, the state is changed to ESelecting, indicating it is the player's turn to select the moves:

```
void CRPSGameAppUi::ConstructL()
    {
    SetState( EStateNull );
    BaseConstructL();
    iAppContainer = new (ELeave) CRPSGameContainer;
    iAppContainer->ConstructL( ClientRect() );
```

```
AddToStackL( iAppContainer );
iSendAppUi = CSendAppUi::NewL(1);

if ( Document().Model().HasChallenge() )
    {
    iState = EStateChallenge;
    iAppContainer->SetOpponentItemsL(
      Document().Model().ChallengeMoves() );
    }
else
    {
    iState = EStateNewGame;
    iAppContainer->SetOpponentItemsL(
      Document().Model().ResponseMoves() );
    iAppContainer->SetMyItemsL(
      Document().Model().ChallengeMoves() );
    }
iAppContainer->SetState ( ESelecting );
}
```

Menus and Resource Files

Playing the RPS game is very simple. An item is browsed and selected with the navigation key and the menu is used to undo the selection, send a challenge, or calculate the result. So, we need to implement three things: a menu, key event handler, and item drawers. Items are drawn in controls, so let us concentrate on the first two things.

The `DynInitMenuPaneL()` function is as follows:

```
void CRPSGameAppUi::DynInitMenuPaneL(TInt aMenuId,
  CEikMenuPane* aMenuPane)
    {
    // Check which menu we are handling (Current version has only 1)
    switch ( aMenuId )
        {
        case R_RPSGAME_OPTIONSMENU:
            {
            if ( iState == EStateNewGame )
                {
                aMenuPane->SetItemDimmed( ERpsGameCmdPlay, ETrue );
                // Are there any items selected?
                if ( iAppContainer->MyItems().Count() == 0 )
                    {
                    // No selected items -> No use for Undo command
                    aMenuPane->SetItemDimmed( ERpsGameCmdUndo, ETrue );
                    // Don't allow challenge command.
                    aMenuPane->SetItemDimmed( ERpsGameCmdChallenge,
                      ETrue );
                    }
                // Are there enough items selected?
                if ( iAppContainer->MyItems().Count() <
                        iAppContainer->MaxCount() )
                    {
                    // Don't allow challenge command.
                    aMenuPane->SetItemDimmed( ERpsGameCmdChallenge,
                      ETrue );
                    }
                }
```

```
        else if ( iState == EStateChallenge )
            {
            aMenuPane->SetItemDimmed( ERpsGameCmdChallenge,
              ETrue );
            // Are there any items selected?
            if ( iAppContainer->MyItems().Count() == 0 )
                {
                // No selected items -> No use for Undo command
                aMenuPane->SetItemDimmed( ERpsGameCmdUndo, ETrue );
                // Don't allow challenge command.
                aMenuPane->SetItemDimmed( ERpsGameCmdPlay, ETrue );
                }
            // Are there enough items selected?
            if ( iAppContainer->MyItems().Count() <
                   iAppContainer->MaxCount() )
                {
                // Don't allow challenge command.
                aMenuPane->SetItemDimmed( ERpsGameCmdPlay, ETrue );
                }
            break;
            }
        else if ( iState == EStateResponse )
            {
            aMenuPane->SetItemDimmed( ERpsGameCmdUndo, ETrue );
            aMenuPane->SetItemDimmed( ERpsGameCmdChallenge,
              ETrue );
            }
        break;
        }
    default:
        {
        break;
        }
    }
}
```

This function is called by the Uikon framework just before a menu pane (`CEikMenuPane`) is displayed. The default implementation of the function is empty. The developer may override the function to set the state of menu items dynamically according to the state of the application data. `CEikMenuPane` is a control derived from `CCoeControl`. We have used its `SetItemDimmed()` function so that menu items, are not shown that are not available to the user. The **dimmed menu items** are identified by commands associated with menu items in the resource file. We will show examples of command definition soon. In the Series 60 Platform, dimmed menu items are not shown at all. In other UI styles, items may be just dimmed or grayed out.

We have only one menu in the RPS game identified by the `R_RPSGAME_OPTIONSMENU` ID in the resource file. In case of a new game, for example, it is not possible to choose challenge or play commands. Commands available in separate states are shown in Figure 8.6.

Figure 8.6 Menu items available in separate states of the rock–paper–scissors game: new game (left), challenge (middle), and response (right)

Resource files are text files used in defining UI components. In Symbian OS, resource files define binary data and text resources only. Use of resource files has three advantages:

- applications are smaller, because it is more efficient to describe UI components in resource files and component behavior in code;
- localization is easier, because it is not necessary to 'walk through' the whole source code;
- it is easier to port applications to a new target, because resource files are separate from application binary.

Resource file definition in the Series 60 Platform can contain a status pane, CBA buttons, menu and menu panes, dialogs, and strings. The first definition in the file is `NAME`, which is up to a four-character-long name of the resource files. The name allows one application to use several resource files. The Uikon as well as Avkon resource file has to begin with three standard resources: `RSS_SIGNATURE`, `TBUF`, and `EIK_APP_INFO`. The signature is left blank. A buffer may be empty as well or it can be used to define a file name that the document class uses. `EIK_APP_INFO` specifies the resources that define the UI components, as shown below:

```
//  RESOURCE IDENTIFIER
NAME    RPSG // 4 letter ID

//  INCLUDES

#include <eikon.rh>
#include <eikon.mbg>
#include <avkon.rsg>
#include <avkon.rh>
#include <avkon.mbg>
#include "rpsgame.hrh"
```

```
#include "rpsgame.loc"
#include "eikon.rsg"

//  RESOURCE DEFINITIONS

RESOURCE RSS_SIGNATURE { }

RESOURCE TBUF { buf="RPSGame"; }

RESOURCE EIK_APP_INFO
    {
    menubar = r_rpsgame_menubar;
    cba = R_AVKON_SOFTKEYS_OPTIONS_BACK;
    }

//----------------------------------------------------
//
//    r_rpsgame_menubar
//    STRUCT MENU_BAR
//    {
//    STRUCT titles[]; // MENU_BAR_ITEMs
//    LLINK extension=0;
//    }
//
//----------------------------------------------------
//
RESOURCE MENU_BAR r_rpsgame_menubar
    {
    titles=
        {
        MENU_TITLE { menu_pane = r_rpsgame_optionsmenu; }
        };
    }

//----------------------------------------------------
//
//    r_rpsgame_menu
//    STRUCT MENU_PANE
//    {
//    STRUCT items[]; // MENU_ITEMs
//    LLINK extension=0;
//    }
//
//----------------------------------------------------
//
RESOURCE MENU_PANE r_rpsgame_optionsmenu
    {
    items=
        {
        MENU_ITEM { command = ERpsGameCmdUndo; txt="Undo"; },
        MENU_ITEM { command = ERpsGameCmdChallenge; txt="Challenge"; },
        MENU_ITEM { command = ERpsGameCmdPlay; txt="Play"; },
        MENU_ITEM { command = EEikCmdExit; txt="Exit"; }
        };
    }

RESOURCE TBUF r_info_msg_challenging { buf="Challenging!"; }
RESOURCE TBUF r_info_msg_not_enough_moves { buf="Not enough moves!"; }
```

There are several kind of files that can be included in the resource file. Files having an extension .rsg are generated resource files. When resources are compiled, there are two output file types. The other is a compiled binary file used by the application in the run time, whereas the other is a generated file used when the application is compiled. For example, avkon.rsg contains identifiers for Avkon resources such as `R_AVKON_SOFTKEYS_OPTIONS_BACK`. Files with an extension .rh contain resource definitions in the form of **resource structures**. Commands are enumerated in the files having an extension .hrh. Finally, it is possible to include ordinary .h files and .loc files. The latter file types contain string definitions in the form of `#define` statements, which are included in the resource file.

All resources are defined in the same way: keyword `RESOURCE`, name of the resource structure, and resource name. For example, `RESOURCE MENU_BAR r_rpsgame_menubar` defines the only menubar of the RPS game. The `EIK_APP_INFO` structure is defined in uikon.rh and is as follows:

```
STRUCT EIK_APP_INFO
    {
    LLINK hotkeys=0;
    LLINK menubar=0;
    LLINK toolbar=0;
    LLINK toolband=0;
    LLINK cba=0;
    LLINK status_pane=0;
    }
```

We have defined only menubar and CBA in the RPS game, but there are a few other possible resource definitions as well. The CBA defines names of the softkeys. `R_AVKON_SOFTKEYS_OPTIONS_BACK` defines that the left softkey is labeled as Options, and the right one as Back. One of the Options combinations should be used. These enable the menu to be shown automatically when the left softkey is pressed.

The menu item resource structure definition is as follows:

```
STRUCT MENU_ITEM
    {
    LONG command=0;
    LLINK cascade=0;
    LONG flags=0;
    LTEXT txt;
    LTEXT extratxt="";
    LTEXT bmpfile="";
    WORD bmpid=0xffff;
    WORD bmpmask=0xffff;
    LLINK extension=0;
    }
```

The command is defined in an .hrh file. A menu item can have other menu items if it is cascaded. Flags may specify, for example, whether there is an item separator between menu items. Txt defines the text label shown in the menu item in the user interface. It is also possible to use bitmaps in menu items in addition to text. Refer to Nokia (2001f) on avkon.rh and uikon.rh for a complete description of resource files.

Commands are enumerated in an .hrh file as in the example below:

```
enum TRPSGameMenuCommands
    {
      ERpsGameCmdUndo = 0x06000,
      ERpsGameCmdChallenge,
      ERpsGameCmdPlay,
      ERpsGameSelect
    };
```

Note that you should start the enumeration from 0x6000, not to get confused with Uikon and Avkon commands. Commands are handled by the `HandleCommandL()` function, explained in the next section.

Command Handling

Commands associated with menu items are handled in the `HandleCommandL()` function. Note that the system shuts down the application by sending a command `EEikCmdExit`. That is why it should be handled in the `HandleCommand()` function, although not used by the application. Its implementation is quite simple, as seen below:

```
void CRPSGameAppUi::HandleCommandL(TInt aCommand)
    {
    switch ( aCommand )
        {
        case EAknSoftkeyBack:
        case EEikCmdExit:
        {
            Exit();
            break;
        }
        case ERpsGameSelect:
        {
            iAppContainer->AddCurrentToSelectedItemsL();
            break;
        }
        case ERpsGameCmdChallenge:
            {
            DoChallengeL();
            break;
            }
```

```
        case ERpsGameCmdPlay:
            {
            DoPlayL();
            break;
            }
        case ERpsGameCmdUndo:
            {
            iAppContainer->UndoLastItem();
            break;
            }
        default:
            break;
        }
    }
```

A specific function is called with respect to the command defined in the .hrh file and provided by the framework. This function implements the basic interactions with the user. Below, we show just the implementation of `DoPlay()`. Other functions are rather straightforward to implement.

`DoPlay()` sets the state, redraws the view, and calls `CheckWinnerL()` to calculate the final results of the game. `CheckWinnerL()` goes through opponent's and player's moves, move by move, and calculates the score according to the following rules: the rock beats the scissors, the scissors beat the paper, and the paper beats the rock. The final result is given as a note to the user:

```
void CRPSGameAppUi::DoPlayL()
    {
    iAppContainer->SetState( EPlaying );
    iAppContainer->DrawNow();
    CheckWinnerL();
    }
void CRPSGameAppUi::CheckWinnerL()
    {
    TInt myScore(0);
    TInt opScore(0);
    if ( iState == EStateChallenge )
        {
        // Challenge message contains the opponent moves and
        // my moves are in container.
        const CRpsMoves& myMoves = iAppContainer->MyItems();
        const CRpsMoves& opMoves = Document().Model().ChallengeMoves();
        for ( TInt i = 0; i < myMoves.Count(); i++ )
            {
            if ( myMoves.Move( i ) > opMoves.Move( i ))
                {
                myScore++;
                }
            if ( myMoves.Move( i ) < opMoves.Move( i ))
                {
                opScore++;
                }
            else
```

```
                {
                // Draw
                }
            }
        }
    else if ( iState == EStateResponse )
        {
        // Other way around.
        const CRpsMoves& opMoves = Document().Model().ResponseMoves();
        const CRpsMoves& myMoves = Document().Model().ChallengeMoves();
        for( TInt i = 0; i < myMoves.Count(); i++ )
            {
            if( myMoves.Move( i ) > opMoves.Move( i ))
                {
                myScore++;
                }
            if( myMoves.Move( i ) < opMoves.Move( i ))
                {
                opScore++;
                }
            else
                {
                // Draw
                }
            }
        }
    if( opScore > myScore )
        {
        InformationNoteL(KLoose);
        }
    else if( opScore < myScore )
        {
        InformationNoteL(KWin);
        }
    else
        {
        InformationNoteL(KDraw);
        }
    }
```

Notes

Notes are dialogs, that may disappear automatically without user interaction. There are several note types in the Series 60 Platform, as we will see in Chapter 11. In the RPS game we have used `CAknInformationNote` to tell whether the player has won or lost the game or whether the final result was a draw. Note how easy it is to use notes. The developer has to provide text to be notified as a descriptor. The note object has to be created, and `ExecuteLD()` shows the object on the screen, after which it deletes the object:

```
void CRPSGameAppUi::InformationNoteL( const TDesC& aText ) const
    {
    CAknInformationNote* note = new (ELeave) CAknInformationNote;
    note->ExecuteLD( aText );
    }
```

8.3.2 Graphics Drawing and Event Handling

The implementation of the `ConstructL()` function of the `CRPSGameContainer` is very typical for a view. Controls may be window-owning or nonwindow-owning. In case of a large number of controls, they should be included in one container owning the window, because window-owning controls tax run-time resources. Ideally, there could be only one top-level window-owning control in the user interface. Nonwindow-owning controls lodge the window of the container to draw themselves. Compound controls (i.e. container controls) reduce the interthread communication (ITC) between the application and the Window server. In addition, they help reducing flickering, because they enable a smaller portion of the screen to be redrawn.

In spite of the benefits of compound controls, we define only one control (`CRPSGameContainer`) in the RPS game. The `ConstructL()` function is as follows:

```
void CRPSGameContainer::ConstructL(const TRect& aRect)
    {
    CreateWindowL();
    iBitmaps = new (ELeave) CArrayPtrFlat<CFbsBitmap>(KMaxNumOfItems);
    iMyItems = CRpsMoves::NewL();
    iOpponentItems  = CRpsMoves::NewL();


    LoadImagesL( KImagePath );


    SetRect(aRect);
    ActivateL();
    }
```

First, we have defined this control to be window-owning. Actually, this has to be, because this is the only control of the application. Window-owning controls call the `CreateWindowL()` function, which creates a new 'child' window in the application's window group. After creating the window, other objects are created. This includes a bitmap array for paper, rock, and scissors bitmaps. `SetRect()` sets the extent (position and size) of the control with respect to the rectangle area provided by the application UI class. `ActivateL()` sets the control ready to be drawn.

The `Draw()` function draws the control. It is also used in redrawing, although it is never called from the application. The framework takes care of calling this function. In case of a redraw, the Window server specifies an invalid area to be redrawn as a parameter to the `Draw()` function. The function may then determine which part of it must actually be redrawn, if any.

Before any drawing, we must get the graphics context. This is a local copy of the system graphics context. Modifications are first done to the local context and then flushed to the Window server taking care of changes on the screen. In the `Draw()` function of the RPS game, there is not much more than bitmap drawing, as shown in the example code below:

```
void CRPSGameContainer::Draw(const TRect& /*aRect*/) const
    {
    CWindowGc& gc = SystemGc();
    gc.SetPenStyle(CGraphicsContext::ENullPen);
    gc.SetBrushColor(KRgbWhite);
    gc.SetBrushStyle(CGraphicsContext::ESolidBrush);
    gc.DrawRect(Rect());

    // Draw the selected images
    if( iState == ESelecting )
        {
        TSize selectedImageSize( 30, 20 );
        TPoint selectedImageTopLeft( 5, 10 );
        for(TInt i = 0; i < iMyItems->Count(); i++)
            {
            TRect rect( selectedImageTopLeft, selectedImageSize );
            TInt imageType = iMyItems->Move( i );
            gc.DrawBitmap( rect, iBitmaps->At( imageType ) );
            selectedImageTopLeft+=TPoint(
              selectedImageSize.iWidth+5, 0 );
            }
        // Draw the big image
        TSize imageSize( 90, 60 );
        TRect clientRect = Rect();
        TPoint topLeft( (clientRect.Width() - imageSize.iWidth)/2, 80 );
        TRect rect( topLeft, imageSize );
        gc.DrawBitmap( rect, iBitmaps->At( iCurrentItemId ) );
        }
    else
        {
        TSize imageSize( 30, 20 );
        TPoint myImageTopLeft( 5, 10 );
        TPoint opponentImageTopLeft( 5, 50 );
        TPoint horizontalAddition( imageSize.iWidth+5, 0 );

        // Draw my items.
        for(TInt i = 0; i < iMyItems->Count(); i++)
            {
            TRect rect( myImageTopLeft, imageSize );
            TInt imageType = iMyItems->Move( i );
            gc.DrawBitmap( rect, iBitmaps->At( imageType ) );
            myImageTopLeft+=horizontalAddition;
            }

        // Draw opponent items.
        for(TInt j = 0; j < iOpponentItems->Count(); j++)
            {
            TRect rect( opponentImageTopLeft, imageSize );
            TInt imageType = iOpponentItems->Move( j );
```

```
            gc.DrawBitmap( rect, iBitmaps->At( imageType ) );
            opponentImageTopLeft += horizontalAddition;
            }
        }
    }
```

ENullPen pen style defines that the pen is not used for drawing. The brush is used to draw the background of the application white. Bitamps are drawn with DrawBitmap(), to which the bitmap location and bitmap file names are given. The selectable item is drawn in the bottom middle of the screen. The use of fixed values such as here is not very good programming. They make, for example, the porting to different UI styles with their own screen resolution difficult. Selected items are shown at the top left of the screen, below which the response (or challenge) moves are drawn, as shown in Figure 8.7.

Options Back

Figure 8.7 Bitmap layout in the rock–paper–scissors game

Event Handling

The CONE environment implements an event loop for handling UI events. The framework automatically creates a session to the Window server from which the events are received. The events may be key-press events, pointer-down or pointer up events (if the smartphone supports a pointer), redraw events or events from other controls. The Window server takes care of drawing to the window using the Font and bitmap server and the graphics device interface (GDI).

The OfferKeyEventL() function below is used for key event handling. It is more complicated than the handling of pointer events, because we know the position where a pointer event occurred. The event is easily delivered to the right control according to the coordinates, but there are no coordinates for key events. In our application this does not cause problems, because we have only one control, which should be responsible for handling any key events:

```
TKeyResponse CRPSGameContainer::OfferKeyEventL( const
  TKeyEvent& aKeyEvent, TEventCode /* aType*/ )
    {
    switch( aKeyEvent.iCode )
        {
        case EKeyLeftArrow:
            break;
        case EKeyRightArrow:
            break;
        case EKeyUpArrow:
            {
            ShowNextItem();
            return EKeyWasConsumed;
            }
            break;
        case EKeyDownArrow:
            {
            ShowPreviousItem();
            return EKeyWasConsumed;
            }
            break;
        case EKeyDevice3:
            {
            AddCurrentToSelectedItemsL();
            return EKeyWasConsumed;
            break;
            }
        default:
            return EKeyWasNotConsumed;
            break;
        }
    return EKeyWasNotConsumed;
  }
```

In general, controls are added into a control stack, when the controls are constructed. A key event from the Window server is received by the window group in the foreground (i.e. the active application). The CONE framework goes through all the controls of the application in the control stack and offers them the event. If the control is added to the stack and it has an implementation of the `OfferKeyEventL()` function, it is capable of processing key events. The return value of this function indicates what happened to the event in the function. If the control is not expecting any key events it does nothing and returns `EKeyWasNotConsumed`. In this case, the CONE framework continues offering the key event to another control in the stack until one is found, which returns `EKeyWasConsumed`.

We have defined interaction for three key events. Items may be browsed by up and down arrows of the navigation key and the item can be selected with `EKeyDevice3`, which corresponds the selection made to the navigation key.

The whole RPS game is available at www.digia.com/books. This chapter and Chapter 5 contain a description of almost every function in the example.

8.4 Summary

The application framework defines a set of base classes from which developers derive their own classes to be used in applications. The base classes of the framework enable the reuse of OS services in a straightforward way. The Uikon framework defines abstract application classes for application creation, UI component drawing, and event handling. In the Series 60 Platform, Avkon is an extension library for the Uikon framework, which provides a concrete look and feel for Uikon GUI classes.

In this chapter we presented a description of the basic base classes for an application: `CAknApplication`, `CAknDocument`, `CAknAppUi`, `CAknViewAppUi`, and `CAknView`. The last two functions are not used in the RPS game example but they plan an important role in view switching, described in Chapter 13. We showed which functions to implement in a minimal GUI application having very simple event-handling requirements. In the following chapters we will go through all the GUI components that can be used in applications. So far, we have used only menus and notes.

9

Standard Panes and Application Windows

In the Series 60 Platform, the term **screen** corresponds to the entire pixel area of the physical screen. **Windows** are components of the screen. One window may fill up the entire screen or a part of it, leaving other parts visible. Each application runs in its own window. **Panes** are subcomponents of windows and are sometimes called subwindows as well. One window may contain many panes, and each pane may contain subpanes, and so on. The **Application window** is the main window, which fills up the entire screen. The application window acts as a parent for the various panes. Typically, an application window owns three panes: **status pane**, **main pane**, and **control pane**.

In this chapter we describe the panes and their usage. We show how to get the whole screen area for the application if that is required. We also explain the structure of panes and application windows having these panes.

9.1 The Status Pane and Its Subcontrols

The status pane is composed of several smaller controls. These are sometimes also referred to as **lodged** or **composite** controls, or just panes. As the Series 60 user interface (UI) style guide notes, a window may contain several panes, and each pane may contain subpanes, and so on; the status pane is just that (Nokia, 2001g).

The function of the status pane is to display information about the device, but mainly about the application and its state. In the implementation of the status pane, these two entities are separated, as we will see later. The status pane occupies quite a large part of the upper window and can in some situations be hidden. The status pane consists of several smaller controls that can be used for various purposes by the application.

(a)

(b)

Figure 9.1 Status pane with (a) normal layout and (b) idle layout

The status pane has two different layouts:

- a normal layout with the **signal**, **context**, **title**, **navigation**, and **universal indicator** panes (Figure 9.1a);
- the idle state layout with the **signal**, **context**, **title**, **navigation**, **battery**, and the **universal indicator** pane, observe here that the universal indicator pane has switched to a sideways orientation and now reveals the hidden battery pane (Figure 9.1b).

Only some of the status pane controls are local to the application. Some general device state indication controls are present on a system server. Thus the `CEikStatusPane` is a client-side view of some of the controls in the status pane. An application has a status pane instance that is created by the application user interface framework upon the application construction. The layout of the status pane is read from the resource files. This stub hooks up with the server-side status pane; its function is to synchronize the server status pane controls with the client. A reference to the status pane can be acquired through the environment. The `CAknAppUi`, `CAknViewAppUi`, and `CAknView` classes have a `StatusPane()` function that can be used to retrieve a handle to the applications status pane instance. In AppUi this is a public function, so when you are not deriving from any of the above classes, you can get the handle through the `CEikonEnv::Static()`; for example, by calling:

```
CEikStatusPane* statusPane = iAvkonAppUi->StatusPane();
```

or

```
CEikStatusPane* statusPane =
  iEikonEnv->AppUiFactory()->StatusPane();
```

The status pane is divided into local-application-owned controls and global-server-owned controls. Applications can interact only with the subpanes they own through the application programming interface (API). Server-side panes can be accessed through the Avkon **notifier** client/server API. The local application owned status pane controls are the:

- title pane
- context pane
- navigation pane

The local default implementations can be replaced with custom controls. These controls are all derived from `CCoeControl`. The server owned controls are:

- battery pane
- signal pane
- indicator pane
- idle layout context pane

Only selected applications have access to the server-side objects through the notifier API.

Retrieving a certain status pane control is done by calling the `ControlL()` function with the status pane identification number ID as a parameter. This call returns a `CCoeControl` that has to be cast into the wanted control; for example, retrieving the navigation pane:

```
iNaviPane = STATIC_CAST<CAknNavigationControlContainer*>
        (StatusPane()->ControlL(TUid::Uid(EEikStatusPaneUidNavi)));
```

In order to be safe when calling this function, it is recommended to ask the pane capabilities whether the required pane is present:

```
if (statusPane && statusPane->PaneCapabilities(paneUid).IsPresent())
    {
    return statusPane->ControlL(paneUid);
    }
```

9.1.1 Title Pane

The default Avkon title pane control is the `CAknTitlePane`. To get a reference to this object you first need to query the status pane capabilities with the `EEikStatusPaneUidTitle` ID. If it is present, a reference to it can be acquired from the status pane with the title pane ID. The Series 60 default implementation supports showing text and bitmaps. The default content of the title pane is the applications

Phonebook

Figure 9.2 Title pane

name taken from the application information file (AIF). The title pain is shown in Figure 9.2.

The text can be set, retrieved, and reset to the default value. When the text is set, a copy of it is made and the ownership of the descriptor remains with the client application. There is also a version that does not leave and takes ownership of the descriptor. In some situations, this can be more than useful – for example, when we have to restore the previous title pane text in the destructor of a class.

Retrieving the text from the title pane does not transfer ownership or the descriptor, so you have to take a copy of it if you need it after assigning a new title. Pictures can be set from bitmaps and Symbian OS bitmap files. Alternatively, the context of the pane can be set from a resource structure. If the resource structure contains both a text and an image, the text is shown primarily:

```
RESOURCE TITLE_PANE r_my_title_pane
{
txt = "Hello world!";
}
```

Most of the time the title pane will contain the application name or a descriptive text about the current view. The text is primarily shown in a large one line font; when a longer text is required an automatic switch to a smaller font is made. In case this is still not enough, two lines are used; after that, the text is truncated from the end.

9.1.2 Context Pane

The context pane shown in Figure 9.3 is used to show the application icon and, in idle state, the clock. The `CAknContextPane` control is the default implementation, which is capable of setting, getting, and swapping bitmaps in the control. When set, the context pane always takes ownership of the image. The context pane default control shows the application icon from the AIF file.

Restoring the default does not access the AIF – an optimization that is very useful. The context pane owns a copy of the default application AIF icon. Swapping the image takes ownership of the new picture but returns ownership of the old image. The resource structure is the most

Figure 9.3 Context pane with an application icon

common way to set the context pane. Here we are using a platform bitmap from the system .mbm file:

```
RESOURCE CONTEXT_PANE r_my_context_pane
    {
    bmpfile = "z:\\system\\data\\avkon.mbm";
    bmpid = EMbmAvkonQgn_menu_unknown_cxt;
    bmpmask = EMbmAvkonQgn_menu_unknown_cxt_mask;
    }
```

9.1.3 Navigation Pane

The navigation pane presented in Figure 9.4 is used to show the user how to navigate in the application and to display information about the state of the application. The default control is empty. It may contain various types of controls:

- tabs
- navigation label
- navigation image
- indicators for volume and editors
- custom controls
- empty

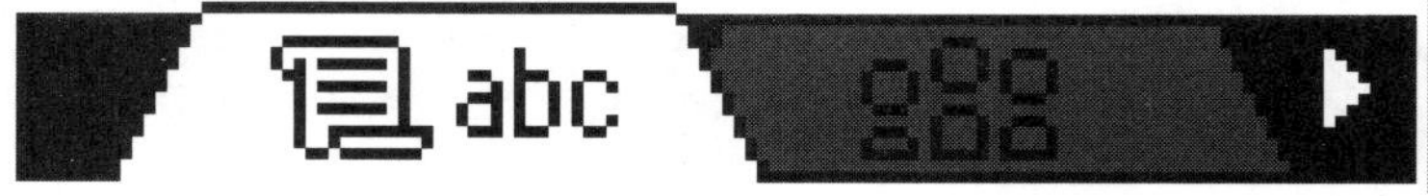

Figure 9.4 Navigation pane with two tabs

The navigation pane can be decorated with the appropriate control. The decorator control container contains the decorated controls. The status pane returns a `CAknNavigationControlContainer` object that is a stack of controls. These controls are pushed on top of the navigation pane stack and are then automatically drawn. Then, again, when popped, the pane below is drawn. An exception to normal stack behavior is the editor indicators, that have higher priority. They are always shown if they exist on the stack.

Any control pushed when an editor indicator is on the stack is inserted right below the indicator control. Also, when the same control is pushed again on the stack it is moved to the topmost position of the stack, so the stack never contains multiple instances of a control.

A control is popped from the stack by deleting it. When there is an editor indicator on the stack and the stack is popped, the topmost non-editor indicator control is popped. If you want to remove a certain non-topmost control from the stack you must use a pop that takes a reference to the decorator; this can be used to remove editor indicator from the stack.

The `CAknNavigationControlContainer` class contains factory functions for creating the various types of decorators. The navigation decorator can also be created from resources using the `NAVI_DECORATOR` structure. This resource structure contains the type that is contained in the structure. For example, if the decorated control is an `ENaviDecoratorControlTabGroup` the structure that is read is a `TAB_GROUP`:

```
RESOURCE NAVI_DECORATOR r_my_navi_decorator
    {
    type = ENaviDecoratorControlTabGroup;
    control = TAB_GROUP
        {
        tab_width = EAknTabWidthWithOneTab;
        active = 0;
        tabs =
            {
            TAB {
                id = EMyAppMainViewId;
                txt = "Main view";
                }
            };
        };
    }
```

The control container does not own the contained controls. Therefore, the programmer is responsible for deleting the controls. To prevent flicker, it is recommended that you first add the new decorator and then delete the old.

Tabs

Tabs can be used to show that there are different types of views from the available data. Example of four navigation tabs in the navigation

Figure 9.5 Four navigation tabs

pane is presented in Figure 9.5. The tabs are created in the navigation pane by first using the factory function in `CAknNavigationControlContainer` to create a tab group. This can be done either from resources or from code. A tab group is a collection of tabs. Here a tab group is created from a resource structure:

```
iCoeEnv->CreateResourceReaderLC(resReader, R_MY_TAB_GROUP);
iNaviDeco = naviPane->CreateTabGroupL(resReader);
```

The resource structure that is loaded is a `TAB_GROUP` and contains information on the configuration of the tab group. The width of the tab group selects the number of tabs shown simultaneously. The index in the tab group of the active tab is followed by a list of `TAB` structures that describe the individual tabs:

```
RESOURCE TAB_GROUP r_my_tab_group
    {
    tab_width = EAknTabWidthWithOneTab;
    active = 0;
    tabs =
        {
        TAB {
            id = EMyMainListViewId;
            }
        };
    }
```

The factory function returns a `CAknNavigationDecorator` object that decorates the contained `CAknTabGroup` object. To reach the contained object, you must cast the return value of the function that returns the contained object to `CAknTabGroup`:

```
CAknTabGroup* tabGroup = static_cast<CAknTabGroup*>
  (NaviDeco->DecoratedControl());
```

Now you can add tabs to the group. They are added in order from left to right.

A tab can contain a bitmap or text. The most common way to add tabs is from resource files, but you can also create them using the bitmaps or text factory functions. Each tab is associated with a tab ID and has a certain index in the tab group. Using the tab ID you can replace or delete a specific tab.

Setting the currently active tab can be done using the tab ID or its index in the group. There are various layouts for tabs: two, three, four small tabs, and three long tabs. The three long tabs are stacked on top of each other so that one is visible. The other layouts have the tabs fully visible. If there are more than four tabs in the tab group they are not all shown but can be navigated to by showing the navigation arrow. In general, applications should not have dynamic tabs or more than six tabs; If these are shown, it will confuse the user and make application navigation difficult (Nokia, 2001g).

The `CAknNavigationDecorator` class has the navigation arrows that are used to show the user that there are tabs available in the equivalent direction. These can be made visible, hidden, highlighted, or dimmed with the corresponding functions. First we must make the arrows visible by calling

```
iNaviDecorator->MakeScrollButtonVisible(ETrue);
```

Then the dimming of the buttons is set with

```
iNaviDecorator->SetScrollButtonDimmed (
  CAknNavigation Decorator::ERightButton, EFalse);
```

Navigation Label

The navigation label can be used when there is a very large number of similar items to be browsed by scrolling horizontally between views. An example of a navigation pane containing a text label is shown in Figure 9.6. The construction is similar to that of tabs in the way that there is a factory function in `CAknNavigationControlContainer` that creates a decorator that contains a `CAknNaviLabel` object. The resource structure used for the resource-loaded version is `NAVI_LABEL`, containing the text for the label:

```
RESOURCE NAVI_LABEL r_my_navigation_label
     {
     txt = "Hello world!";
     }
```

This provides setting and retrieval of the text in the navigation label. The factory function in the control container can be used to set a default text.

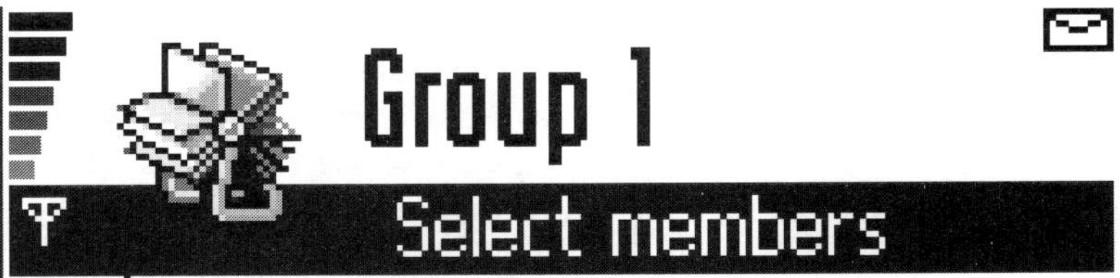

Figure 9.6 Navigation pane containing a text label

Navigation Image

An image can be set on the navigation pane. This decorator-contained image is a `CEikImage` that is constructed from a bitmap and its masked with the factory function in `CAknNavigationControlContainer`. Also, a resource file factory function is available. This uses the `NAVI_IMAGE` structure. The architecture is similar to the one described in connection with tabs:

```
RESOURCE NAVI_IMAGE r_my_navigation_image
    {
    bmpfile = "z:\\system\\data\\avkon.mbm";
    bmpid = EMbmAvkonQgn_indi_attach_audio;
    bmpmask = EMbmAvkonQgn_indi_attach_audio_mask;
    }
```

Indicators

The navigation pane can be used to show various indicators. The `CAknNavigationControlContainer` can create a volume indicator and editor indicator decorators. An example of navigation pane containing an editor indicator is shown in Figure 9.7. The volume indicator is constructed from a resource structure. The `CAknVolumeControl` class use left and right arrow keys for the volume adjustment.

The behavior of the `CAknNavigationControlContainer` toward editor indicators was discussed at the beginning of this section on the navigation pane. They always have the highest priority on the control stack. Thus, any other non-editor indicator

Figure 9.7 Navigation pane with an editor indicator

control that is pushed is not the topmost control on the stack. The control that is decorated in the navigation pane control container is `CAknIndicatorContainer`. From there we can set the message length indicator and adjust its state.

Only the front end processor (FEP) is supposed to alter the editor indicators. The responsibility of the FEP is to keep track of the current editor and its input mode and to adjust the editor indicators accordingly. The FEP owns the control container editor indicators. If the application has to set its own editor indicators to the pane, a reference to the editor indicator container can be acquired, when the focus is on the editor, by calling

```
iAvkonEnv->EditingStateIndicator()->IndicatorContainer().
```

Custom Controls

To make a custom navigation pane control you can use `NewL` in `CAknNavigationDecorator` to decorate your custom control. This takes a reference to the `CAknNavigationControlContainer` and a `CCoeControl`-derived control that is decorated for the container. Now you can push your own control on the navigation pane.

Empty

The navigation pane can also be left empty by the application. It is easiest to do this by pushing an empty control on the navigation control container stack. An empty navigation pane is shown in Figure 9.8.

Figure 9.8 An empty navigation pane

9.1.4 Signal Pane

The signal pane shown in Figure 9.9 is a server-side status pane control. There is only a single instance of this object at any time in the device. Only system applications can communicate with this, through the `RNotifier`. The interface to this can be found in the notify server's `AknNotifier` interface. It communicates with the

Figure 9.9 Signal pane

Eikon notify server's `AknNotifier` plug-in. The Eikon notifier server forwards its state changes to the signal control.

9.1.5 Battery and Universal Indicator Panes

The battery pane is another server-side status pane control. Like the signal pane, there is only a single instance of this in the device, and only system applications can change the value of this pane through the same `AknNotifier` server interface. Its state changes are forwarded to the battery control.

When the battery pane is not visible the universal indicator pane is shown. The battery pane is only visible when the device is in the idle state. The universal indicator pane contains systemwide indicators. It is also a server-side control. Indicators, for example, a new inbox item, the system in-call, or the infrared transmission is active.

The available controls are shown in a priority queue. If there is not enough room for all the controls, only the highest-priority indicators are shown. The state changes are handled in the same way as they are in the signal pane and battery pane. An active object can be used to monitor certain events. The active object gets the event of the state change and uses the notifier interface to change the pane indicators. There are two layouts for the universal indicator pane: a horizontal and a vertical orientation. The universal indicator pane is shown in Figure 9.10.

Figure 9.10 Universal indicator pane with a new message indicator

Currently, there is no way for third-party developers to add their own controls. The class used is `CAknSmallIndicator`; it does not have a public API but contains a notice/promise of the upcoming feature to add one's own indicators.

The system application handles the small indicator by creating one with the wanted indicator ID as parameter:

```
iEnvelopeInd = CAknSmallIndicator::NewL(KAknSmallIndicatorEnvelopeUid);
```

The indicator's state is changed by calling:

```
iEnvelopeInd->SetIndicatorState(EAknIndicatorStateOn);
```

The possible indicator states are `EAknIndicatorStateOff`, `EAknIndicatorStateOn`, and `EAknIndicatorStateAnimate`. After use, it is deleted.

9.1.6 Full-screen Mode

In some cases you might wish to use the whole screen for your application as shown in Figure 9.11 (e.g. for games and certain types of viewers). The status pane can be hidden, but you have to be very careful that if a leave occurs the status pane is restored. The system does not take care of this for you. You can use the cleanup stack with a `TCleanupItem`. This is achieved by calling the status pane class `MakeVisible` function. For example:

Figure 9.11 An application in full-screen mode

```
CEikStatusPane* statusPane = StatusPane();
statusPane->MakeVisible(EFalse);
```

9.1.7 Custom Control

Local controls on the status pane can be replaced. To replace one of the controls in the status pane with a custom control you need to construct a new control, then swap the default implementation with the new control by calling the `SwapControlL` function in the `CEikStatusPane` base class `CEikStatusPaneBase`. The swap function takes the pane ID as a parameter and returns the old control. The calling application is responsible for deleting the old control. The ownership of the new control is transferred to `CEikStatusPane`.

The whole status pane content can be set automatically from the resources when the application is started. The application's `EIK_APP_INFO` structure contains a status pane section. This section is a link to a `STATUS_PANE_APP_MODEL` structure, which contains `SPANE_PANE` resources. The overridden panes are defined there.

9.2 Control Pane

The control pane holds the left and right softkeys and the scroll indicator. The scroll indicator is present whenever there is a list that can be scrolled vertically. The corresponding keys are usually situated just below the control pane labels. The softkeys are implemented using the Uikon command button array (CBA). The array contains the labels for the softkeys and the scroll indicator frame. Any container of control can own a CBA.

9.2.1 Softkey Labels

The softkeys are bound to command IDs. This can be achieved easily with the CBA resource structure. The labels support two modes: long and short. This is described in resources by a tab character (/t) limiter. The long version is the first in the resource label, followed by the limiter, and then the short version. The long version is used when the scroll indicator is invisible.

Usually, the softkey label is short enough to fit the available space, but the long and short versions might have to be used if the software is ported to other user interfaces. The platform defines many different softkey combinations in the `Avkon.Rsg` file. They are defined as, for example, `R_AVKON_SOFTKEYS_OK_EMPTY`; here, `OK` is the left softkey and `EMPTY` the right. The `EMPTY` text means the right softkey is left

Ok Cancel

Figure 9.12 Control pane with Ok and Cancel softkeys

without a text and a command ID. An example control pane with Ok and Cancel softkeys is shown in Figure 9.12.

If the suitable softkey combination is not available in the platform you will have to make your own softkeys. The button array contains two `CBA_BUTTON` structures; the first will be the left softkey, and the second the right softkey:

```
RESOURCE CBA r_softkeys_ok_cancel
     {
     buttons =
            {
            CBA_BUTTON { id = ESoftKeyOK; txt = "OK"; },
            CBA_BUTTON { id = ESoftKeyCancel; txt = "Cancel"; }
            };
     }
```

Either button structure can be left empty to indicate an empty softkey.

The softkeys can be changed using the handle retrieved through the `CAknView` classes `Cba()` function. The `CEikButtonGroup-Container` has functions to control the softkey buttons. For example, new softkeys can be set using

```
Cba()->SetCommandSetL(R_AVKON_SOFTKEYS_OPTIONS_EXIT);
```

In a dialog, the CBA can be retrieved through the `CEikDialog` class `ButtonGroupContainer()` function.

9.2.2 Scroll Indicator

Traditionally, the scroll indicators are shown at the sides on the window, as shown in Figure 9.13. In the Series 60 Platform, the vertical scroll indicator is on the control pane.

Figure 9.13 Scroll indicators between the two softkeys

When a component that needs a scroll indicator is constructed, it creates an instance of `CEikScollBarFrame`. Using the Object provider mechanism, the client code must ensure that a suitable CBA is made known to the object provider so that the control will get the correct CBA. The component does not create its own local scrollbars.

Object Provider Mechanism

There are several situations where a control needs to know about another control but cannot get a reference to it. The object provider is a generic mechanism to solve this problem. For example, you have a listbox inside a dialog that needs access to the scroll indicator in the CBA. The two do not know about each other. To get around this, the participants derive from the `MObjectProvider` interface. `CCoeControl`, `CEikAppUi` and `CAknView` already do this, and set the Object provider parent in

```
CreateWindowL(const CCoeControl* aParent)
```

and `SetContainerWindowL()` automatically. So, what is left for the programmer is to set the top-level control or view. These must set the application user interface or view as their object provider. These know the CBA and the menu bar, so a handle can be acquired. This is achieved by calling the

```
SetMopParent(MObjectProvider* aParent)
```

of the control.

The object-provider mechanism can be used to provide objects to subobjects, or, more generally parent to child controls. All that is needed is to override the `MopSupplyObject(TTypeUid aId)`. The ID corresponds to the requested object. If the control corresponds to the requested ID, the object should be returned by calling `return aId.MakePtr(object)`, otherwise return `TTypeUid::Null()`. This way, the subcontrols request forwards all the way to the object that can provide the requested control.

The objects that need to be provided are `CEikButtonGroupContainer`, `CEikCba`, and `CEikMenuBar`. In `Eikmop.h` there is a utility function that helps to return any of these. The ID of the object provided is declared in the header with the `DECLARE_TYPE_ID` macro. To access an object though the object provider, you need to call the `T* MopGetObject(T*& aPtr)` function in the `MObjectProvider`-derived class. The requested object reference is set in the pointer, passed to the function.

9.3 Main Pane

The main pane is the principal area of the screen where an application can display its data. The main pane can be used freely by the applications to draw whatever needed. In that case, the responsibility of the look and feel rests on the application's designer. However, the main pane will usually include one of a number of standard components:

- list: there are several standard list types to choose from; list types are described in Chapter 10;
- grid: there are several types available; see Chapger 10;
- find pane: this is used together with a list, allowing the user to search list items alphabetically;
- status indicators: there are present only in the idle state, immediately below the navi pane, displaying status indicators; few indicators appear also in the universal indicator pane (when not in idle); others exist only in the status indicator pane;
- soft indicators: these are present only in the idle state.

9.4 Summary

In this chapter we have provided a description of panes. Panes are subcomponents of windows and they may themselves contain subpanes. There are three panes dividing the screen of a smartphone: status pane, main pane, and control pane. The status pane shows the application caption and icon, if these are defined in addition to battery and signal strength. The main pane is available for the application, although it is possible to allocate the whole screen for the application. The control pane shows the labels of the softkeys.

10

Lists and List Types

Lists are a common way to represent data in the Series 60 Platform. Lists can be separated roughly into **vertical** and **two-dimensional grids**, but there are many different layout types for lists. Many applications in Series 60 use vertical lists, but in some situations two-dimensional lists are very convenient. At the end of this chapter, there are examples on how to use lists. This chapter concentrates mainly on vertical lists.

10.1 List Architecture

Basic architecture for lists in Symbian OS is shown in Figure 10.1. To understand how lists work, it is helpful to go through the classes related to `CEikListBox`. The list framework follows the Model–View–Controller (MVC) design pattern presented in Chapter 3. `CEikListBox` is the controller, `CListBoxView` the view, and `MListBoxModel` the model. The abstract class `MEikListBoxObserver` can be used to attach an observer.

`CEikListBox` is the base class for all lists in the Series 60 Platform. `CListBoxView` displays items currently visible in a list, and `CListItemDrawer` draws individual items for `CListBoxView`. All the events are reported to the observer.

To show data, lists use `MListBoxModel`, which offers data in a format suitable for lists. In the Series 60 Platform, every list uses `CTextListBoxModel`, whether it is a grid or a vertical list. This model makes the use of lists very flexible, because the data are offered through the abstract class `MDesCArray`, which is one of the key elements used with lists. This array can be changed dynamically, or it is possible to use a default array, created by `CTextListBoxModel`. This is not recommended, because the actual type of this array is not known.

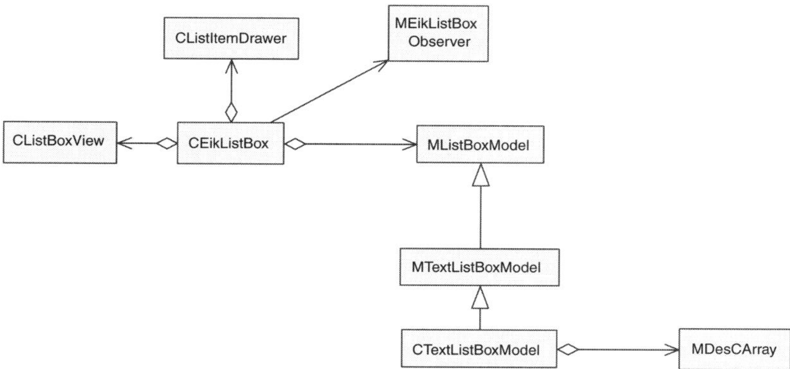

Figure 10.1 Basic list architecture

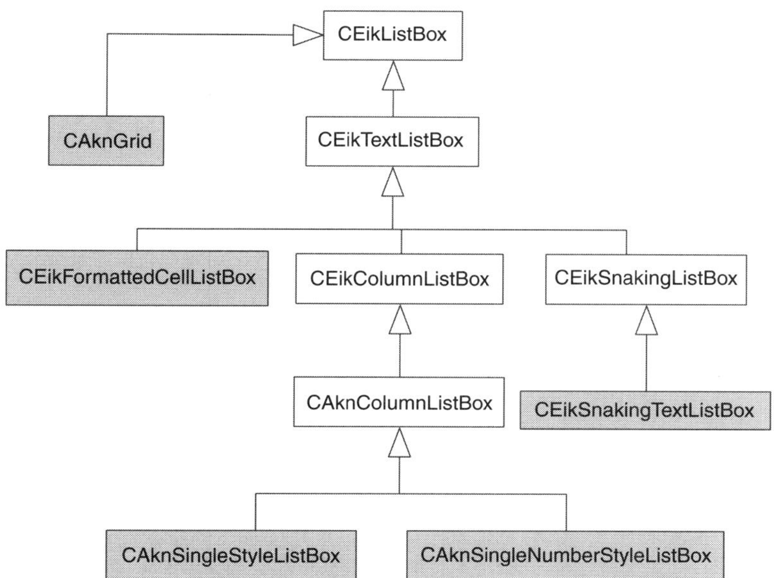

Figure 10.2 Various types of lists in Series 60. Note: gray indicates a 'concrete' class in Avkon

Figure 10.2 shows the architecture of various types of lists in the Series 60 Platform. There are four main categories: `CAknGrid` is the base class for two-dimensional lists, and `CEikTextListBox` is the base class for all vertical lists. Vertical lists are further divided in three different list types, based on their layout. Classes presented in gray boxes are concrete classes in Avkon (Nokia, 2001f).

10.2 List Types

List types are divided based on their look and use; further information can be found in the Nokia Series 60 UI Style Guide (Nokia, 2001g).

10.2.1 Menu List

Menu lists are used to select one item from a list. Those are generally used in popup windows (see Figure 10.3).

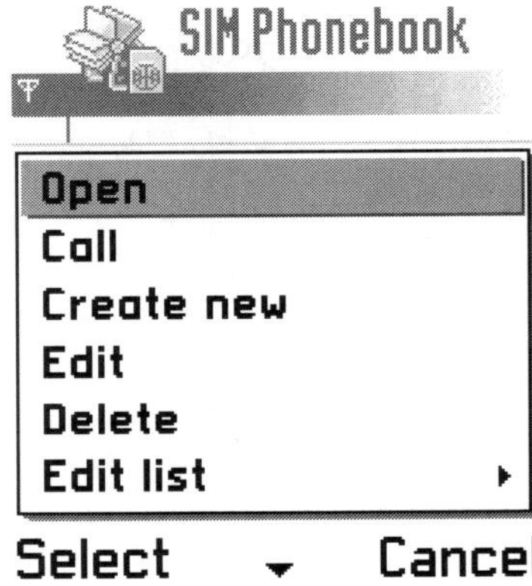

Figure 10.3 Menu list

10.2.2 Selection List

Selection lists are a common way of displaying and accessing data in applications, shown in the main pane. In Figure 10.4, the selection opens a more detailed view.

10.2.3 Markable List

A markable list is the same as a selection list, except that a marking feature has been added. With markable lists it is possible to mark any number of items and execute a single command for all of them. Marking is shown in the right-hand corner of a list item (see Figure 10.5). Selection can be done in the options menu or with the simultaneous use of shift and selection keys.

10.2.4 Multiselection List

Multiselection lists are used when it is emphasized that several items can be selected from a list (see Figure 10.6), for instance when the

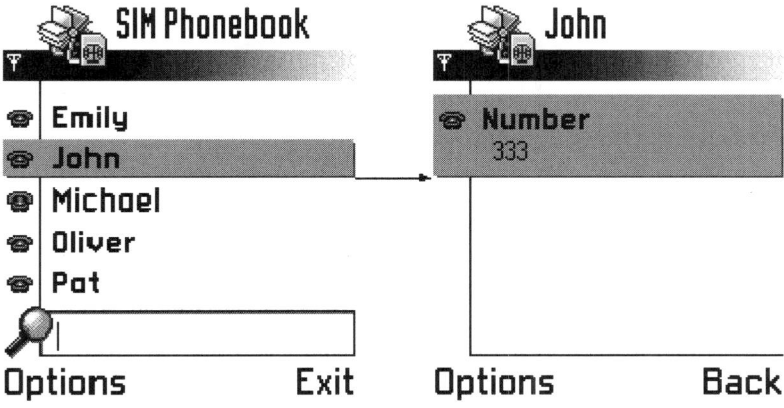

Figure 10.4 Selection lists

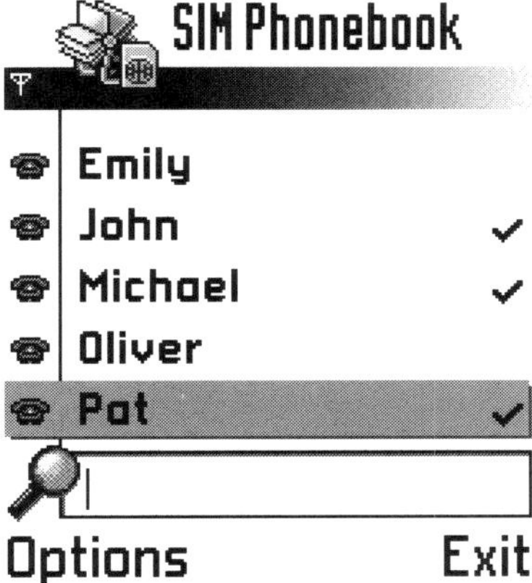

Figure 10.5 A markable list

same short message is sent to several people. The selection differs from markable lists. In multiselection lists, the selection is done only with the selection key, and acceptance is given with the Ok softkey.

10.2.5 Setting List

The setting list shows settings that can be adjusted. For instance, the application shown in Figure 10.7 uses a setting list. The setting items are covered in more detail in Chapter 11, on other user interface (UI) components.

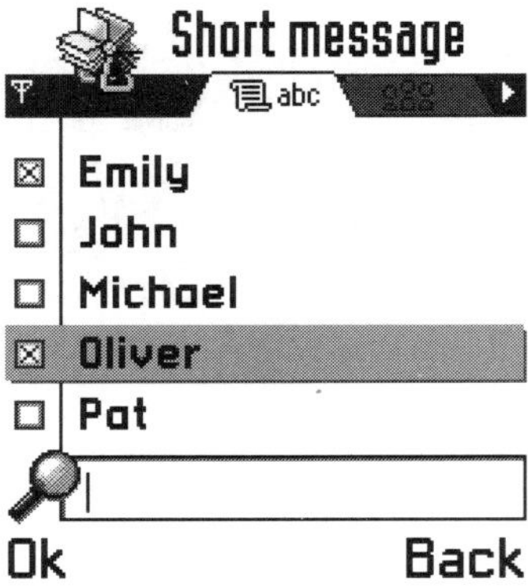

Figure 10.6 Multiselection list

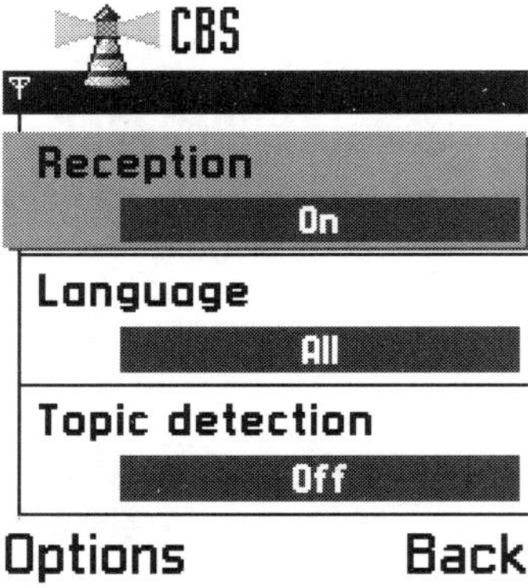

Figure 10.7 A setting list

10.2.6 List Layouts

Vertical list items can in general contain more elements than grids, and their layout structure has been defined quite accurately. The following layout rules apply to all vertical lists (Nokia, 2001g):

- list items have equal height;
- the same column structure is used with all list items (e.g., the list item can have a single column or several columns, but not a combination of these).

Column Structure of Lists

All the list items are divided into three different columns starting from the left: A, B, and C (see Figure 10.8). It is not necessary to keep these separate, so combinations are possible. The list items can have combinations: AB, BC, or ABC, but you cannot mix different layouts within a list. A dynamic column still exists after C, where additional icon indicators can be added. Additional icons can be added to every list item dynamically and therefore it is not the same kind of column as the other three (Nokia, 2001g).

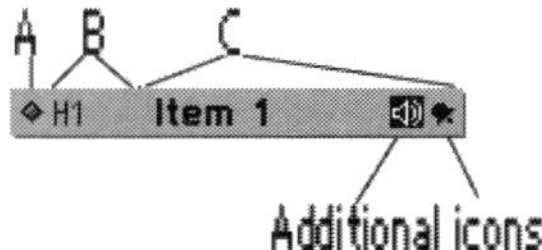

Figure 10.8 Column structure

Standard list item columns are as follows:

- column A:
  - small graphic (icon) (Nokia, 2001g),
  - item number (Nokia, 2001g);
- column B:
  - heading (title or attribute of the item) (Nokia, 2001g);
- column AB:
  - heading (title or attribute of the item) (Nokia, 2001g),
  - large graphic (e.g. icon or image thumbnail) (Nokia, 2001g);
- column C/BC/ABC:
  - main text of the item (Nokia, 2001g).

List Item String Formatting

In implementation, different columns are divided with tab characters (\t), and icons are indicated with numbers. If the list type is graphical, then column A contains an icon, and there has to be a number to the icon array before the tab (see Figure 10.9a). If the list type is a number, then the first number is presented as a number not as an icon (see Figure 10.9b). Additional icons are indicated with numbers on the icon array after the tab, as seen in Figure 10.9.

A collection of lists and list item string formats can be found in Table 10.1.

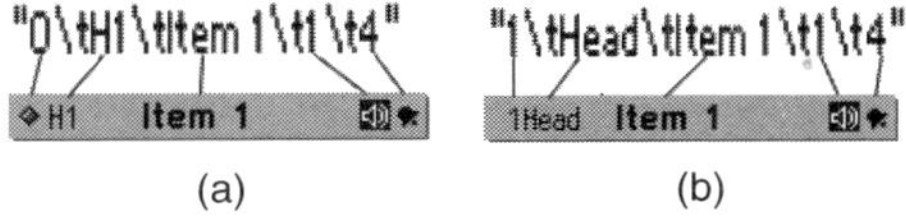

Figure 10.9 List styles: (a) graphic-style list and (b) number-style list

Table 10.1 Vertical list layouts (from aknlists.h)

List item types	Example picture	Class name	List item string format
Single-line item	TextLabel	`CAknSingleStyleListBox`	`"\tTextLabel\t0\t1"`
Single-line item with number	1 TextLabel	`CAknSingleNumberStyleListBox`	`"2\tTextLabel\t0\t1"`
Single-line item with heading	Heading TextLabel	`CAknSingleHeadingStyleListBox`	`"Heading\tTextLabel\t0\t1"`
Single-line item with graphic	TextLabel	`CAknSingleGraphicStyleListBox`	`"0\tTextLabel\t1\t2"`
Single-line item with graphic and heading	Headi... TextLabel	`CAknSingleGraphicHeadingStyleListBox`	`"0\tHeading\tTextLabel\t1\t2"`
Single-line item with number and heading	1Headi... TextLabel	`CAknSingleNumberHeadingStyleListBox`	`"1\tHeading\tTextLabel\t2\t3"`
Single-line item with large graphic	TextLabel	`CAknSingleLargeStyleListBox`	`"1\tTextLabel\t0"`
Double item	FirstLabel SecondLabel	`CAknDoubleStyleListBox`	`"\tFirstLabel\tSecondLabel\t0"`
Two-line item	This is a double 2 line list box	`CAknDoubleStyle2ListBox`	`"\tFirstLongLabel\t\t0"`

Double item with number	1 FirstLabel SecondLabel	CAknDoubleNumberStyleListBox	"1\tFirstLabel\t\t0" and "1\tFirstLabel\tSecondLabel\t0"
Double item with time	13:00 FirstLabel SecondLabel	CAknDoubleTimeStyleListBox	"Time\tPM\tFirstLabel\tSecondLabel" Time has to consist of numbers and could be separated with. or:
Double item with large graphic	FirstLabel SecondLabel	CAknDoubleLargeStyleListBox	"1\tFirstLabel\tSecondLabel\t0" Recommended image sizes: 30 × 40, 36 × 44, 40 × 30, 40 × 48, 44 × 36.
Double item with graphic	FirstLabel SecondLabel	CAknDoubleGraphicStyleListBox	"0\tFirstLabel\tSecondLabel\t0" Where 0 is index for icon array
Double item with graphic	FirstLabel SecondLabel	CAknFormDoubleGraphicStyleListBox	"1\tFirstLabel\tSecondLabel\t0"
Double item	FirstLabel SecondLabel	CAknFormDoubleStyleListBox	"\tFirstLabel\tSecondLabel\t0"
Setting item	FirstLabel ValueText	CAknSettingStyleListBox	"\tFirstLabel\t\tValueText"

(continued overleaf)

Table 10.1 *(continued)*

List item types	Example picture	Class name	List item string format
Setting item	FirstLabel	CAknSettingStyleListBox	"\tFirstLabel\t0\t"
Setting item	FirstLabel ValueText	CAknSettingStyleListBox	"\tFirstLabel\t\tValueText\t*"
Setting item	FirstLabel SecondLabel	CAknSettingStyleListBox	"\tFirstLabel\t\t\t\tSecondLabel"
Setting item with number	1 FirstLabel ValueText	CAknSettingNumberStyleListBox	Same as above four entries except that there is a number before the first item: "1\tFirstLabel\t\tValueText"
Single-line item	FirstLabel	CAknSinglePopupMenuStyleListBox	"FirstLabel\t0"
Single-line item with graphic	Label	CAknSingleGraphicPopupMenuStyleListBox	"0\tLabel"
Single-line item with graphic	Label	CAknSingleGraphicBtPopupMenuStyleListBox	"0\tLabel"

Single-line item with heading	Heading Label	CAknSingleHeadingPopupMenuStyleListBox	"Heading\tLabel"
Single-line item with graphic and heading	Headi... Label	CAknSingleGraphicHeadingPopupMenuStyleListBox	"0\tHeading\tLabel"
Double item	FirstLabel SecondLabel	CAknDoublePopupMenuStyleListBox	"FirstLabel\tSecondLabel"
Double item with large graphic	FirstLabel SecondLabel	CAknDoubleLargeGraphicPopupMenuStyleListBox	"0\tFirstLabel\tSecondLabel"
Single item with graphic	ShortLabel LongLabel	CAknFormGraphicStyleListBox	"1\tShortLabel" and "\t\tLongLabel"

10.3 Use of Lists

We now present a few examples of how to use lists. First, we give basic construction examples; the last example will be little bit more complex. The main focus in the last example is to show a proper way to offer and formulate data for the lists. In the last example you can also see how lists are used in view-based applications (see Chapter 12 for more information on view architecture).

10.3.1 Creating Lists Manually

First, we choose a layout for the list. For now, we are using a single-line item list. Before we can use the Series 60 Platform lists, we have to include the following files into our cpp file:

```
#include <aknlists.h>
#include <avkon.hrh>
```

We should make a member variable for the list, and `CEik-TextListBox` is a base class for all vertical lists, so we can use that:

```
CEikTextListBox* iListBox;
```

Then we create an instance of this list, and construct it with a flag that defines what kind of list it will be (see Table 10.2):

```
iListBox = new (ELeave) CAknSingleStyleListBox;
iListbox->SetContainerWindowL(*this);
iListBox->ConstructL(this, EAknListBoxSelectionList);
```

Table 10.2 Construct flags for vertical lists (from avkon.hrh)

Flag	Type
`EAknListBoxSelectionList`	Selection list
`EAknListBoxMenuList`	Menu list
`EAknListBoxMarkableList`	Markable list
`EAknListBoxMultiselectionList`	Multiselection list
`EAknListBoxLoopScrolling`	List can be looped from first to last, and from last to first. This cannot be used alone, but it can be appended with or operator (e.g. `EAknListBoxSelectionList | EAknListBoxLoopScrolling`)

For offering data for the list, we can make an instance of, for example, `CDesCArrayFlat`, and format items correctly before offering those to the list. It is also possible to use the default array of lists, but, as previously mentioned, it is not recommended. When we set a new array to the list model, the default array will be destroyed. The ownership type of the array can be changed with `SetOwnershipType()`, if it is preferable that the creator owns the array:

```
CDesCArray* array = new (ELeave) CDesCArrayFlat(5);
CleanupStack::PushL(array);
array->AppendL(_L("\tItem 1"));
array->AppendL(_L("\tItem 2"));
CleanupStack::Pop();     // array
iListBox->Model()->SetItemTextArray(array);
```

Adding Items

If we want to add items dynamically, we have to ask for the item text array from the list model. The function `ItemTextArray()` is returning `MDesCArray`, so we have to know what kind of array it really is and change the type to that, to append more items to the array. After appending the item, the list has to be informed of item addition, so that it knows to redraw itself. Of course, it is possible to keep the pointer to array as a member variable to use it directly, but here the array is asked in the list box model:

```
CDesCArray* array = static_cast<CDesCArray*>(iListBox->
  Model()-> ItemTextArray());
array->AppendL(_L("\tNew item"));
iListBox->HandleItemAdditionL();
```

Removing Items

In some cases we need to remove items from the list. Again, we have to get the array from the list model. In the example below we remove a currently focused item. After deletion we have to set the current item index again, because of the highlight. The highlight means that the currently focused item is emphasized with some color (e.g. see 'Pat' in Figure 10.5). In case the deleted item was the last, we have to set the current item to some other place; here, the last item will be set:

```
TInt currentItem( iListBox->CurrentItemIndex() );
if ( currentItem >= 0 )
    {
    CDesCArray* array = static_cast<CDesCArray*>(iListBox->
      Model()->ItemTextArray());
    array->Delete(currentItem);
    if ( currentItem >= array->Count() )     // last item
        {
        currentItem = array->Count();
        }
```

```
    if ( currentItem >= 0 )
        {
        iListBox->SetCurrentItemIndex(currentItem);
        }
    iListBox->HandleItemRemovalL();
    }
```

Scrollbars are needed if there are more items than one display can accommodate; they indicate the position of the currently focused item in a list. Creation of scrollbars can be done immediately after list creation, because the list shows scrollbars automatically if the number of items exceeds the display:

```
iListBox->CreateScrollBarFrameL(ETrue);
iListBox->ScrollBarFrame()->SetScrollBarVisibilityL(
  CEikScrollBarFrame::EOff, CEikScrollBarFrame::EAuto);
```

Note: remember to offer your list in `ComponentControl()` and in `CountComponentControls()` and pass key events on the list in `OfferKeyEventL()` function. To be able to see the list, we still have to give the area where it is going to be drawn. The usual place for this is `SizeChanged()` function, where it could be done in the following way:

```
if ( iListBox )
    {
    iListBox->SetRect( Rect() );
    }
```

10.3.2 Creating a List from the Resource File

One of the simplest ways to create a list is to create it from the resource file. This is especially convenient if the data in the array is static. The resource structure for the lists is as follows:

```
STRUCT LISTBOX
    {
    BYTE version=0;      // do not change
    WORD flags = 0;      // what kind of list will be created,
                         // see Table 10.2
    WORD height = 5;      // in items
    WORD width = 10;      // in chars
    LLINK array_id = 0;      // link to array definition
    }
```

In this resource example, we define the list in the following way:

```
RESOURCE LISTBOX r_list_box_resource
    {
```

```
    array_id = r_list_box_array;
    flags = EAknListBoxMarkableList;
    }

RESOURCE ARRAY r_list_box_array
    {
    items =
        {
        LBUF
            {
            txt = "1\tFirstLabel 1\tSecondLabel 1";
            },
        LBUF
            {
            txt = "2\tFirstLabel 2\tSecondLabel 2";
            }
        };
    }
```

After these resource definitions, the implementation would proceed as follows:

```
iListBox = new (ELeave) CAknDoubleNumberStyleListBox;
iListBox->SetContainerWindowL(*this);

TResourceReader rr;
iCoeEnv->CreateResourceReaderLC(rr, R_LIST_BOX_RESOURCE);
iListBox->ConstructFromResourceL(rr);
CleanupStack::PopAndDestroy();     // rr
```

10.3.3 Use of Lists Inside Dialogs

Lists can also be used inside a dialog. There are two special classes for this: `CAknSelectionListDialog` and `CAknMarkableList-Dialog`. Those can be found from `aknselectionlist.h` (Nokia, 2001f). `CEikDialog` can also be used, but it would be difficult to determine which item or items were selected. If `CEikDialog` were derived, then `OkToExitL()` could be overridden to retrieve the selections.

Another advantage of using the Avkon dialogs is that there is the possibility of adding a find box inside the dialog. These are the reasons why these two different types of dialogs have been made. These dialogs are constructed from a resource file.

In the next example, we make a `CAknSelectionListDialog`. The resource definition looks like this:

```
RESOURCE DIALOG r_selection_list_dialog
    {
    flags = EAknDialogSelectionList;
```

```
buttons = R_AVKON_SOFTKEYS_OK_BACK;
items =
    {
     DLG_LINE
        {
        type = EAknCtDoubleNumberListBox;
        // see Table 10.3
        id = ESelectionListControl;
        control = LISTBOX
            {
            flags=EAknListBoxSelectionList|
              EAknListBoxLoopScrolling;
            array_id = r_list_box_array;
            };
        }
    }
}
```

Table 10.3 List enumerations for dialogs (from avkon.hrh)

Type	Class
EAknCtSingleListBox	CAknSingleStyleListBox
EAknCtSingleNumberListBox	CAknSingleNumberStyleListBox
EAknCtSingleHeadingListBox	CAknSingleHeadingStyleListBox
EAknCtSingleGraphicListBox	CAknSingleGraphicStyleListBox
EAknCtSingleGraphicHeadingListBox	CAknSingleGraphicHeadingStyleListBox
EAknCtSingleNumberHeadingListBox	CAknSingleNumberHeadingStyleListBox
EAknCtSingleLargeListBox	CAknSingleLargeStyleListBox
EAknCtDoubleListBox	CAknDoubleStyleListBox
EAknCtDoubleNumberListBox	CAknDoubleNumberStyleListBox
EAknCtDoubleTimeListBox	CAknDoubleTimeStyleListBox
EAknCtDoubleLargeListBox	CAknDoubleLargeStyleListBox
EAknCtDoubleGraphicListBox	CAknDoubleGraphicStyleListBox

The `NewL()` function for `CAknSelectionListDialog` looks like this:

```
CAknSelectionListDialog* NewL(TInt &aOpenedItem,
  MDesCArray* aArray, TInt aMenuBarResourceId,
  MEikCommandObserver* aCommand = 0 );
```

The first parameter is used to return the item which was selected and the second parameter is a pointer to a list item array. In this example, the item array is created from the resource file, so we can ignore this parameter.

Parameter `aMenuBarResourceId` is a resource identifier (ID) for the menu bar, and the last parameter is used if one wishes to catch the option menu commands. The option menu is actually a menu bar, which contains menu panes (e.g. see the picture on the option menu in Figure 10.3). For these special dialogs, the menu bar has to consist of at least one menu pane and should be either:

```
R_AVKON_MENUPANE_SELECTION_LIST
```

or

```
R_AVKON_MENUPANE_SELECTION_LIST_WITH_FIND_POPUP
```

It is possible to add your own menu panes into the menu bar. In this example, we use only `R_AVKON_MENUPANE_SELECTION_LIST`, which is empty, and that is why we are defining buttons to `R_AVKON_SOFTKEYS_OK_BACK`. The resource definition for the menu bar is:

```
RESOURCE MENU_BAR r_menubar
    {
    titles =
        {
         MENU_TITLE
            {
            menu_pane = R_AVKON_MENUPANE_SELECTION_LIST;
            }
        };
    }
```

For implementation, we have:

```
TInt selectedItem(0);
CAknSelectionListDialog* dlg = CAknSelectionListDialog::NewL(
  selectedItem, NULL, R_MENUBAR);
if (dlg->ExecuteLD(R_SELECTION_LIST_DIALOG))
     { ...use selectedItem here... }
else { ...canceled... }
```

The use of `CAknMarkableListDialog` is very similar to this, except that one can define an array for the selection items (see the correct use in `aknselectionlist.h`).

10.3.4 Use of Icons

All lists are able to show icons, it just depends on the layout type where, of what size, and how many of these can be added (see Table 10.1). There are no default icons for the lists; everything should be set by the programmer. For instance, if we are using a markable list, then we have to set some icons for the list or else it is not able show any marking icons. In udeb (unicode debug) builds, it even panics if icons are not set.

Note that the first icon from the icon array is used as the marking icon (see Figure 10.5, where selections are indicated with an icon on the right-hand side of the list item):

```
iListBox->ConstructL(this, EAknListBoxMarkableList);
```

First, we have to create a pointer array for icons:

```
CArrayPtr<CGulIcon>* icons = new(ELeave)
  CArrayPtrFlat<CGulIcon>(2);
```

How to set the icon array for the lists depends on the base class. If the base class is `CEikFormattedCellListBox`, icons are set with

```
iListBox->ItemDrawer()->FormattedCellData()->
  SetIconArray(icons);
```

If the base class is `CEikColumnListBox`, the icon array is set with

```
iListBox->ItemDrawer()->ColumnData()->SetIconArray(icons);
```

Then we just append icons in the way shown below:

```
icons->AppendL(iEikonEnv->CreateIconL(
  _L("Z:\\system\\data\\avkon.mbm"), EMbmAvkonQgn_indi_marked_add,
  EMbmAvkonQgn_indi_marked_add_mask));
```

Notice that the function `SetIconArray()` changes ownership, and therefore we do not have to push the array on the cleanup stack. But if the ownership change is done later, remember to push the icon array to the cleanup stack before appending any items on it and pop it after appending.

It is possible to append items to the list that is using icons. Actually, it does not matter whether you set an item text array before the icon array; just do not activate the container before setting the icon array.

Note: example icons are used from avkon.mbm, so avkon.mbg should be included.

10.3.5 Popup Menu Example

To create a popup menu, we have to use a special class called `CAkn-PopupList`. It is an assistant class for creating a popup window for lists with an optional title. Only special lists are allowed to be used within this class. Their class names contain the word `PopupMenuStyle`

Table 10.4 Standard popup note layouts (from aknpopuplayout.h)

AknPopupLayouts	Class used with the type
EMenuWindow	CAknSinglePopupMenuStyleListBox
EmenuGraphicWindow	CAknSingleGraphicPopupMenuStyleListBox
EMenuGraphicHeadingWindow	CAknSingleGraphicHeadingPopupMenuStyleListBox
EmenuDoubleWindow	CAknDoublePopupMenuStyleListBox
EMenuDoubleLargeGraphicWindow	CAknDoubleLargeGraphicPopupMenuStyleListBox
EPopupSNotePopupWindow	CAknSinglePopupMenuStyleListBox

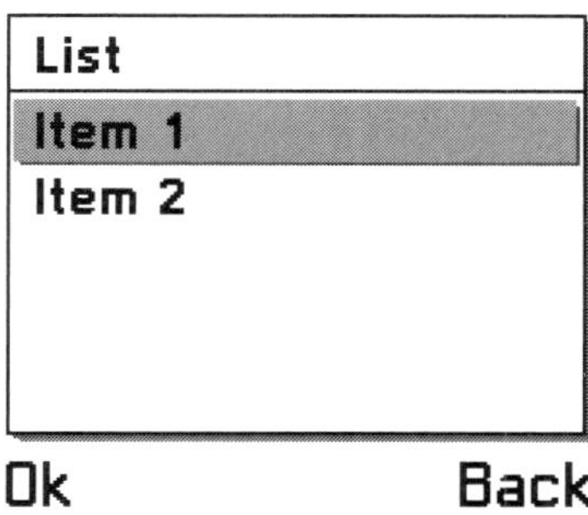

Figure 10.10 Popup menu example

(e.g. `CAknSinglePopupMenuStyleListBox`). They are listed in Table 10.4. A popup menu example is shown in Figure 10.10.

To be able to use `CAknPopupList` class we have to include

```
#include <aknPopup.h>

CEikFormattedCellListBox* listBox = new (ELeave)
                               CAknSinglePopupMenuStyleListBox;
CleanupStack::PushL(listBox);
```

Before we can continue list creation we have to create a popup list. When creating a popup list we have to offer a list, soft keys, and a layout type in the `NewL()` function. Layout types can be found in Table 10.4 and in Table 10.7:

```
CAknPopupList* popupList = CAknPopupList::NewL(
                    listBox, R_AVKON_SOFTKEYS_OK_BACK,
                    AknPopupLayouts::EPopupSNotePopupWindow);
CleanupStack::PushL(popupList);
popupList->SetTitleL(_L("List"));
```

After the popup list, we can continue list creation. In `ConstructL()`, we offer a popupList as a parent for the list. Explanations for the flags can be found in Table 10.1:

```
listBox->ConstructL(popupList,
          EAknListBoxSelectionList|EAknListBoxLoopScrolling);
```

Then we continue with normal list construction, like setting scroll indicators and items for list model. After this, popup list is ready to be launched:

```
CleanupStack::Pop();      // popupList
if (popupList->ExecuteLD())
    {
    TInt index(listBox->CurrentItemIndex());
    // do something with the current item index
    }
else
    {
    // popup list canceled
    }
CleanupStack::PopAndDestroy();      // listbox
```

10.3.6 Grids

Grids are not as developed in the Series 60 Platform as the vertical lists, so the application developer has to do more work than with vertical lists. Series 60 provides a common base class for all grids, which is then customized for applications. There are a few almost complete grids, and these are `CAknPinbStyleGrid`, `CAknQdialStyleGrid`, and `CAknCaleMonthStyleGrid`. Item string formatting for those can be found in Table 10.5. For construction, these grids are almost the same as vertical lists. You create, for example, an instance of `CaknPinbStyleGrid`, and the only thing that differs is enumeration `EAknListBoxSelectionGrid` and how to set the empty text:

```
iGrid = new (ELeave) CAknPinbStyleGrid;
iGrid->ConstructL(this, EAknListBoxSelectionGrid);
iGrid->SetContainerWindowL(*this);
iGrid->SetEmptyGridTextL(_L("Empty grid\n Two lines of text"));
```

Table 10.5 Construct flags for grids (from avkon.hrh)

Flag	Type
`EAknListBoxSelectionGrid`	Selection grid
`EAknListBoxMarkableGrid`	Markable grid
`EAknListBoxMenuGrid`	Menu grid

Table 10.6 Grid layouts (from aknlists.hrh)

Style name	Example picture	Class name	
CellPinbPaneStyle		`CAknPinbStyleGrid`	`"1\t2\t3"`, `"1\t2"`, `"1"` and `"1\t\t3"`

CellQDialPaneStyle	Txt1 Txt2 Txt3 / Txt1 Txt2 Txt3	`CAknQdialStyleGrid`	`"\t\t\t\t\t0\t0\t\t\t\t\t\t\t\t\t0"`,
			`"\t\t\t\t\t\t\t0\t0\t\t\t\t\t\t\t0"`,
			`"\t\t\t\t\t\t\t\t\t0\t0\t\t\t\t\t0"`,
			`"\t\t\t\t\t\t\t\t\t\t\t0\t0\t\t\t0"`,
			`"\t\t\t\t\t\t\t\t\t\t\t\t\t0\t0\t\t0"`,
			`"Txt1\t\t\tTxt2\tTxt3\t\t\t\t\t\t\t\t\t\t\t\t0"`
			`"\t0\tTxt1\tTxt2\tTxt3\t\t\t\t\t\t\t\t\t\t\t\t\t0"`
			and `"\t\t\t\t\t\t\t\t\t\t\t\t\t\t\t0\t0"`

CellCalcMonthPaneStyle		`CAknCaleMonthStyleGrid`	`"1\t2\tTxt"`

After this, you should set an icon array

```
iGrid->ItemDrawer()->FormattedCellData()->SetIconArray(icons);
```

and an item array:

```
iGrid->Model()->SetItemTextArray(array);
```

However, you may need to make your own custom grid. This can be achieved in several ways, but here is one way to do it. You need to inherit a `CAknGrid` and override the `SizeChanged()` and `MinimumSize()` functions, which are virtual functions defined in `CEikListBox`. In the next example, we make a 4 × 4 grid of custom width, because there is no grid for that size (see Figure 10.11).

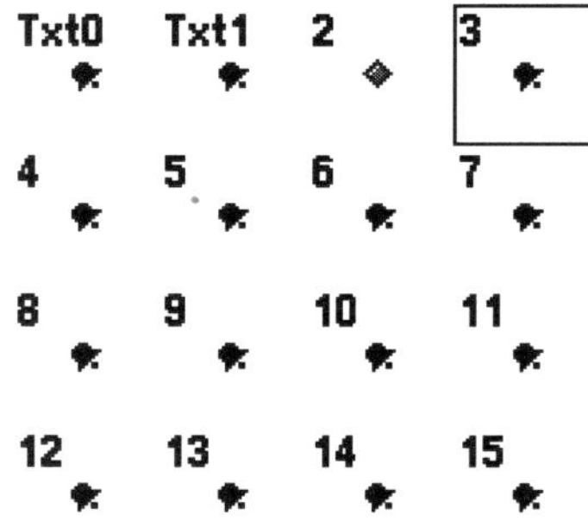

Figure 10.11 A custom grid

Table 10.7 Standard popup note layouts without resizing (from aknpopuplayout.h)

AknPopupLayouts	Class used with the type
`EDynMenuWindow`	`CAknSinglePopupMenuStyleListBox`
`EDynMenuGraphicWindow`	`CAknSingleGraphicPopupMenuStyleListBox`
`EDynMenuGraphicHeadingWindow`	`CAknSingleGraphicHeadingPopupMenuStyleListBox`
`EdynMenuDoubleWindow`	`CAknDoublePopupMenuStyleListBox`
`EDynMenuDoubleLargeGraphicWindow`	`CAknDoubleLargeGraphicPopupMenuStyleListBox`

To achieve a 4 × 4 grid we have to set layout with `SetLayoutL()` (see akngrid.h):

```
void SetLayoutL(TBool aVerticalOrientation,
  TBool aLeftToRight, TBool aTopToBottom,
  TInt aNumOfItemsInPrimaryOrient,
  TInt aNumOfItemsInSecondaryOrient,
  TSize aSizeOfItems, TInt aWidthOfSpaceBetweenItems=0,
  TInt aHeightOfSpaceBetweenItems=0);
```

This can also be done in the resource with `SetLayoutFromResourceL()`. The resource structures used with the grid are as follows:

```
STRUCT GRID
    {
    BYTE version = 0;      // do not change
    WORD flags = 0;      // type of grid
    WORD height = 5;     // in items
    WORD width = 10;     // in chars
    LLINK array_id = 0;     // link to item array
    LTEXT emptytext = "";     // what is shown if grid
                              // is empty
    LLINK style = 0;     // GRID_STYLE
    }

STRUCT GRID_STYLE
    {
    WORD layoutflags = 0;     // items layout, see Table 10.8
    WORD primaryscroll=0;     // primary scroll behavior,
                              // see Table 10.9
    WORD secondaryscroll=0;     // secondary scroll behavior,
                                // see Table 10.9
    WORD itemsinprimaryorient = 0;     // how many items
                                       // primary oriently
    WORD itemsinsecondaryorient=0;     // how many items
                                       // secondary oriently
    WORD height = 0;     // cell height
    WORD width = 0;     // cell width
    WORD gapwidth = 0;     // gap between cells
    WORD gapheight = 0;     // gap between cells
    }
```

Table 10.8 Grid layout flags (from avkon.hrh)

Flag	Description
`EAknGridHorizontalOrientation`	Items are appended horizontally and, if items do not fit on the screen, the grid is expanded vertically
`EAknGridVerticalOrientation`	Items are appended vertically and, if items do no fit on the screen, the grid is expanded horizontally
`EAknGridRightToLeft`	Items are appended from right to left
`EAknGridLeftToRight`	Items are appended from left to right
`EAknGridBottomToTop`	Items are appended from bottom to top
`EAknGridTopToBottom`	Items are appended from top to bottom

Our example structures look like this:

```
RESOURCE GRID r_grid
    {
    emptytext = "Empty grid";
    style = r_grid_style;
    }
```

Table 10.9 Grid scroll flags (from avkon.hrh)

Primary scroll	Description
`EAknGridFollowsItemsAndStops`	Focus follows grid and stops when edge of grid has been reached
`EAknGridFollowsItemsAndLoops`	Focus follows grid and loops back to beginning when edge of grid has been reached

With layout flags, we define our primary orientation. The orientation can be horizontal or vertical, and primary and secondary scrolling follows that style:

```
RESOURCE GRID_STYLE r_grid_style
    {
    layoutflags = EAknGridHorizontalOrientation |
      EAknGridLeftToRight | EAknGridTopToBottom;
    primaryscroll = EAknGridFollowsItemsAndStops;
    secondaryscroll = EAknGridStops;
    itemsinprimaryorient = 4;
    itemsinsecondaryorient = 4;
    height = 44;     // this doesn't affect, because
                     // it is overridden when
                     // creating sub cells
    width = 44;      // same thing, but these
                     // have to be defined
    gapwidth = 0;
    gapheight = 0;
    }
```

The `SizeChanged()` function determines the layout of the grid; here, We set what the cells are going to look like and what they will contain. The cell in Figure 10.12 consists of two subcells: icon and text.

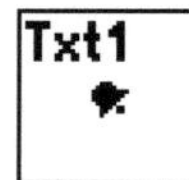

Figure 10.12 Subcell for icon and for text

To define the subcells in the cell, we can use a utility class called `AknListBoxLayouts` (see aknlists.h); some calculation is needed before setting the start and end positions of subcells:

```
void CMyOwnGrid::SizeChanged()
   {
   CAknGrid::SizeChanged();
```

The `SizeChanged()` function cannot leave, so we have to trap the creation of subcells. We have to get a pointer to the item drawer to set up the subcells:

```
TRAPD(error,
  {
  CAknGrid& aListBox = *this;
  CFormattedCellListBoxItemDrawer* itemDrawer =
      aListBox.ItemDrawer();
```

First, we set the default foreground, background colors, and grid position. Position values should be constant, so they can easily be changed and understood. For instance, the `const TInt KMainCellWidth=44` and `TInt KMainCellHeight=44`, or they can be calculated if you wish to make a very flexible and customizable grid:

```
  AknListBoxLayouts::SetupStandardGrid(aListBox);
  AknListBoxLayouts::SetupGridPos(aListBox, 0, 0, -1,
        -1, KMainCellWidth, KMainCellHeight);
```

For grids, we can set up a single graphic subcell:

```
  AknListBoxLayouts::SetupGridFormGfxCell(aListBox,
        itemDrawer, 0, 0, 0, -1, -1, KMainCellWidth,
        KMainCellHeight, TPoint(0, 0),
        TPoint(KMainCellWidth, KMainCellHeight));
```

The text subcell is defined in the following way:

```
  const TInt KFontColor = 215;
  const TInt KLeftMargin = 2;
  const TInt KNotUsed = 0;
  const TInt KBaseLine = 12;
  const TInt KWidth = 2;

        AknListBoxLayouts::SetupFormAntiFlickerTextCell(
  aListBox, itemDrawer, 1,
  LatinBold12(), KFontColor, KLeftMargin,
  KNotUsed, KBaseLine, KWidth,  CGraphicsContext::ELeft,
  TPoint(0, 0), TPoint(EMainCellWidth,EMainCellHeigth));
```

Of course, there should be exact calculations for the base line and for all the position values. Transparent cells can also be created, so that they do not overlap with some other lower-level cell:

```
  itemDrawer->FormattedCellData()->
        SetTransparentSubCellL(1, ETrue);
  } ); // TRAPD ends
} // SizeChanged ends
```

At the same time that subcells are created, we define our customized way of formatting items in a grid. Our first subcell is an icon, and the second is a text, hence the format is:

```
array->AppendL(_L("1\\tTxt"));
```

where number 1 is an index for the icon array. Then we can construct our example grid from the resource file, and items can also be set from the resource file, or they could be set later. However, the icon array has to be set manually and not from the resource file.

10.3.7 List with One's Own Model

In the following example, the intention is to show the user contact data, which consists of a name and a number. In Figure 10.13, one contact datum is separated into two rows. Application consists of three different parts, which is common for Symbian OS programs. There is an engine, a view, and an application. These are not separated into different dynamic link libraries (DLLs), because the size of the program is so small.

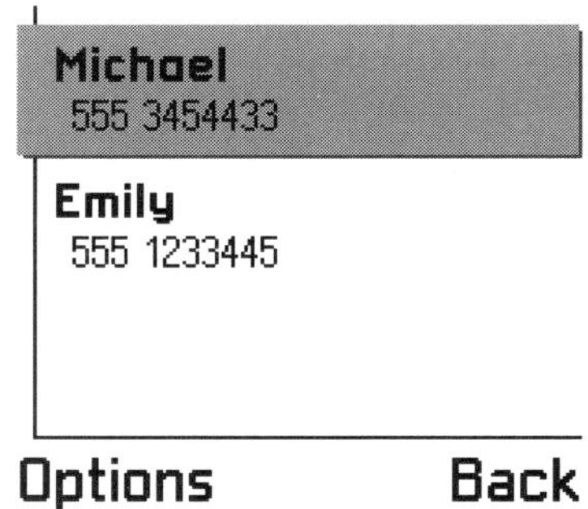

Figure 10.13 List with one item separated into two rows

Engine

`CMyEngine` provides a very simple contact database to which it is possible to add contacts. It uses its own data class, called `CMyContact`, where it is possible to store a name and a number (see Figure 10.14). For this type of data, it is convenient to create an array of pointers to `CMyContact`.

```
CArrayPtr<CMyContact>* iArray;
```

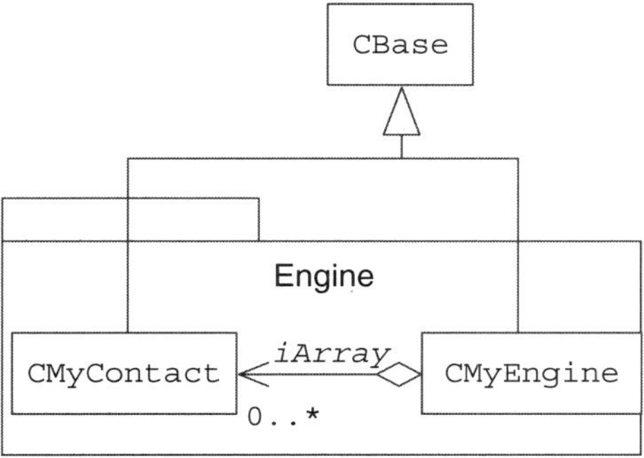

Figure 10.14 Engine

View

Unfortunately, it is not possible to place an engine array straight in the list, because the base class is not `MDesCArray`. That is why it is practical to use the adapter design pattern (Gamma *et al.*, 2000) to adapt the engine data to `MDesCArray`. This is done in the `CMyListBoxModel` of the view. This is the reason why this model has to be inherited from `MDesCArray`, as seen in Figure 10.15. `CMyListBoxModel` also formats data, so that the list is able to show the data correctly.

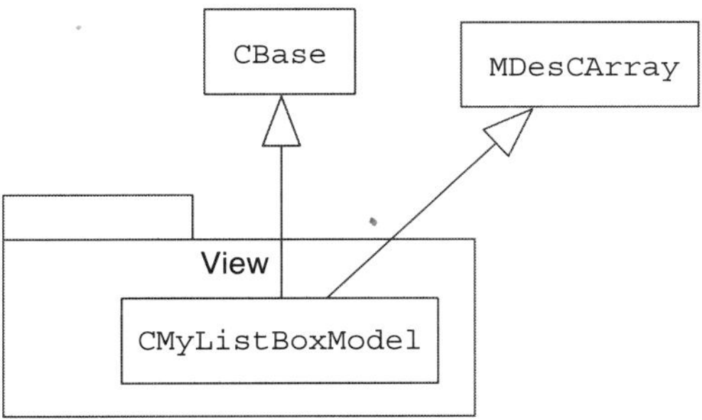

Figure 10.15 View

Application

In Figure 10.16 the application architecture is shown as a whole. In the application package, `CListBoxView` owns a container, which has `CEikFormattedCellListBox`. This list is actually instantiated to `CAknDoubleStyleListBox`, which shows one item separated into two rows, as seen in Figure 10.13.

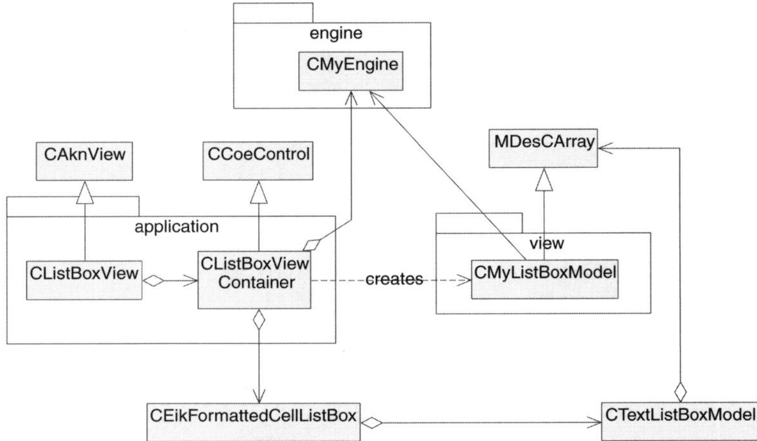

Figure 10.16 The whole application

A code example will further clarify this application. In the next subsection we will go through the key elements of an example.

Code Analysis

`CListBoxViewContainer` creates a list in its `ConstructL()` function. The actual instance of the class will be `CAknDoubleStyle-ListBox`. `CMyEngine` and `CMyListBoxModel` are also created in `CListBoxViewContainer`. A reference to the engine is given to the model, which enables access to contacts. The model is given to `CTextListBoxModel`, which uses `CMyListBoxModel` as an `MDesCArray` to provide data to be shown in the list. It is possible to give ownership to the list model or to keep it in the application. Here, the ownership is changed with `SetItemTextArray()` to `CTextListBoxModel`:

```
CMyListBoxModel* model = CMyListBoxModel::NewL(*iEngine);
iListBox->Model()->SetItemTextArray(model);
  // changes ownership
```

Before we can show data stored in the engine, we have to implement the model to offer the data in the correct format for the list. `CMyListBoxModel` is derived from `MDesCArray`, so we have to implement two functions that are defined as pure virtual in `MDesCAr-ray` (see bamdesca.h). First, we implement `MdcaCount()`, which is very simple; it just returns the count of items in the engine:

```
return iEngine.Count();
```

In order to implement `MdcaPoint`, we have to do a little bit more work. When the list draws visible list items in the screen, these are asked for from `MDesCArray` through `MdcaPoint`. `MdcaPoint` is always used when the list draws something where the list does not copy or use pointer/reference when showing items.

Visible items are only in the screen buffer, not as a text, but as pixels. For example, if you press down the arrow to move in the list, the viewable items in the list are all redrawn and therefore asked from the model. The implementation depends on whether the array is static or dynamic. If data in the array do not change and the size is quite small, the best way is probably to do a static array in the resource file. If the array is dynamic and the size can be quite large, it is advisable format each item separately. Since the lists do not copy or use reference for list items, it allows the possibility of using a single buffer in the model to format items for the list.

Implementing the buffer could be done with, for instance, an automatic class variable **mutable TBuf<256> iBuffer**. This type of buffer works very well in the Series 60 Platform, because Series 60 is not designed for wide screens. That is why 256 characters is sufficiently long to show the needed list items with formatting parameters. Mutable is used, because `MdcaPoint` is constant and cannot therefore change the member data, but mutable allows mutating (i.e. modifying) even constant data.

In this example, a dynamic buffer is used with an initial size of 50. This size is chosen to ensure that the model is always able to offer something for the list. If reallocation fails, we can still use the buffer that was initially created. Of course, with every application, it should be considered whether the user should be informed about potential data loss. This means that some data could be missing at the end of the list item, which could be critical in some stockmarket applications, for example. Below, we show the implementation for `MdcaPoint` and `CMyListBoxModel`, which uses `HBufC* iBuffer` for formatting data.

In the example, the data are first fetched from the engine. If the current buffer size is not sufficiently long, a new allocation is needed. `MdcaPoint` cannot leave, and that is why it reallocates with `ReAlloc()` and checks the return value whether the reallocation was successful. If the reallocation was successful, we have to make `iBuffer` to point to the new data area, because `HBufC` is always allocated as a single cell and the previous address may be to an invalid area in the memory – where the `iBuffer` used to reside. In case of failure, we keep using the old buffer:

```
const CMyContact* contact = iEngine.GetContact(aIndex);

TPtrC name( contact->Name() );
```

```
TPtrC number( contact->Number() );

// there is going to be two tabulatures, because of formatting
TInt length( name.Length() + number.Length() +
  (KTab().Length() * 2) );
if ( iBuffer->Des().MaxLength() < length )
    {
    HBufC* tmp = iBuffer->ReAlloc(length);
    // Do not destroy the old buffer if allocation has failed!
    if (tmp)
        {
        iBuffer = tmp;
        }
    }
```

We then simply append the data to the buffer with label separators, but before we actually add the original data to the buffer we have to remove possible tabs (\t) from the data, otherwise, the list would show items incorrectly:

```
TPtr buffer( iBuffer->Des() );
buffer.Zero();
buffer.Append(KTab);
AppendAndRemoveTabs(buffer, name);
buffer.Append(KTab);
AppendAndRemoveTabs(buffer, number);

return *iBuffer;
```

10.4 Summary

Now we have concluded the lists chapter and know that there are many different layouts and many ways to construct lists. The use is challenging, but the basic idea is the same with all lists. Some container needs to own an instance of a list and for that, item icon arrays must be set. The abstract class `MDesCArray` makes it possible to create flexible systems that can hide complicated data structures and makes using them easy.

11

Other User Interface Components

In this chapter, a large number of Series 60 Platform user interface (UI) controls are described. These vary from user **notification** controls to **setting pages**. There are not that many controls, but every one has some variations that are used for different situations. To assist in the selection of the proper UI controls for particular situations, refer to the Series 60 UI style guide (Nokia, 2001g). This book is not intended to override that document, but is an aid to constructing the control chosen from the style guide.

11.1 Dialogs

In the Series 60 Platform, the dialog architecture has some new features. The dialogs are capable of having menus. This is implemented in the `CAknDialog` class, which derives from `CEikDialog`. Applications should use dialogs derived from this class if they need dialogs with a menu. If they do not, they can still derive from `CEikDialog` (e.g., if one needs only the softkey buttons for selection). An example dialog is shown in Figure 11.1.

The `CAknDialog` uses the same dialog resource, `DIALOG`, as in the previous platform versions, except that now a `MENU_BAR` resource is needed for the menu. The menu bar and dialog resources are defined separately. The `ConstructL()` function of `CAknDialog` takes the menu bar resource identifier (ID) and loads it from the resource file. The derived class can implement the `DynInitMenuPaneL()` function, which is called just before the menu is launched. In this function, the menu may be formatted according to the current application or focus state:

```
void CMyAknDialog::DynInitMenuPaneL (TInt aResourceId,
  CEikMenuPane* aMenuPane)
```

Figure 11.1 Dialog: the appearance of a dialog can be similar to that of a view

```
{
if (aResourceId == R_MY_DIALOG_MENUPANE)
     {
     if (iMyControl->NumberOfItems() == 0)
          {
          aMenuPane->SetItemDimmed(EMyCmdDeleteItem, ETrue);
          }
     }
}
```

In the Series 60 menu, items are dimmed instead of deleted. A dimmed item is not shown in the menu; there are no dimmed menu items at all. This is done to make scrolling of the menus easier to use. The menus are faster to use and can contain more items when there are no dimmed items.

Make sure the correct menu pane is in question when dimming items; this is done by comparing `aResourceId` against the intended menu pane. Every menu pane present in the menu bar receives its own `DynInitMenuPaneL()` call. In our experience, the best practice is to include all possible menu items in the resource and use dimming as the only strategy to alter the menu context.

Also, the derived class must implement the `ProcessCommandL()` function. Here, selected menu items are handled. The unhandled commands are forwarded to the application user interface:

```
iEikonEnv->EikAppUi()->HandleCommandL(aCommandId);
```

The `OkToExitL()` function is called when a softkey is pushed. The dialog state can be verified and then return to `ETrue` if the dialog can be closed.

11.2 Forms

Forms are special types of dialogs, as shown in Figure 11.2. They can be in either **view** or **edit** state. The item layout and functionality can differ between the two states. In the view state, the fields are not editable; it looks and acts like a listbox. The listbox focus can be moved, and items can be selected. In the edit state, the form fields can be edited; it may contain text fields, popup fields, and sliders. The state can be switched by using a menu item. In some cases, the view state is not used at all and the form is always in the editable state. The menu can also be disabled if it seems inappropriate to the situation.

Figure 11.2 A double line form in edit state

The editable form may be in one- or two-line format. In the one-line style, the optional icon is the left-most, followed by the label, editor, and second icon. The forms also use `DIALOG` resource structures. In the following example, the last item of the resource structure, `form`, is used as a link to the `FORM` structure:

```
RESOURCE FORM r_my_dynamic_form
      {
      flags = EEikFormShowBitmaps | EEikFormUseDoubleSpacedFormat;
      }
```

The `FORM` structure contains two fields: the flags, and items for the form. The flag field sets the modes of the form. The left-most icon is enabled with the `EEikFormShowBitmaps` flag and the two-lined form state with the `EEikFormUseDoubleSpacedFormat` flag. In a static form, the form items can be defined in this resource as `DLG_LINE`s. If the content of the form is dynamic the items section can be left blank.

The form items are constructed dynamically in the form `PreLayoutDynInitL()` function, using the `CreateLineByTypeL()` function.

The form can contain various controls. Below, we will discuss the editor controls individually. The editors can expand to multiple lines dynamically, depending on the input length. 'Up' and 'down' arrows are used to navigate between fields in the form. In some situations, where we want to group certain controls together, a separator can be used.

The `CAknForm` class has several virtual functions that can be used to implement specific features into the Series 60 form control. The `SaveFormDataL()` function can be overridden to implement specific save logic to a form. For example, a note could be shown if saving is not possible. The `DoNotSaveFormDataL()` can be used to reset the editor values to their previous values. The `QuerySaveChangesL()` function default implementation presents a save confirmation query. This can also be overridden to customize the query. The virtual `EditCurrentLabelL()` function displays the old label as a caption, and the editor is used to retrieve the new label. The `DeleteCurrentItemL()` presents a query about deleting the currently focused form field and then deletes the form line depending on this result. `AddItemL()` is used to add a field to the form.

11.3 Editors

There are several new editor modes in the Series 60 user interface. These range from telephone number editors to secret editors for passwords and such. With the possibility of having a predictive text-entry system, extra care has to be taken when setting the editor flags. These can depend on the input type for the retrieved data. The mode of the predictive text-entry system has to be controlled and the special character set verified for each editor.

These flags can be set through a resource structure that has been added to support the above. The edwin, global text and rich text editor resource structures that the Series 60 editors use are extended to include an `AKN_EDITOR_EXTENSIONS` structure. This contains flags for the default case, allowed case modes, numeric key map, allowed input modes, default input mode, special character table, and a flags field. The definitions are in the `UIKON.hrh` file.

11.3.1 Editor Cases

The first two flags – default case and allowed case mode – control the case type of the editor. The default case flag defines the mode the editor will be in when constructed. The allowed case mode defines the case modes the editor can acquire during its lifetime. The allowed case modes are listed in Table 11.1.

Table 11.1 Allowed editor cases

Allowed case mode	Description
EAknEditorUpperCase	Upper-case input only
EAknEditorLowerCase	Lower-case input only
EAknEditorTextCase	Automatic change from upper case to lower case; first letter is in upper case.
EAknEditorAllCaseModes	User is able to switch between upper-case and lower-case input

When the editor is constructed dynamically the `SetAknEditorCase()` and `SetAknEditorPermittedCaseModes()` functions can be used to manipulate the case mode, as shown in Figure 11.3.

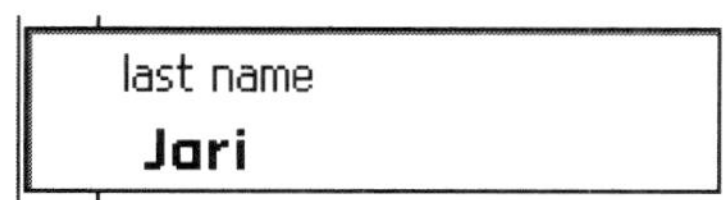

Figure 11.3 A text editor in text-case mode

11.3.2 Numeric Keymap

This flag defines the behavior of the * and # keys while in a numeric input mode. The most important ones are listed in Table 11.2.

Table 11.2 Keymap flags and interpretation of keys

Keymap	Type
EAknEditorStandardNumberModeKeymap	Pressing * key inputs *, +, p, or w; Pressing # key inputs #
EAknEditorPlainNumberModeKeymap	No functionality
EAknEditorCalculatorNumberModeKeymap	Pressing * key inputs +, −, *, or / Pressing # key inputs.(the decimal separator)
EAknEditorConverterNumberModeKeymap	Pressing * key inputs +, −, and E Pressing # key inputs.(the decimal separator)
EAknEditorToFieldNumberModeKeymap	Pressing * key inputs + Pressing # key inputs;
EAknEditorFixedDiallingNumberModeKeymap	Same as EAknEditorStandardNumberModeKeymap

`SetAknEditorNumericKeymap()` function can be used when the editor is constructed dynamically. In Figure 11.4 a number editor is presented in a two-line form.

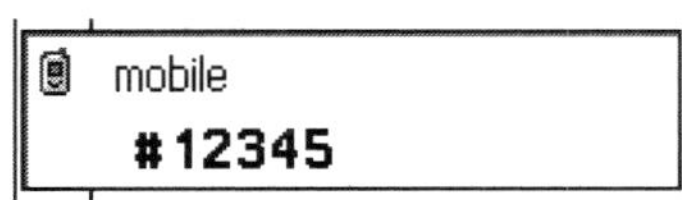

Figure 11.4 A number editor in a two-line form

11.3.3 Input Modes

The allowed input and default input modes define the use of the keypad and predictive text-entry system mode. Remember that from the same keypad, numbers and text are input. In some cases, you must disable some modes in order not to confuse the user. The allowed input mode flag defines all the modes the editor can operate during its lifetime, and the default input mode defines the mode the editor is in when constructed.

An attempt to place the editor in a mode not defined in the allowed input mode will result in a panic. The `SetAknEditorInputMode()` and `SetAknEditorAllowedInputModes()` functions can be used to set the state of the editor when it is constructed dynamically.

11.3.4 Special Character Table

Special character table is a link to a special character resource structure used by the editor (see Figure 11.5). When it is set to value 0 no special character table is used for the editor. Also, any other valid special character table can be used. The link is to a `DIALOG` structure where one dialog line has a `SPECIAL_CHAR_TABLE` structure as the control. This in turn has a `SCT_CHAR_SET`, which defines a string of characters that are shown in a table control. The `chars` field maximum length is 255.

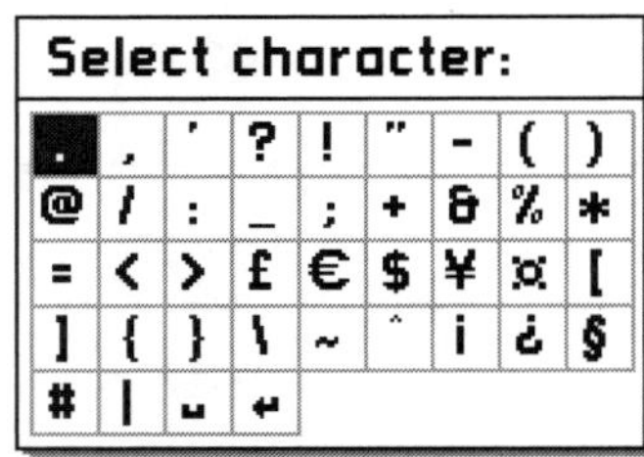

Figure 11.5 The default special character table

The `SetAknEditorSpecialCharacterTable()` function can be used when the editor is constructed dynamically, but you still have to have a resource for the table. Avkon resources contain the special character tables shown in Table 11.3. The content of these can vary depending on the active character case and locale.

11.3.5 Flags

This field allows the setting of various attributes to the editors. The value of these flags can also be set by using the `SetAknEditorFlags()` function. The flags are listed in Table 11.4.

Table 11.3 Special character tables

Special character table resource id	Description
`r_avkon_special_character_table_dialog`	The default special character table
`r_avkon_url_special_character_table_dialog`	Contains the special characters that are available in an URL address field
`r_avkon_email_addr_special_character_table_dialog`	Contains all possible special characters available in an e-mail address field
`r_avkon_currency_name_special_character_table_dialog`	The special characters that are used in the currency name editor; they must be 7-bit ASCII characters.

Table 11.4 Editor flags

Flag	Description
`EAknEditorFlagDefault`	Resets the flag bits (it is defined as `0x000`)
`EAknEditorFlagFixedCase`	Disables the case change possibility.
`EAknEditorFlagNoT9`	Disables the predictive text input (T9) system
`EAknEditorFlagNoEditIndicators`	Disables the editor indicator
`EAknEditorFlagNoLRNavigation`	Disables left-key and right-key navigation in the editor
`EAknEditorFlagSupressShiftMenu`	Disables the use of the special character table from the shift key (ABC)
`EAknEditorFlagEnableScrollBars`	Enables the scrollbar indicator for the editor
`EAknEditorFlagMTAutoOverwrite`	Enables automatic overwriting of text.
`EAknEditorFlagUseSCTNumericCharmap`	Enables the numeric character map
`EAknEditorFlagLatinInputModesOnly`	Enables only Latin character input modes

11.4 Notifications

Notes are a feedback component that informs the user about the current situation. They contain text and a graphical icon that classifies the notes to the following categories. They do not have softkeys except for the wait note, which can contain a cancellation in some situations.

11.4.1 Confirmation Notes

A confirmation note informs the user about a successfully completed process. The duration of the note is short, accompanied with a subtle tone. An example of a confirmation note is shown in Figure 11.6.

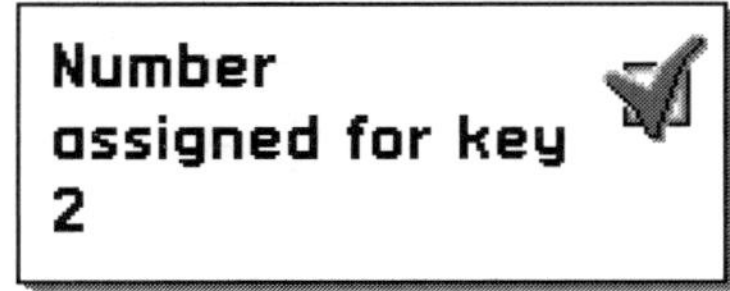

Figure 11.6 Confirmation note

There are two ways to show a confirmation note. First, using the `CAknNoteDialog` class with the fully defined `DIALOG` resource:

```
CAknNoteDialog* dlg = new (ELeave) CAknNoteDialog(
CAknNoteDialog::EConfirmationTone,
CAknNoteDialog::EShortTimeout);
dlg->ExecuteLD(R_MY_CONFIRMATION_NOTE);
```

You must give the correct tone and time-out parameters for each of the note types. The resource definition for the note is shown in the example below:

```
RESOURCE DIALOG r_my_confirmation_note
    {
    flags = EAknConfirmationNoteFlags;
    buttons = R_AVKON_SOFTKEYS_EMPTY;
    items =
        {
        DLG_LINE
            {
            type = EAknCtNote;
            id = EConfirmationNote;
            control = AVERELL_NOTE
                {
                layout = EGeneralLayout;
                singular_label = "note text";
                imagefile = "z:\\system\\data\\avkon.mbm";
                imageid = EMbmAvkonQgn_note_ok;
                imagemask = EMbmAvkonQgn_note_ok_mask;
                animation = R_QGN_NOTE_OK_ANIM;
                };
            }
        };
    }
```

There is also a possibility for having a plural label for the note text. The label text can be set dynamically using the `SetTextL()` function. The plural mode can be set by using the `SetTextPluralityL()` function with `ETrue` as the parameter. The `SetTextNumberL()` function sets the number of the plural texts to the one indicated by the parameter. This function works for resource and dynamically set texts. The number is inserted where the `\%d` string is in the descriptor.

For example: "Delete \%d items" would become "Delete 30 items". If you have multiple formatting needs in your string, you can use the string loader component and then set the text.

Notice how all the static note type information has to be defined over and over again for different notes. This can become a very 'heavy' updating procedure. Combined with localization, it can cause considerable binary size bloat if your application has many notes. The second method of note construction is far more resource-friendly. Use the above method only if you need to divert from the standard note types or need some special features. In other cases, it is strongly recommended that note wrappers, described below, be used.

The second method for making a note is use of the note wrapper component. The class CAknConfirmationNote has all the tones and icons already set on a central resource structure:

```
CAknConfirmationNote* note = new(ELeave) CAknConfirmationNote;
note->ExecuteLD(_L("prompt"));
```

The ExecuteLD takes the prompt text as a parameter, so no resource definition is needed, only the prompt text. This can be loaded from a resource buffer and then formatted, if needed. Using this will save you some space in your resource file, and you do not have to worry about the correct tone and icon parameters.

11.4.2 Information Notes

Information notes are used to inform the user about an unexpected situation. The duration of the note is slightly longer and the tone is a bit more noticeable than the confirmation note. They are used to inform the user about an error that is not too serious. An example information note is shown in Figure 11.7.

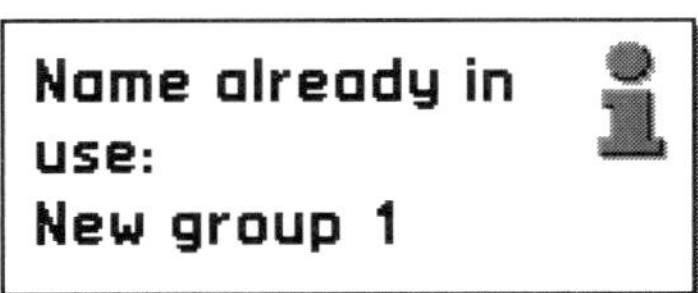

Figure 11.7 Information note

The first method needs the correct parameters for the note. Refer to Section 11.4.1, on confirmation notes, on how to create the CAknNoteDialog and the resource structure for the note. For the tone, the parameter is EConfirmationTone and ELongTimeout time for the length. The resource structure must have the ID EAknInformationNoteFlags and the correct icon for the information note.

The note wrapper class for this type of note is CAknInformationNote:

```
CAknInformationNote* note = new(ELeave) CAknInformationNote;
note->ExecuteLD(_L("prompt"));
```

The use is similar to that for the confirmation note. This is one advantage when using the wrappers: you have only to change the created object in order to change the note type:

11.4.3 Warning Notes

Warning notes are used to notify the user of a situation that may require an action. A common example would be a battery-low situation. The tone can be heard even when the phone is in one's pocket, and the duration is long enough to be noticeable. An example of a warning note is given in Figure 11.8.

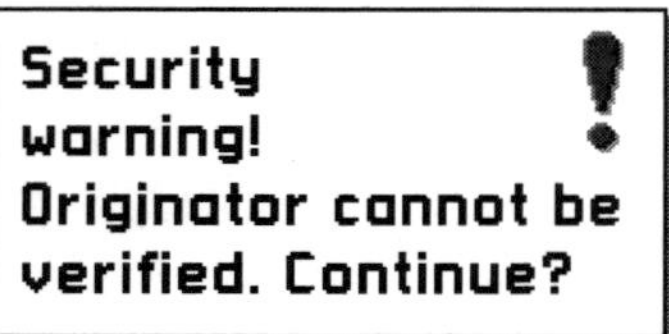

Figure 11.8 Warning note

The first method needs the correct parameters for the note. Refer to Section 11.4.1, on confirmation notes, on how to create the CAknNoteDialog and the resource structure for the note dialog. For the tone, the parameter is EErrorTone and ELongTimeout time for the length. The resource structure must have the correct ID EAknWarningNoteFlags and the icon for the information note.

The wrapper class for the warning note is CAknWarningNote:

```
CAknWarningNote* note = new(ELeave) CAknWarningNote;
note->ExecuteLD(_L("prompt"));
```

11.4.4 Error Notes

Error notes are used in situations when the user has done something seriously (e.g. when he or she has typed the wrong password).

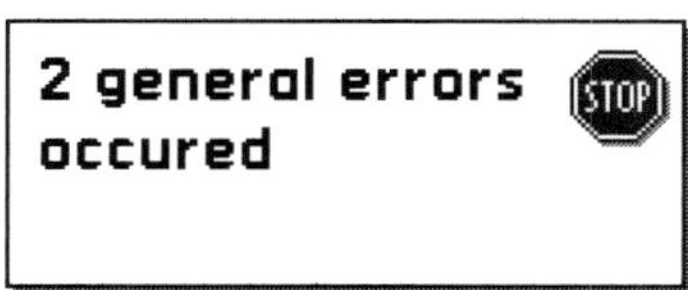

Figure 11.9 Error note

Information notes are always used when there is no irreversible error. Error notes should be used very wisely. An example error note is shown in Figure 11.9.

The first method needs the correct parameters for the note. Refer to Section 11.4.1, on confirmation notes, on how to create the `CAknNoteDialog` and the resource structure for the note. For the tone, the parameter is `EErrorTone` and `ELongTimeout` time for the length. The resource structure must have the correct ID `EAknErrorNoteFlags` and the icon for the information note.

The wrapper class is `CAknErrorNote` for the error note:

```
CAknErrorNote* note = new(ELeave) CAknErrorNote;
note->ExecuteLD(_L("prompt"));
```

11.4.5 Permanent Notes

The permanent note has no timeout value so it stays on the screen until the indicated action is taken or until the software dismisses it. The class `CAknStaticNoteDialog` combined with the `DIALOG` resource structure lets us define a static note:

```
RESOURCE DIALOG r_my_notification
     {
     flags = EEikDialogFlagNoDrag | EEikDialogFlagNoTitleBar
       | EEikDialogFlagCbaButtons | EEikDialogFlagNoShadow;
     buttons = r_avkon_softkeys_show_cancel;
     items =
         {
         DLG_LINE
             {
             type = EAknCtNote;
             id = EMyNotification;
             control = AVKON_NOTE
                 {
                 layout = ETextualNotificationLayout;
                 singular_label = "You have 1 message";
                 plural_label = "You have %d messages";
                 };
             }
         };
     }
```

Using the above resource as a parameter for the `PrepareLC()` function, and then calling `RunLD()` executes the static note:

```
CAknStaticNoteDialog* dlg = new(ELeave) CAknStaticNoteDialog;
dlg->PrepareLC(R_MY_NOTIFICATION);
dlg->SetNumberOfBorders(4);
if (dlg->RunLD())
    {
    <<positive response>>
    }
```

The `RunLD` returns a positive value if the response is accepted and a negative value if declined. The `SetNumberOfBorders()` function sets the number of borders visible to the user. It can be used to indicate that there are additional unhandled soft notifications in the queue.

11.4.6 Wait and Progress Notes

Wait and progress notes are used for long-running processes. An example is shown in Figure 11.10. The progress animation shows an estimate of how long the process will take. The wait animation is used when the length or progress of the process cannot be estimated. The user should be able to stop this kind of process, so the right softkey is used as a cancel. A small delay on the presentation and timeout are used to avoid quickly flashing notes to the user.

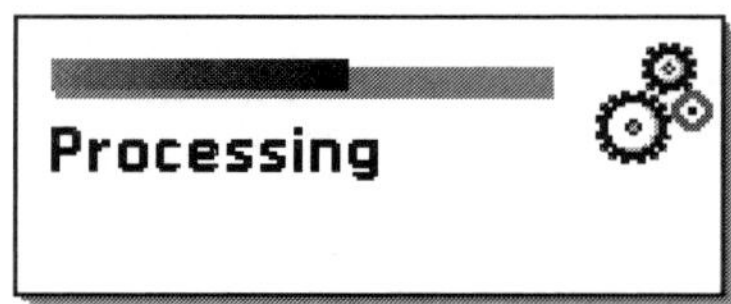

Figure 11.10 Progress note

If a long-running operation is run synchronously with an application, the application freezes for the duration of the operation. This can be very confusing to the user; the application would appear to have stalled. In order to solve this problem the user is presented with a wait or progress note while the long process is run. When the process is completed, the wait note is dismissed.

In order to get a wait note on the screen, you need to make an active object that has a wait note dialog and that then runs the long, synchronous, operation in small portions. The long operation is divided into smaller units of execution. It is not possible to issue the wait note and then start the long synchronous operation and, when that returns, dismiss the wait note without an active object. The wait note is never drawn on the screen! Again, there are two methods for doing this. An example of a wait note is shown in Figure 11.11.

Figure 11.11 Wait note

The basic method is to make the active object wait note oneself. This has to be done when the long-running operation contains another service request one would have to wait for (e.g. when one would have to wait for a notification of a data structure being ready for use).

The `CAknWaitDialog` and `CAknProgressDialog` classes implement the wait note and progress note components. The `CAknWaitDialog` derives from `CAknProgressDialog`, which has most of the application programming interface (API) for the functionality. In order for clients to receive notification when the user has dismissed the note, the `DialogDismissedL` function of the `MProgressDialogCallback` interface must be implemented; the parameter is the button ID of the user-pressed key. To set the client to receive this call, the `SetCallback()` function must be used:

```
iWaitDialog = new(ELeave) CAknWaitDialog(
  REINTERPRET_CAST (CEikDialog**, &iWaitDialog));
iWaitDialog->SetCallback(this);
iWaitDialog->ExecuteLD(R_MY_WAIT_NOTE);
```

The resource structure for the wait dialog is a `DIALOG`:

```
RESOURCE DIALOG r_my_wait_note
     {
     flags = EAknWaitNoteFlags;
     buttons = R_AVKON_SOFTKEYS_CANCEL;
     items =
         {
         DLG_LINE
             {
             type = EAknCtNote;
             id = EWaitNote;
             control = AVERELL_NOTE
```

```
                {
                layout = EWaitLayout;
                singular_label = "Copying";
                imagefile = "z:\\system\\data\\avkon.mbm";
                imageid = EMbmAvkonQgn_note_copy;
                imagemask = EMbmAvkonQgn_note_copy_mask;
                animation = R_QGN_GRAF_WAIT_BAR_ANIM;
                };
            }
        };
    }
```

The `CAknProgressDialog` constructor takes the final value, increment, interval, and a self-pointer as parameters. The `CEikProgressInfo` class controls the progress bar; a handle to this can be retrieved from the progress dialog `GetProgressInfoL()` function. Through this class, the increment of the progress bar can be controlled by using the `IncrementAndDraw()` function. When the process is finished, the progress dialog `ProcessFinishedL()` function must be called. The progress dialog also uses the `DIALOG` resource structure:

```
RESOURCE DIALOG r_my_progress_note
    {
    flags = EAknProgressNoteFlags;
    buttons = R_AVKON_SOFTKEYS_CANCEL;
    items =
        {
        DLG_LINE
            {
            type = EAknCtNote;
            id = EProgressNote;
            control = AVERELL_NOTE
                {
                layout = EProgressLayout;
                singular_label = "deleting items";
                imagefile = "z:\\system\\data\\avkon.mbm";
                imageid = EMbmAvkonQgn_note_progress;
                imagemask = EMbmAvkonQgn_note_progress_mask;
                };
            }
        };
    }
```

The second way to do this is by using a wait note wrapper. The above situation is so similar and common that a wait note runner can eliminate most of the hardest active object handling work. The `CAknWaitNoteWrapper` class is an active object that already owns a wait note control. What the client needs to implement is the `MAknBackgroundProcess` interface that then makes the callbacks to the wait note wrapper client. The `CAknWaitNoteWrapper` class

ExecuteL function takes the wait note resource ID, the interface implementer object, visibility delay, tone setting, and, optionally, the prompt text parameters.

When the wrapper is executed, an active object service request is made. When the active scheduler services the wait note request, the MAknBackgroundProcess implementer object IsProcess-Done() function is called. This function determines whether the client process is done. If it returns EFalse, a single step of the process is executed by calling the StepL() function, and another service request is made. If it returns to Etrue, the ProcessFinished() function is called; this signals the client that the process is done. The sequence diagram of the wait note wrapper is shown in Figure 11.12.

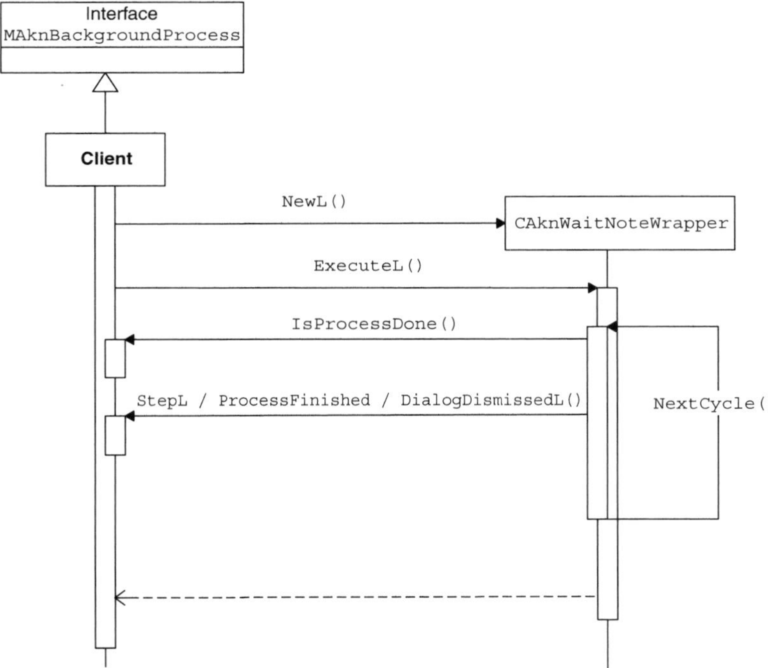

Figure 11.12 Wait note wrapper sequence diagram

If the StepL() function leaves, the client CycleError() function is called with the error code as a parameter. If the client is able to handle this error and the process can logically continue it should return to KErrNone. All other errors are escalated to the active scheduler. If the user dismisses the wait note, the client DialogDismissedL() function is called with the softkey button ID as a parameter:

```
iWaitNoteWrapper = CAknWaitNoteWrapper::NewL();
iWaitNoteWrapper->ExecuteL(R_MY_WAIT_NOTE, *this);
```

The `R_MY_WAIT_NOTE` resource structure is similar to the wait note structure used above.

11.4.7 Soft Notifications

Soft notifications are reminders to the user that a certain event has taken place (e.g. that an incoming text message has arrived). The user can respond to these with the softkeys. The left softkey is used to proceed with the appropriate action, and the right to dismiss the notification event.

Soft notifications are shown only in the idle state. This means that when an event occurs and the user is using an application and chooses to ignore the event, such as an incoming call, the notification is shown when he or she returns the idle state of the phone.

There are two different appearance styles for soft notifications: ungrouped and grouped. Ungrouped notifications contain a single notification, and they use the same control as notes. Grouped notifications can contain many different notification items. The control used in this situation is a list query, where the user can choose an item from the list. In the case where many ungrouped notifications are active, they are stacked on top of each other. After the top-most is handled, the one behind appears. Notifications have a priority that determines the order of the stack.

The application that issued the soft notification can discard it without the user being involved. Notifications should remain active as long as the event is pending. The notification can be discarded when a user reacts to the event, even though he or she has not seen the soft notification.

At the moment there is no way or any need for third-party applications to issue soft notifications. Though the `CAknSoftNotifier` class is the interface for the notifications, you might want to check if all the functions are in fact private and thus cannot be used for your client code. All the classes that can use soft notifications are friends to this class. If a public interface for it appears, applications can issue soft notifications.

11.5 Queries

Queries are used in situations where the user needs to select a way to proceed or requires user input, or validation of the selected action. There are several different types of queries for these situations.

The Series 60 Platform contains local and global queries. Local queries are local to a certain application. For instance, switching to another application hides a local query, but is again shown when the user switches back to the original application. Global queries are systemwide queries, and are shown on top of everything; so, switching to a different application still shows a global query. The following types of queries are defined and described below in detail:

- confirmation query
- data query
- list query
- multiselection list query

11.5.1 Local Queries

A confirmation query is a local query used to request a selection or a confirmation from the user. An example of a confirmation query is shown in Figure 11.13. There are one or two possible selections to be made. The two softkeys are used for the selection. One softkey is used as the way to proceed with the operation and the other softkey to cancel the operation. If there is only one way to proceed with the action then only the first softkey is used. The selection key maps to the first softkey. In most situations, the left softkey is the positive operation key.

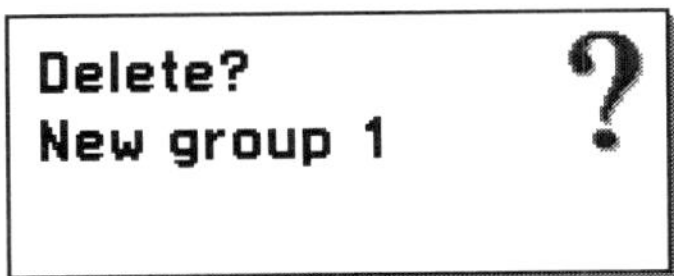

Figure 11.13 Confirmation query

Confirmation queries are used to confirm an action from the user or to answer a yes–no type of question. For example, before deleting an item, the query is shown to confirm that the user does not accidentally delete something important. The layout of the query is the same as the note window, with an optional graphic icon.

As seen from the following example, the `CAknQueryDialog` class is used with the dialog resource for the confirmation query:

```
CAknQueryDialog* dlg = CAknQueryDialog::NewL();
dlg->SetPromptL(*prompt);
if(dlg->ExecuteLD(R_MY_GENERAL_CONFIRMATION_QUERY))
    {
    }
```

The SetPromptL method is used to set the prompt text for the query. The R_MY_GENERAL_CONFIRMATION_QUERY resource is standard DIALOG resource structure:

```
RESOURCE DIALOG r_my_general_confirmation_query
     {
     flags = EGeneralQueryFlags;
     buttons = R_AVKON_SOFTKEYS_YES_NO;
     items =
           {
           DLG_LINE
                 {
                 type = EAknCtQuery;
                 id = EGeneralQuery;
                 control = AVKON_CONFIRMATION_QUERY
                       {
                       layout = EConfirmationQueryLayout;
                       };
                 }
           };
     }
```

When defined, the flags for the dialog are usually the EGeneralQueryFlags and the softkeys that are used for the query. The EGeneralQueryFlags sets these flags, as shown in Table 11.5.

Table 11.5 Dialog flags

Flag	Description
EEikDialogFlagWait	Makes this dialog modal; the ExecuteLD returns when the dialog is dismissed
EEikDialogFlagNoDrag	The dialog window cannot be moved
EEikDialogFlagNoTitleBar	The title bar of the dialog is not used
EEikDialogFlagCbaButtons	The CBA buttons are enabled; in this case they are the softkeys

The confirmation dialog contains one item of type EAknCtQuery with the EGeneralQuery ID. The control contains the confirmation query resource structure. This defines the layout of the confirmation query and the static label for the control.

Data Query

A data query contains a prompt text and an input editor (see Figure 11.14). Data queries are used in some situations where an application requires user input. The editors can accept alphanumeric

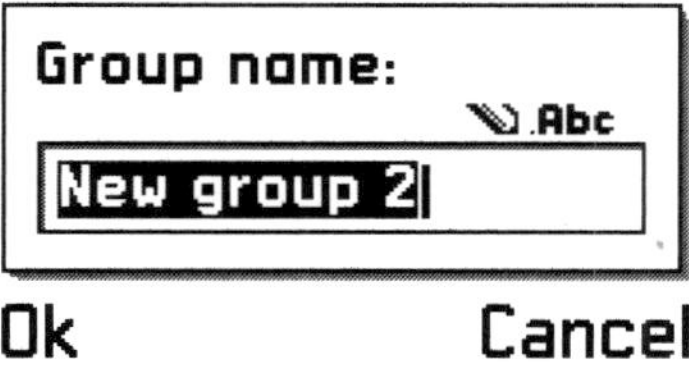

Figure 11.14 Text query

or numeric information, depending of the type of the data query. There are several possible data query types for different situations.

The `CAknTextQueryDialog` is used for querying alphanumeric text, secret text, a phone number, or the PIN (personal identification number) code from the user:

```
CAknTextQueryDialog* dlg = CAknTextQueryDialog::NewL
  (editText, *prompt);
dlg->SetMaxLength(aMaxLength);
dlg->ExecuteLD(R_MY_TEXT_QUERY);
```

The `editText` is the descriptor that is used for the editor; prompt is the descriptor that contains the prompt text for the query. The prompt text can also be described in the resource structure of the query. `SetMaxLength()` sets the maximum length of the text editor. The parameter for the `ExecuteLD()` is the resource structure for the query. The used structure is a standard `DIALOG` resource structure:

```
RESOURCE DIALOG r_my_text_query
    {
    flags = EAknGeneralQueryFlags;
    buttons = R_AVKON_SOFTKEYS_OK_CANCEL;
    items =
        {
        DLG_LINE
            {
            type = EAknCtQuery;
            id = EGeneralQuery;
            control = AVKON_DATA_QUERY
                {
                layout = EDataLayout;
                control = EDWIN
                    {
                    };
                };
            }
        };
    }
```

The structure is similar to confirmation query resource structure. The only difference is the contained control structure, which is an `AVKON_DATA_QUERY` resource. This structure again contains the layout of the query, the static label, and the control used in the query. The layout and control mappings for the different editors are listed in Table 11.6.

Table 11.6 Layout and control mappings for editors

Type	Layout	Control
Alphanumeric text	`EDataLayout`	EDWIN
Secret editor	`ECodeLayout`	SECRETED
Phone number	`EPhoneLayout`	EDWIN
PIN code	`EPinLayout`	SECRETED

The resource-defined prompt text is static, so you cannot format its run-time. In most cases, this is just what is needed. To do this, you need to insert a second `DLG_LINE` before the editor dialog line. This time, the type of the line is an `EAknCtPopupHeadingPane` and the ID `EAknMessageQueryHeaderId`. The control structure is an `AVKON_HEADING` that contains a label field where you insert your prompt text. Also, the optional bitmap icons are defined here. The file that contains the bitmap is put to `bmpfile` and the ID of the bitmap and its mask is set to `bmpid` and `bmpmask`.

`CAknNumberQueryDialog`, shown in Figure 11.15, is used when a number input is needed. The API is very similar to that of a data query. The `NewL` constructor takes a reference to the integer typed parameter that is altered to contain the user input. The `SetMinimumAndMaximum()` function allows one to set the minimum and maximum values for the editor. The resource structure is similar to that of a data query, except that the contained EDWIN control is replaced by a `NUMBER_EDITOR` structure, containing the compulsory initial minimum and maximum values. When using the `SetMinimumAndMaximum()` function, one must note that the parameter values must be between the initial minimum and maximum values.

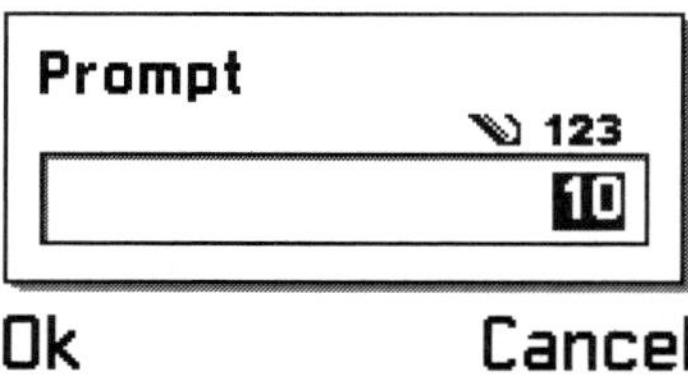

Figure 11.15 Number query

CAknTimeQueryDialog, shown in Figure 11.16, is used for requesting a time or a date. The NewL() constructor takes a reference to a TTime object that is altered by the user. The SetMinimumAndMaximum() function acts in the same way as in the CAknNumberQueryDialog. The resource structure is similar to that of a data query except for the control, which has a DATE_EDITOR structure. It initializes the minimum and maximum dates for the editor. When used for querying the time, the control structure is replaced by a TIME_EDITOR that contains the minimum and maximum time values.

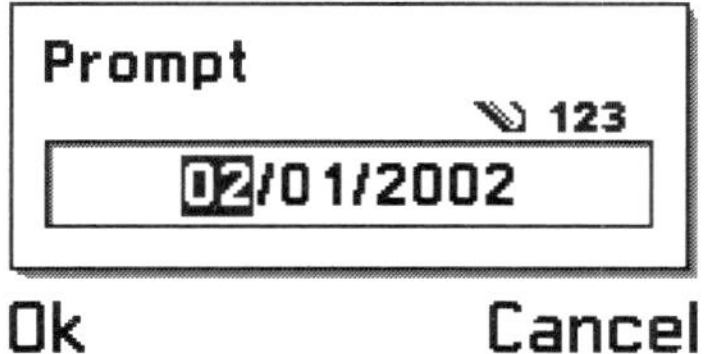

Figure 11.16 Time query

CAknDurationQueryDialog, shown in Figure 11.17, is used for getting the time duration. The NewL() constructor parameter is a reference to a TTimeIntervalSeconds object. The resource structure is similar to that of a data query, except that the layout is an EDurationLayout and the control is DURATION_EDITOR. This structure describes the minimum and maximum duration for the editor.

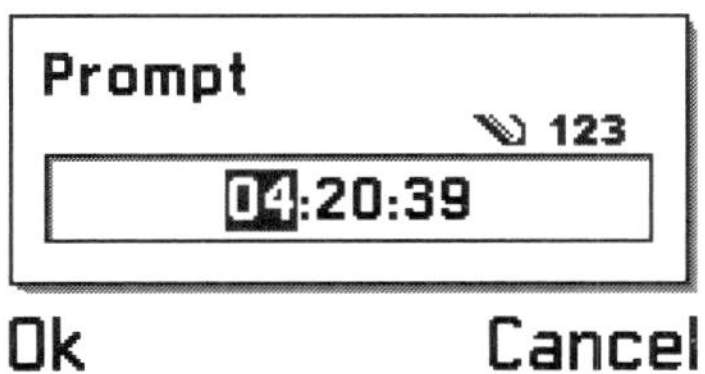

Figure 11.17 Duration query

CAknFloatingPointQueryDialog, shown in Figure 11.18, is used for floating point input. The NewL takes a TReal reference that is modified by the user input. The resource layout is an EFloatingPointLayout and the controls resource structure is a FLPTED. The maximum and minimum values and the length of the number can be initialized here.

CAknMultiLineDataQueryDialog, shown in Figure 11.19, is used for queries that have multiple editors. At the moment it is two at the maximum. There are varying sets of NewL() constructors for possible input situations. These all take reference parameters for the editors.

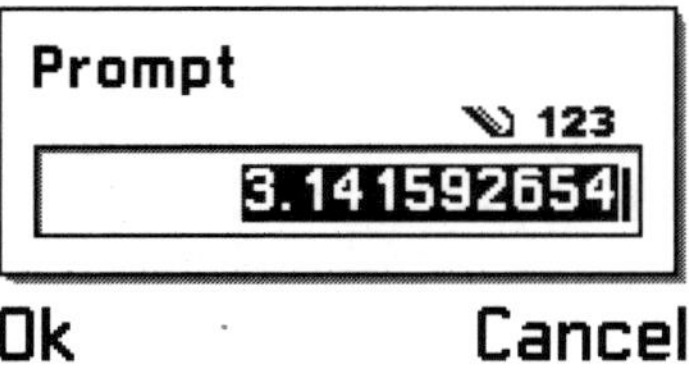

Figure 11.18 Floating point query

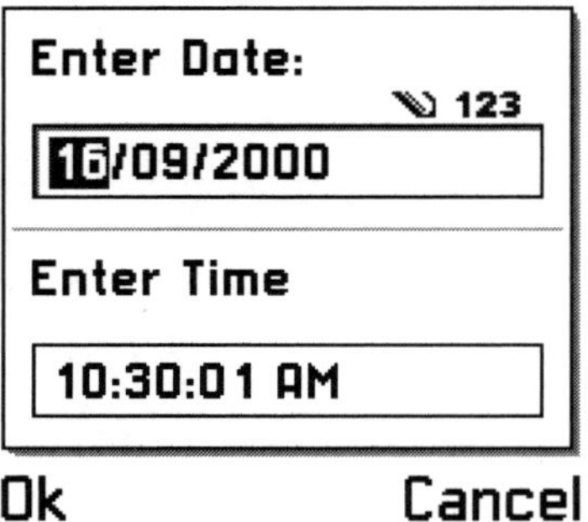

Figure 11.19 Multiline query with date and time lines

The resource structure now contains two dialog lines instead of one. A date and time multiline data query resource structure would be:

```
RESOURCE DIALOG r_date_and_time_multiline_query
     {
     flags = EGeneralQueryFlags;
     buttons = R_AVKON_SOFTKEYS_OK_CANCEL;
     items =
           {
           DLG_LINE
                 {
                 type = EAknCtMultilineQuery;
                 id = EMultilineFirstLine;
                 control = AVERELL_DATA_QUERY
                       {
                       layout = EMultiDataFirstDateEd;
                       label = "Enter Date:";
                       control = DATE_EDITOR
                             {
                             minDate = DATE {year=1986;};
                             maxDate = DATE {year=2060;};
                             flags=0;
                             };
                       };
                 },
           DLG_LINE
                 {
                 type = EAknCtMultilineQuery;
                 id = EMultilineSecondLine;
```

```
                control = AVERELL_DATA_QUERY
                      {
                           layout = EMultiDataSecondTimeEd;
                           label = "Enter Time";
                           control = TIME_EDITOR
                           {
                           minTime = TIME {second=0;minute=0;hour=0;};
                           maxTime = TIME
                             {second=59;minute=59;hour=23;};
                           flags=0;
                           };
                      };
                }
          };
    }
```

The dialog line type is `EAknCtMultilineQuery` for both of the lines. The ID determines the multiline line number; for the first line it is `EMultilineFirstLine` and for the second it is `EMultilineSecondLine`. Then the controls are as for a the general data query.

List Query

The list query, an example of which is shown in Figure 11.20, is a list of possible selections, of which the user must select one. It can contain a heading at the top of the popup window, if chosen. Below, there is a list of possible choices; the handling of the softkeys and the selection key is similar to the confirmation query. Also, a grid component can be used instead of a list:

```
TInt index = 0;
CAknListQueryDialog* dlg = new(ELeave) CAknListQueryDialog(&index);
if(dlg->ExecuteLD(R_MY_LIST_QUERY))
    {
    \\ ok pressed
    }
```

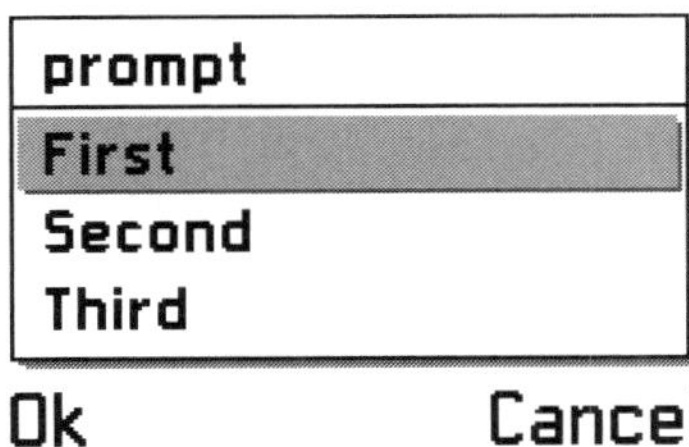

Figure 11.20 List query with three items

The `CAknListQueryDialog` class takes the integer pointer parameter that contains the index of the selection after the list has been executed. When the dialog is dismissed, ExecuteLD() returns the softkey ID that was pressed.

The list query uses an AVKON_LIST_QUERY resource structure:

```
RESOURCE AVKON_LIST_QUERY r_my_list_query
    {
    flags = EGeneralQueryFlags;
    softkeys = R_AVKON_SOFTKEYS_OK_CANCEL;
    items =
        {
        AVKON_LIST_QUERY_DLG_LINE
            {
            control = AVKON_LIST_QUERY_CONTROL
                {
                listtype = EAknCtSinglePopupMenuListBox;
                listbox = AVKON_LIST_QUERY_LIST
                    {
                    array_id = r_listbox_item_list_array;
                    };
                heading = "prompt";
                };
            }
        };
    }
```

In the source code example above we have an AVKON_LIST_QUERY_DLG_LINE, containing an AVKON_LIST_QUERY_CONTROL that defines the listbox type. The AVKON_LIST_QUERY_LIST contains a resource link to the array of items in the list:

```
RESOURCE ARRAY r_listbox_item_list_array
    {
    items =
        {
        LBUF {txt = "1\tFirst"; },
        LBUF {txt = "1\tSecond"; },
        LBUF {txt = "1\tThird"; }
        };
    }
```

A dynamic list query can be constructed with a CAknPopupList class that takes an existing listbox control and makes it a popup frame with an optional heading. The NewL constructor takes any CEikListBox derived listbox, the command button array (CBA) resource ID, and a popup layout type as a parameter. Even though this might appear to be a dialog and has an ExecuteLD() function as well, it is not a dialog.

The ExecuteLD() returns after the popup has been dismissed; the return value is true if it is accepted. The user is responsible for deleting the popup list object. The `SetTitleL()` function can be used to set the title of the popup window. Only the following styles of listboxes work with the popup list:

- `CAknSinglePopupMenuStyleListBox`
- `CAknSingleGraphicPopupMenuStyleListBox`
- `CAknSingleGraphicHeadingPopupMenuStyleListBox`
- `CAknSingleHeadingPopupMenuStyleListBox`
- `CAknDoublePopupMenuStyleListBox`
- `CAknDoubleLargeGraphicPopupMenuStyleListBox`

Multiselection List Query

A multiselection list shown in Figure 11.21 is similar to the list query described above, but now the user can select multiple items from the list. The selected items are passed to the client via a special array of indices of the items shown.

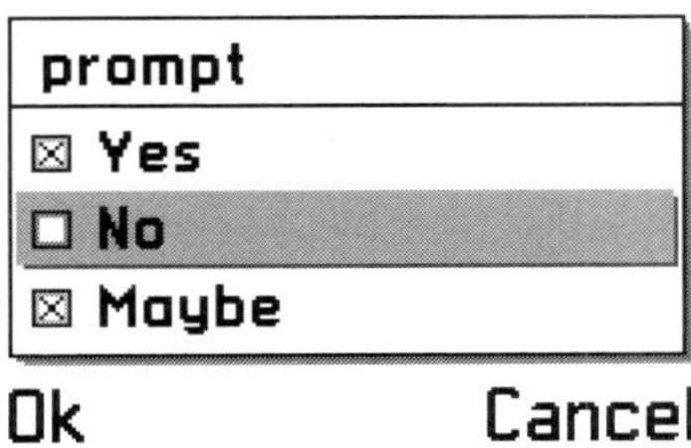

Figure 11.21 Multiselection list query with two selected items

The `CAknListQueryDialog` constructor takes a `CListBoxView::CSelectionIndexArray` pointer that will contain the selected items indices after the query has executed. This `CSelectionIndexArray` can be any `CArrayFix` deriving class. In the example below it is a `CArrayFixFlat`; what you must watch out for here is the cleanup stack – the listbox does not take ownership of it, so it must be on the cleanup stack:

```
CArrayFixFlat<TInt>* array = new(ELeave)CArrayFixFlat<TInt>(3);
CleanupStack::PushL(array);
CAknListQueryDialog* dlg = new(ELeave) CAknListQueryDialog(array);
if(dlg->ExecuteLD(R_MY_MULTI_SELECTION_LIST_QUERY))
    {
    // ok pressed
    }
CleanupStack::PopAndDestroy(array);
```

The resource structure is somewhat similar to that of a list query, but it has its own structures and types:

```
RESOURCE AVKON_MULTISELECTION_LIST_QUERY
  r_my_msl_query
    {
    flags = EGeneralQueryFlags;
    softkeys = R_AVKON_SOFTKEYS_OK_CANCEL;
    items =
        {
        AVKON_MULTISELECTION_LIST_QUERY_DLG_LINE
            {
            control = AVKON_LIST_QUERY_CONTROL
                {
                listtype = EAknCtSingleGraphicPopupMenuListBox;
                listbox = AVKON_MULTISELECTION_LIST_QUERY_LIST
                    {
                    array_id = r_listbox_item_list_array_icons;
                    };
                heading = "prompt";
                };
            }
        };
    }
```

The multiselection list needs a listbox array that has tabs (the correct tab listing is given in Chapter 10):

```
RESOURCE ARRAY r_listbox_item_list_array_icons
    {
    items =
        {
        LBUF {txt = "1\tYes"; },
        LBUF {txt = "1\tNo"; },
        LBUF {txt = "1\tMaybe"; }
        };
    }
```

11.5.2 Global Queries

There are three types of global query. These use the notification framework to show the queries. The user must create an active object, start it, and then pass the `TRequestStatus` to the query execution function. The user selection is contained in the request status object. All the parameters passed to the global query have to be class members; they must exist when the server executes.

Global Confirmation Query

The confirmation query `ShowConfirmationQueryL()` function has many parameters: the request status, prompt, softkeys, animation, image information, tone, and a dismissal flag:

```
if (iAOObserver)
    {
    iAOObserver->Cancel();
    delete iAOObserver;
    }
iAOObserver = new(ELeave) CMyGlobalConfirmationObserver(iEikonEnv);
iAOObserver->Start(); //SetActive
iGlobalConfirmationQuery->ShowConfirmationQueryL(iAOObserver->iStatus,
  iPrompt, R_AVKON_SOFTKEYS_OK_CANCEL, R_QUERY_NOTE_ANIMATION);
```

One must be extra careful that the prompt text object exists when the query is shown. On completion, the request status will contain the pressed softkey ID. The `CMyGlobalConfirmationObserver` object `RunL` function is executed when the query is dismissed.

Global List Query

`ShowListQueryL()` of the list query takes an array of list elements, the request status, and the initially focused index in the list. The request status will hold the index of the selected item on completion; if the list is canceled it will contain the −1 value.

Global Message Query

`ShowMsgQueryL()` function of the message query takes the request status, message text, tone, header text, and its image file ID and mask. The request status will contain the softkey ID on completion.

11.6 Setting Views

All application-specific settings should be collected to a special settings page. The setting pages are specialized screens for the manipulation of certain controls. These are presented in a list to the user. If the list is long it can be split into different views by using tabs, or into a hierarchical settings tree. The setting structure is presented top-down, starting from the setting listbox. The settings are collected and presented in specialized setting item listboxes. Some of these setting items are modified through editors in specific setting pages.

11.6.1 Setting Item List

The setting item list shown in Figure 11.22 displays a number of setting items that are not yet editable, although some items with a binary state can be switched in the listbox. These setting item listboxes can be used

Figure 11.22 Setting item list

in various places, such as in a control in a view architecture view, or as part of a multipage dialog.

The content of the setting list is likely to be very stable, and the best place to load the list is in the resources. The resource structure used is an AVKON_SETTING_ITEM_LIST. This structure contains the title and an array of AVKON_SETTING_ITEM structures for the contained items. An example of a setting list resource is as follows:

```
RESOURCE AVKON_SETTING_ITEM_LIST r_my_setting_item_list
    {
    title = "My settings";
    initial_number = 2;
    items =
        {
        AVKON_SETTING_ITEM
            {
            identifier = EMyDateSettingId;
            setting_page_resource = r_my_date_setting_page;
            name = "Date";
            },
        AVKON_SETTING_ITEM
            {
            identifier = EMyNumbersSettingId;
            name = "Number";
            type = EEikCtEdwin;
            setting_editor_resource = r_my_numbers_edwin;
            compulsory_ind_string = "*";
            }
        };
    }
```

The identifier is a unique ID for the setting item. The setting_page_resource is a link to an AVKON_SETTING_PAGE resource that determines the appearance of the setting page. The `setting_editor_resource` overrides the previous resource by defining a link to an editor; when this is used, the type of the editor must also be specified. The `compulsory_ind_string` is used to show a special index character on the setting item list; in this case, it is a * character.

You must derive your own setting list class from `CAknSettingItemList`. To load the setting list resource you should call the `ConstructFromResourceL()` function of the `CAknSettingItemList` class. The base class has a `CreateSettingItemL()` function that needs to be overridden. The purpose of this function is to construct a setting item that corresponds to a specific ID. When the setting item is constructed, the item constructor usually requires the editable value to be passed. This is then shown to the user in the list. The setting list owns a `CAknSettingItemArray`.

In some situations, you will want to create the setting listbox dynamically. This is done by constructing the wanted setting listbox (e.g. a `CAknSettingStyleListBox`). This listbox is then filled and formatted with the setting texts, such as a standard listbox. The view is made an observer of the listbox. The listbox generates events, for example, when a selection is made, and then calls the client `HandleListBoxEventL()`. From this function, using the current listbox index, we launch the correct setting page.

11.6.2 Setting Items

The setting items are contained in the setting item listbox. The setting item can construct the correct setting page by calling the virtual `EditItemL()` function. If it is needed, you can derive from any of these classes to implement your own setting item and page. All the different `CAknSettingItem` deriving classes override this function to construct their corresponding setting page.

The setting item base class provides a common interface to all the setting item classes that make up the array of setting values. Each setting item has both a `LoadL()` and a `StoreL()` function that can internalize and externalize the editor value. The virtual `LoadL()` is called when the item is constructed. Again, the virtual `StoreL()` must be called by the client, though the setting item list `StoreSettingsL()` function calls them to all the contained items. A list of standard setting items and the pages they create are shown in Table 11.7.

11.6.3 Setting Pages

Selecting a setting item in the setting item listbox and pressing the selection key, or choosing a menu item to open the setting for editing, should be translated into a function call that creates the setting

Table 11.7 Setting items and the pages they create

Setting item class	Corresponding setting page class
CAknTextSettingItem	CAknTextSettingPage
CAknIntegerSettingItem	CAknIntegerSettingPage
CAknPasswordSettingItem	CAknAlphaPasswordSettingPage or CAknNumericPasswordSettingPage, depending on the password mode
CAknVolumeSettingItem	CAknVolumeSettingPage
CAknSliderSettingItem	CAknSliderSettingPage can be overridden with CreateAndExecuteSettingPageL and CreateSettingPageL
CAknTimeOrDateSettingItem	CAknTimeSettingPage or CAknDateSettingPage, depending on the mode.
CAknIpFieldSettingItem	CAknIpFieldSettingPage
CAknEnumeratedTextPopupSettingItem	CAknPopupSettingPage
CAknBinaryPopupSettingItem	CAknPopupSettingPage

page. The method for constructing and executing the setting page is described below.

The CAknSettingPage is a base class for all the different setting page types. These deriving setting pages then own one of the controls mentioned below. Possible controls and the setting page classes that contain them are listed in Table 11.8. In this example case, the owner is a CAknVolumeSettingPage. For example the volume setting is as in Figure 11.23.

Table 11.8 Setting page classes and controls

Setting page class	Control
CAknSliderSettingPage	CAknSlider
CAknVolumeSettingPage	CAknVolumeControl
CAknTextSettingPage	CEikEdwin
CAknIntegerSettingPage	CAknIntegerEdwin
CAknTimeSettingPage	CEikTimeEditor
CAknDateSettingPage	CEikDateEditor
CAknDurationSettingPage	CEikDurationEditor
CAknAlphaPasswordSettingPage	CEikSecretEditor
CAknNumericPasswordSettingPage	CAknNumericSecretEditor
CAknCheckBoxSettingPage	CAknSetStyleListBox
CAknPopupSettingPage	
CAknRadioButtonSettingPage	

The slider, volume, radio button, and text setting pages manipulate a single value, but the checkbox allows for multiple selected items. The resource structure used for the setting pages is an AVKON_SETTING_PAGE. This is used for all the setting page types

Figure 11.23 Volume setting page

by changing the values of the structure. For instance, a volume setting page resource could be:

```
RESOURCE AVKON_SETTING_PAGE r_my_volume_setting_page
    {
    label = "Volume";
    softkey_resource = R_AVKON_SOFTKEYS_OK_BACK;
    type = EAknCtVolumeControl;
    editor_resource_id = r_my_volume_control;
    }
```

The `editor_resource_id` is a link to the setting page control of the type defined by the `type` line above it. The control would then be:

```
RESOURCE VOLUME r_my_volume_control
    {
    flags = EMyVolumeControl;
    value = 10;
    }
```

A setting page may have its own menu. This is defined by the `menubar` link, which works similarly to dialog menus. In this case, you will also want to change the `softkey_resource`. Menus are generally used with settings lists. Also, some optional features such as a setting item number, setting text description, and a navigation pane hint text are supported. The mapping of the different control types and the resource structures they use is given in Table 11.9.

The `CAknSettingPage` class has set and get functions for overriding the resource-defined functions. This can be used for context-sensitive texts and for setting the index. The various setting page deriving classes also have constructors that take the needed

Table 11.9 Control types and their resource structures

Setting page type	Control type	Resource structure
Slider	EAknCtSlider	SLIDER
Volume	EAknCtVolumeControl	VOLUME
Text	EEikCtEdwin	EDWIN
Integer	EAknCtIntegerEdwin	AVKON_INTEGER_EDWIN
Time	EEikCtTimeEditor	TIME_EDITOR
Date	EEikCtDateEditor	DATE_EDITOR
Duration	EEikCtDurationEditor	DURATION_EDITOR
Radio button list, Check Box	EAknSetListBox	LISTBOX
Numeric secret editor	EAknCtNumericSecretEditor	NUMSECRETED
IP address editor	EAknCtIpFieldEditor	IP_FIELD_EDITOR
Popup list	EAknCtPopupSettingList	POPUP_SETTING_LIST

data as parameters. These can be used instead of using resource-defined structures. In most cases, the control needs to be defined in a resource anyway.

The specific setting page constructor usually takes the setting value as a reference parameter. This reference is updated according to the `ExecuteLD()` parameter. The check box setting page uses a particular `CSelectionItemList` class to handle the text and their selection states. It is basically an array that contains `CSelectableItem` objects. The items contain the item text and its selection status, as shown in Figure 11.24.

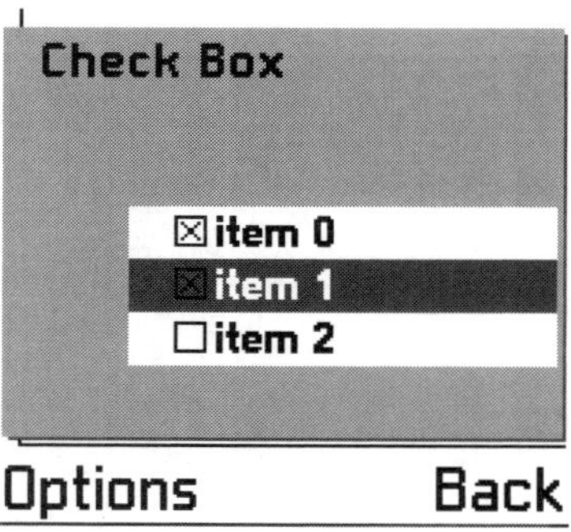

Figure 11.24 Check Box setting page

Setting pages are like modal dialogs, the `ExecuteLD()` sticks, but events may still occur and run to other parts of the application code. The parameter determines whether the parameter passed by the reference is changed whenever the editor is changed or only when the page is dismissed. The setting page is constructed and executed in the setting list according to the user selection:

```
CAknVolumeSettingPage* dlg = new (ELeave)CAknVolumeSettingPage(
  R_MY_VOLUME_SETTING_PAGE, volume);
if (dlg->ExecuteLD(CAknSettingPage::EUpdateWhenChanged))
    {
    //Accepted, set system volume to new value
    }
```

11.6.4 Custom Setting Page

To implement your custom setting page you can derive from the most suitable settings page and then override some of the virtual functions. The `CAknSettingPage` declares these virtual functions, and they are called in various situations during the lifetime of the page:

- The `DynamicInitL()` is called from the `ExecuteLD()` just prior to activating and adding the dialog to the control stack.
- `UpdateSettingL()` is called when the client value needs to be changed. This is called only when the `ExecuteLD()` parameter has been set to `EUpdateWhenChanged`.
- The `OkToExitL()` function is called when the user attempts to exit the setting page. This can be used to check that the data are valid, etc. It works just as in the dialog architecture.
- The `AcceptSettingL()` is called when the user accepts the setting and the page is about to be dismissed. The value in the setting control is written to the client variable. `RestoreOriginalSettingL()` is called when the setting is rejected; some restoring may be done here. Both are called after calling the `OkToExitL()` function.
- A setting page may have an observer that reacts to changes done to the editable value, calling the `SetSettingPageObserver()` function with a `MAknSettingPageObserver` deriving class as the parameter sets the observer. The deriving class implements the `HandleSettingPageEventL()` function that is called every time an event occurs.

11.7 Summary

The reader should now be familiar, visually and codewise, with the available controls that can be used in applications. All the frustrating work of finding which enumerations and flags belong to which type of control should be made a little easier with use of the relevant tables in this chapter. It is hoped this chapter will help developers implement their applications with fewer bugs and in less time.

12

View Architecture

To aid maintainability and flexibility, Series 60 Platform applications are commonly split into two distinct parts, the engine and the user interface. In this chapter, the focus will be on how the user interfaces from separate applications can cooperate together to provide the end user with a seamless way of navigating the functionality in the platform.

12.1 View Deployment

In essence, application views are fairly self-contained and independent user interfaces with their own controls and event handling. Use of the **view architecture** is straightforward once the menu and the key event handling code is physically separated into the individual view classes.

The actual systemwide management of views, foreground and background states as well as **fading** (or dimming) of views is handled by the **View Server** component. This server is not directly accessible to the applications, but certain important aspects of systemwide behavior are explained later in this chapter View run-time behavior.

12.1.1 Views and User Interface Controls

For the application to function properly, the main user interface (UI) class has to be derived from `CAknViewAppUi` instead of the classes higher in the hierarchy such as the `CEikAppUi` familiar from earlier versions of Symbian OS. If the application logic can be clearly separated into individual views, the main UI class does not need to contain much code at all. A compulsory view relates to the `ConstructL()` method that creates and registers the application views to be available to the rest of the system and sets one of the views as the default view. This view will be activated automatically when the application is first launched by the user. An example implementation is shown below:

```
void CMyAppUi::ConstructL()
    {
    // Initialize base class
    BaseConstructL();
    // Create an application view
    CAknView* view;
    view = CMyFirstView::NewL();
    // Register the view with the framework
    CleanupStack::PushL( view );
    AddViewL( view );
    CleanupStack::Pop();
    // Set this as the default view
    ActivateLocalViewL( view->Id() );
    // Create and register another view
    view = CMyOtherView::NewL();
    CleanupStack::PushL( view );
    AddViewL( view );
    CleanupStack::Pop();
    }
```

A view is identified by the system with a unique identifier (UID) value that is unique within the application. Each of the view classes should implement an identifier method that returns the view UID. As the class `CAknView` is derived from the class `CBase`, it is instantiated with the conventional Symbian construction process. However, since only *one* view is active within an application at a time, it is best to use all UI-specific construction and destruction code in the `DoActivateL()` and `DoDeactivate()` methods. It is important to note that one should always check within the `DoActivateL()` method whether a view is already active, to avoid resource leaks. This is advisable, since by having only a single view active, the memory consumption and system resource usage is significantly decreased. Finally, a `HandleCommandL()` method is needed in the view to handle user-originated events. An example implementation of these methods is shown below; as views do not own any pieces of the screen themselves, each view must own at least one window-owning control:

```
// Construct a view. This method is called only once during the
// lifetime of the object.

void CMyFirstView:ConstructL()
    {
    // Load view-specific data from resource file
    BaseConstructL( R_MYFIRSTVIEW );
    }
// DoActivateL is called each time when a view is activated
// For example, this occurs when the user starts a new application
// or the active view in the foreground is changed by the system itself
void CMyFirstView::DoActivateL( const TVwsViewId& /* aPrevViewId */,
  TUid /* aCustomMessageId */, const TDesC8& /* aCustomMessage */ )
```

```
    {
    // Create view container control that will then create
    // all child controls
    if( !iContainer )
        {
        iContainer = new ( ELeave ) CMyFirstViewContainer( *this );
        iContainer->ConstructL( AppUi()->ClientRect() );
        iContainer->SetMopParent( this );
        // Add the container to the control stack
        // So that it and its children can receive key events
        AppUi()->AddToViewStackL( *this, iContainer );
        // Complete the initialization
        iContainer->ActivateL();
        }
    }

// DoDeactivate is called whenever a view is deactivated

void CMyFirstView::DoDeactivate()
    {
    // The controls are not needed anymore. Remove them to save
    // memory.
    if( iContainer )
        {
        AppUi->RemoveFromViewStack( *this, iContainer );
        delete iContainer;
        iContainer = NULL;
        }
    }
```

It is customary for views to have a command button array (CBA) and a menu. These can be specified within the application resource file using a resource structure of type AVKON_VIEW. The name of the resource has to be passed as a parameter in the method `BaseConstructL()` during view initialization. An example of the resource is shown below:

```
// This resource is located in the application resource file.
//
RESOURCE AVKON_VIEW r_myfirstview
    {
    // Define the softkeys. Certain commonly used softkeys are
    // pre-defined in avkon.hrh
    cba = R_AVKON_SOFTKEYS_OPTIONS_EXIT;
    // Define the contents of the options menu. This should be
    // present in the same resource file as this view resource.
    menubar = r_myfirst_view_menubar;
    }
```

The relationships of the various view-related classes are shown in Figure 12.1.

12.1.2 Implementing View Transitions

Owing to the way that the Series 60 applications interact with each other, it is often necessary for applications to launch views from other

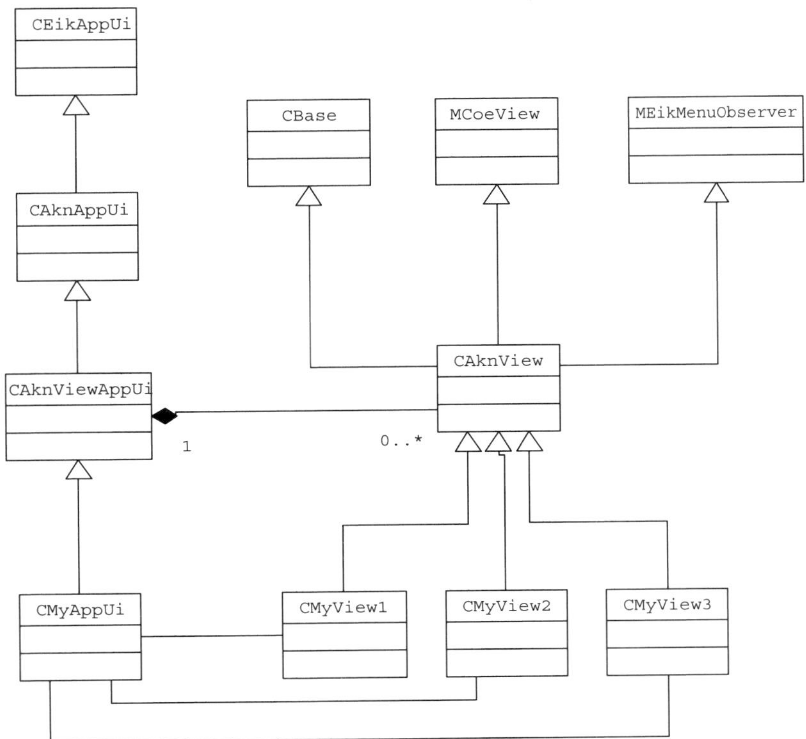

Figure 12.1 Relationships between various application-side classes in a view-based application

applications in the system. Any application that supports views can easily be launched if the view UID and application UID are known. For the view switches within the context of a single application, a method `ActivateLocalViewL()` of the class `CAknViewAppUi` is used. For interapplication transitions, the same class provides the method `ActivateViewL()`. An example of the use of each of these is shown below:

```
// Switch to a new view within the same application
// The information needed is the view UID.
// The UIDs for all views should preferably be located in a
// single header file.

AppUi->ActivateLocalViewL( KMyOtherViewUID );

// Switch to a view within another application.
// The necessary information is the application UID and view UID.
// This is a method of the class CAknView, unlike the above,
// which comes from CAknViewAppUi. The helper class
// TVwsViewId is used to store
// the UID values.
```

```
TVwsViewId targetView( KTargetAppUid, KTargetViewUid );
ActivateViewL( targetView );
```

For more complex applications, it is necessary to parameterize the views. An example of this may be a day view of a calendar application. The content of the view is naturally parameterized by a specific date. For applications launching this view, a mechanism is necessary to pass this extra information to the calendar day view class. This is made possible by the interview messaging mechanism. A view that needs to be parameterized can receive a special, custom-defined message ID and message string parameters for use in the `DoActivateL()` method at the time of activation. Other applications can then pass the message and the ID to the view activation calls that were discussed above. An example of this more complex interaction is as follows:

```
// An example implementation of the parameterized view
CMyCalendarDayView::DoActivateL( const TVwsViewId&
/* aPrevViewId */, TUid aCustomMessageId,
   const TDesC8& aCustomMessage )
    {
    // Create view container control that will then create
    // all child controls
    if( !iContainer )
        {
        if( aCustomMessageId == KStartDayView )
            {
            TUint8 day = aCustomMessage[0];
            iContainer = new ( ELeave ) CDayViewContainer( *this );
            iContainer->SetMopParent( this );
            iContainer->ConstructL( day, AppUi()->ClientRect() );
            // Add the container to the control stack
            // So that it and its children can receive key events
            AppUi()->AddToViewStackL( *this, iContainer );
            // Complete the initialization
           iContainer->ActivateL();
           }
        }
    }

// Within application code, parameterized views can be launched
// as follows
TVwsViewId targetView( KTargetAppUid, KTargetViewUid );
// Create the message buffer
TBufC8<1> dayBuffer;
dayBuffer[0] = 15;
// Launch the view
ActivateViewL( targetView, KStartDayView, dayBuffer );
```

If you want your application views to be accessible to other applications, it is recommended to document the usage. At least view, message UID, and syntax should be defined in the documentation. In practice, this is accomplished by including the information in a separate header file.

12.2 View Runtime Behavior

As mentioned above, the Series 60 Platform supports one active view per running application. Although it is simple for the programmer to implement switching between active views in different applications, the runtime behavior must be taken into account when designing the view structure.

12.2.1 Activation and Deactivation

A view switch is initiated by the application calling the method `ActivateViewL()`. The application acts here as the view server client, though this interaction is hidden within the methods of the classes provided by the view architecture. If the application to which the target view belongs is not running, the view server will execute it first and then signal the view to activate. Following this, the application user interface and the target view cooperate together to perform the view activation. The client application and the view server are notified of this, and continue their processing. The whole sequence of operations is completed by deactivating the previously active view of the target application.

12.2.2 Exception Handling

During any view activation, a leave can occur for a number of reasons (e.g. out-of-memory situations, resource shortage, or general error conditions). The basic rule is as follows: if the leave occurs (within the application/view framework) during a view switch, the leave is propagated back to the application that initiated the operation. This will result in the same view still being active and an error note on the screen for the respective leave code.

However, the cleanup operation is more complex if the leave occurs during, for instance, the construction or activation of the target view. By then, the system will already have deactivated the previously active view within the target application. To cope with the situation, the view that caused the leave is first deactivated and an appropriate error message is displayed to the user. Since the target application has now no active view, the previously active view is reactivated, or the application is simply closed if reactivated. The application can also be closed if there was no previously active view. Finally, the original view in the calling application is brought to the foreground. As a whole, this behavior ensures that the framework, applications, and the view server are always left in a consistent state, no matter how the transitions may fail.

Owing to the way in which the current view server is implemented, crashing applications or badly programmed applications with long UI delays can pose a problem for the overall usability of the platform. The view server can terminate a currently active application if it has 'hung' or remains otherwise unresponsive for a prolonged period of time. In this manner, other applications in the platform eventually become accessible. However, the developers are responsible for guaranteeing that no major loops or delays exist within applications.

12.3 Summary

The view architecture makes it possible to create solid applications that can interact easily with each other. By implementing a view-based application, the developer can make the software more maintainable, reduce the memory footprint, and take advantage of reusable views.

13

Audio

In this chapter we discuss how to use different sounds in own applications. The Series 60 Platform currently supports three different audio formats: basic monophonic sinewave tones, WAV format sound clips, and MIDI format polyphonic tunes. In this chapter, we also describe how to record and stream WAV format audio.

When programming audio software for the Series 60 Platform, an important point to be noted is that every sound utility has a priority value. By that value, it is possible, for example, to let the phone ring when a phone call comes or short message arrives.

Sound utilities, playing, recording, and so on are asynchronous operations. Handling these operations is done by a callback interface. The sound utility classes have their own 'M' classes that describe the appropriate callback methods for the class to take care of asynchronous operations. An observer has to exist for every audio client. A single observer can observe more than one player or recorder.

This chapter includes example implementation for methods that are essential for enabling the use of sounds in own applications. The class diagram of the audio components is given in Figure 13.1. `CMyTonePlayer`, `CMyAudioPlayer`, `CmyRecorder`, and `CMyStreamer` are the classes used in the example codes.

13.1 Playing

The Series 60 Platform includes three player classes for different purposes: one for playing sinewave tones, one for playing WAV sound clips and other supported formats, and one for streaming audio. The support for additional audio formats is done by media server plug-ins. Currently, in Series 60 there is a plug-in for polyphonic MIDI songs.

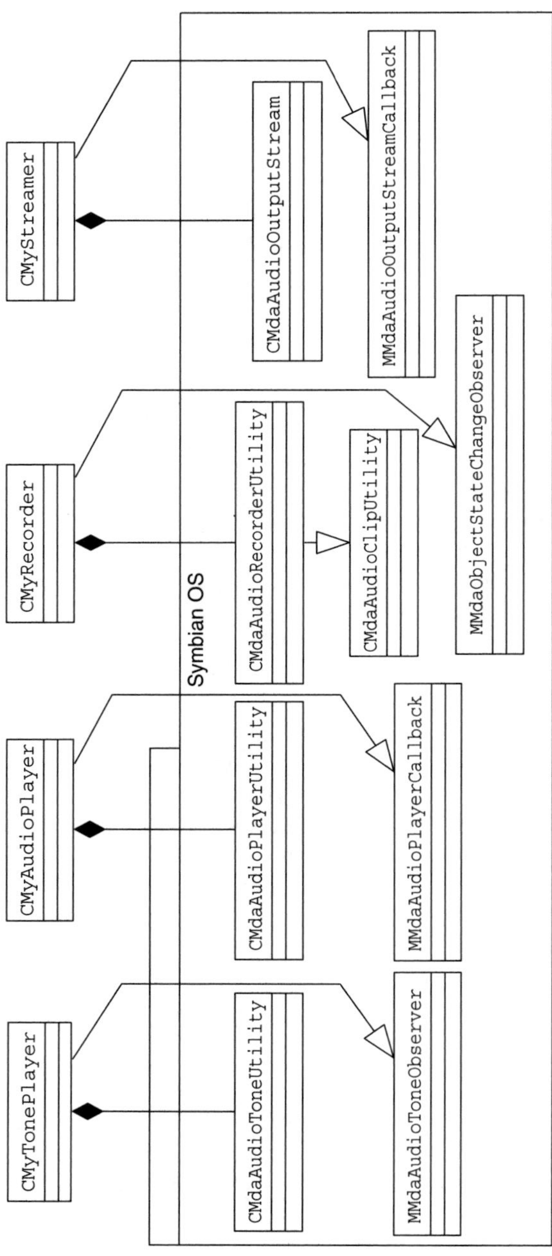

CMyTonePlayer
CMyAudioPlayer
CMyRecorder
CMyStreamer
Symbian OS
CMdaAudioToneUtility
MMdaAudioToneObserver
CMdaAudioPlayerUtility
MMdaAudioPlayerCallback
CMdaAudioRecorderUtility
CMdaAudioClipUtility
MMdaObjectStateChangeObserver
CMdaAudioOutputStream
MMdaAudioOutputStreamCallback

Figure 13.1 Audio class diagram

13.1.1 Priorities

All playing requests have to have a **priority**. Through the priority value, Media Server chooses the client that has the highest priority to play and gives the playing shift for that. The priority value range is from −100 to 100. The default priority value is zero.

The priority value also has a priority preference. By that value, the audio driver decides what to do to the playing process in the lower-priority **player**. This decision is made when another client makes a playing request with a higher priority. The priority values and preferences are described in Avkon.hrh. It is very important to set a priority value and preference to every playing request. If such values are not set, the player client's request will receive the previous request's priority and preference.

13.1.2 Sinewave Tones

Sinewave tones are used, for instance, in basic monophonic ringing tones, information notes, and so on. The class for playing sinewave tones supports playing just one tone containing a note frequency and duration, playing a number of notes from the descriptor or a file, and playing a dual-tone multifrequency (DTMF) from a descriptor. These DTMF tones can be used during a phone call to simulate phone number key presses, every number, star, and pound mark has its own multifrequency tone.

```
class CMdaAudioToneUtility : public CBase
     {
public:
     IMPORT_C static CMdaAudioToneUtility* NewL(
       MMdaAudioToneObserver& aObserver,
       CMdaServer* aServer=NULL);
     //
     virtual TMdaAudioToneUtilityState State()=0;
     //
     virtual TInt MaxVolume()=0;
     virtual TInt Volume()=0;
     virtual void SetVolume(TInt aVolume)=0;
     virtual void SetPriority(TInt aPriority,
       TMdaPriorityPreference aPref)=0;
     virtual void SetDTMFLengths(
       TTimeIntervalMicroSeconds32 aToneOnLength,
       TTimeIntervalMicroSeconds32aToneOffLength,
       TTimeIntervalMicroSeconds32aPauseLength)=0;
     virtual void SetRepeats(TInt aRepeatNumberOfTimes,
       const TTimeIntervalMicroSeconds& aTrailingSilence)=0;
     virtual void SetVolumeRamp(const TTimeIntervalMicroSeconds&
       aRampDuration)=0;
     virtual TInt FixedSequenceCount()=0;
     virtual const TDesC& FixedSequenceName(TInt aSequenceNumber)=0;
```

```
    //
    virtual void PrepareToPlayTone(TInt aFrequency,
      const TTimeIntervalMicroSeconds& aDuration)=0;
    virtual void PrepareToPlayDTMFString(const TDesC& aDTMF)=0;
    virtual void PrepareToPlayDesSequence(const TDesC8& aSequence)=0;
    virtual void PrepareToPlayFileSequence(const TDesC& aFilename)=0;
    virtual void PrepareToPlayFixedSequence(TInt aSequenceNumber)=0;
    virtual void CancelPrepare()=0;
    //
    virtual void Play()=0;
    virtual void CancelPlay()=0;
    };

class MMdaAudioToneObserver
    {
public:
    virtual void MatoPrepareComplete(TInt aError)=0;
    virtual void MatoPlayComplete(TInt aError)=0;
    };
```

When adding a sinewave player to your own application, there has to be an observer class derived from the appropriate M class. The observer class also has to have implementations for methods described in the M class. Calling `NewL()` method creates an instance of the player class. This method takes a reference to an observer as a parameter. The observer's methods are called when the preparation is completed or when the playing has ended. Both of those methods could be called with an error value. If everything has gone fine, the parameter has `KErrNone` as a value. If there is an error it should be handled by the player class developer.

Below, we present an example implementation of initializing the player utility and starting playback. The `CMyTonePlayer` class is derived from `MMdaAudioToneObserver`, so it also implements the callback methods. In this case, the audio utility is prepared to play a file-based ringing tone. The format of the file sequence is described in the Smart messaging specification (Nokia, 2000). The playing volume and the priority of the utility can only be set when the player has been prepared successfully, otherwise a panic is raised. The implementation is as follows:

```
Void CMyTonePlayer::ConstructL()
    {
    iMyToneUtility = CMdaAudioToneUtility::NewL(*this);

    // KFile has complete path and filename of a ringing tone.
    iMyToneUtility->PrepareToPlayFileSequence(KFile);
    }

void CMyTonePlayer::Play()
    {
    iMyToneUtility->Play();
    }
```

```
void CMyTonePlayer::MatoPrepareComplete(TInt aError)
    {
    // If error occurred things below cannot be done.
    if (!aError)
        {
        iMyToneUtility->SetVolume(iMdaAudioToneUtility->MaxVolume());
        iMyToneUtility->SetPriority(0, EMdaPriorityPreferenceTime);
        }
    else
        {
        // error handling.
        }
    }
void CMyTonePlayer::MatoPlayComplete (TInt aError)
    {
    If (aError)
        {
        // error handling.
        }
    }
```

13.1.3 Audio Clips

An audio clip player can be used to play, for instance, MIDI songs and single WAV clips in one's own applications. Audio playing other than the playing of sinewave tone audio formats is included in this player through plug-ins. The device includes some plug-ins, for instance, for MIDI. The plug-ins can also be installed afterwards. The audio clips can be played from a file or from a descriptor. The audio format is recognized from the header of the data. If the format is not supported or recognized, the `init` callback function is called with an `KErrNotSupported` error code:

```
class CMdaAudioPlayerUtility : public CBase
    {
public:
    IMPORT_C static CMdaAudioPlayerUtility* NewFilePlayerL(
    const TDesC& aFileName,
    MMdaAudioPlayerCallback& aCallback,
    TInt aPriority=EMdaPriorityNormal,
    TMdaPriorityPreference aPref= EMdaPriorityPreferenceTimeAndQuality,
    CMdaServer* aServer=NULL);
    IMPORT_C static CMdaAudioPlayerUtility* NewDesPlayerL(
    TDes8& aData, MMdaAudioPlayerCallback& aCallback,
    TInt aPriority= EMdaPriorityNormal,
    TMdaPriorityPreference aPref= EMdaPriorityPreferenceTimeAndQuality,
    CMdaServer* aServer=NULL);
    IMPORT_C static CMdaAudioPlayerUtility* NewDesPlayerReadOnlyL(
    const TDesC8& aData, MMdaAudioPlayerCallback& aCallback,
    TInt aPriority=EMdaPriorityNormal,
    TMdaPriorityPreference aPref= EMdaPriorityPreferenceTimeAndQuality,
    CMdaServer* aServer=NULL);

    // Control functions
    virtual void Play()=0;
```

```
    virtual void Stop()=0;
    virtual void SetVolume(TInt aVolume)=0;
    virtual void SetRepeats(TInt aRepeatNumberOfTimes,
      const TTimeIntervalMicroSeconds& aTrailingSilence)=0;
    virtual void SetVolumeRamp(const TTimeIntervalMicroSeconds
      aRampDuration)=0;

    // Query functions
    virtual const TTimeIntervalMicroSeconds& Duration()=0;
    virtual TInt MaxVolume()=0;
    };

class MMdaAudioPlayerCallback
    {
public:
    virtual void MapcInitComplete(TInt aError,
      const TTimeIntervalMicroSeconds& aDuration)=0;
    virtual void MapcPlayComplete(TInt aError)=0;
    };
```

Differences between the audio clip player and the tone player are that the priority value and preferences are given to the audio clip player when it is created. These cannot be changed later, for instance, with the tone player. Control functions and query functions cannot be used before the player utility has been successfully initialized. If an error occurs during preparation and query or control methods are called, it causes the player to panic. The `init` callback function has the length of the initialized audio clip in microseconds as a parameter to check whether if the initialization has gone successfully:

```
void CMyAudioPlayer::ConstructL(const TDesC& aFileName)
    {
    // KFile has complete path and filename of WAV file
    iMdaAudioPlayerUtility =
    CMdaAudioPlayerUtility::NewFilePlayerL(KFile, *this);
    }

void CMyAudioPlayer::MapcInitComplete(TInt aError,
  const TTimeIntervalMicroSeconds& aDuration)
    {
    // If error occurred things below cannot be done.
    if (!aError)
        {
        iPlayer->SetVolume(iPlayer->MaxVolume());
        iPlayer->Play();
        }
    else
        {
        // error handling.
        }
    }

void CMyAudioPlayer::MapcPlayComplete(TInt aError)
    {
    If (aError)
```

```
        {
        // error handling.
        }
    }
```

The player class is derived from the `MMdaAudioPlayerCallback` class; the player class also implements callback methods. There is no need to define the audio format before use. The recognition of the format and codec is done from the header of the audio clip. Playing of the example is started right after initialization.

13.1.4 Streaming

An audio stream can be used when continuous audio data play is required (e.g. for many wave clips after each other, or if the audio data come continuously from another device or application). Currently, the streaming supports WAV format audio that has been recorded with sample rate 8 kHz or 16 kHz and 16 bits per sample. The audio stream is a media server plug-in:

```
class CMdaAudioOutputStream : public CBase
     {
public:
     IMPORT_C static CMdaAudioOutputStream*
       NewL(MMdaAudioOutputStreamCallback& aCallBack,
         CMdaServer* aServer = NULL);
     //
     virtual void SetAudioPropertiesL(TInt aSampleRate,
       TInt aChannels) = 0;
     virtual void Open(TMdaPackage* aSettings) = 0;
     virtual TInt MaxVolume() = 0;
     virtual TInt Volume() = 0;
     virtual void SetVolume(const TInt aNewVolume) = 0;
     virtual void SetPriority(TInt aPriority,
       TMdaPriorityPreference aPref) = 0;
     virtual void WriteL(const TDesC8& aData) = 0;
     virtual void Stop() = 0;
     virtual const TTimeIntervalMicroSeconds& Position() = 0;
protected:
     CMdaAudioOutputStream();
     };

class MMdaAudioOutputStreamCallback
     {
public:
     virtual void MaoscOpenComplete(TInt aError) = 0;
     virtual void MaoscBufferCopied(TInt aError,
       const TDesC8& aBuffer) = 0;
     virtual void MaoscPlayComplete(TInt aError) = 0;
     };
```

The audio stream should be created by the `NewL()` method, which takes the appropriate M-class observer as a parameter. The initializing is done by calling the stream's open method, which must be called before use. If the opening of the stream has been done successfully, the `MaosOpenComplete()` callback method is called with `KErrNone` value. After successful callback, the properties of the audio, playing volume, and priority can be set and audio writing to the player plug-in buffer can be started. The callback method `MaosBufferCopied()` is called when the buffer has been copied to the server.

Playback will be started when the first buffer has been copied to the player plug-in, or when the player plug-in's own buffer is full. When the playing position reaches its end in the player plug-in's buffer, and there is no audio data waiting to be copied in the plug-in buffer, the callback method `MaosPlayComplete()` is called with error code `KErrUnderFlow`.

If the playing is terminated by calling the `Stop` method, the callback method `MaosBufferCopied()` is called with KErrAbort error code for every buffer that has not yet copied the player plug-in's buffer. The callback method `MaosPlayComplete()` is also called with error code `KErrCancel`. The `WriteL()` method can be called for copying the audio data to plug-in at anytime and many times after successful initialization of the stream. There is no need to wait for callback of the buffer copy.

A wave audio sound clip (load from a file) is used to describe the stream usage is the example below:

```
void CMyStreamer::ConstructL()
    {

    // reading wave to buffer
    TInt size(0);
    TInt headerBytes(44);
    iFs.Connect();

    // KStreamFile has complete path and filename of WAV file
    iFile.Open(iFs, KStreamFile, EFileRead);
    iFile.Size(size);

    // member variable HBufC8* iBuffer
    iBuffer1 = HBufC8::NewL(size - headerBytes);
    // size of audio data = whole size - headers (44 bytes in PCM
    // wave)
    TPtr8 ptr1(iBuffer1->Des());

    // read the data after headers
    iFile.Read(headerBytes, ptr1);
    iFile.Close();

    // creating an Instance of stream.
    iMyAudioOutputStream = CMdaAudioOutputStream::NewL(*this);
    TMdaAudioDataSettings settings;
    settings.Query();
```

```
    settings.iSampleRate = TMdaAudioDataSettings::ESampleRate8000Hz;
    settings.iChannels = TMdaAudioDataSettings::EChannelsMono;

    //Open the stream.
    iMyAudioOutputStream->Open(&settings);
    }

void CMyStreamer::PlayL()
    {
    // same buffer is sent twice to player plug-in.
    iMyAudioOutputStream->WriteL(*iBuffer1);
    iMyAudioOutputStream->WriteL(*iBuffer1);
    }

void CMyStreamer::MaoscOpenComplete(TInt aError)
    {
    if (aError == KErrNone)
        {
        TRAPD(err,
        iMyAudioOutputStream->SetAudioPropertiesL(
        TMdaAudioDataSettings::ESampleRate8000Hz,
        TMdaAudioDataSettings::EChannelsMono));
    if (err)
        {
        // error handling
        }
        iMyAudioOutputStream->SetVolume(
        iMyAudioOutputStream->MaxVolume() );
        }
    }

void CMyStreamer::MaoscBufferCopied(TInt aError, const TDesC8& aBuffer)
    {
    if(aError != KErrUnderflow && aError !=KErrAbort)
        {
        // error handling
        }
    }

void CMyStreamer::MaoscPlayComplete(TInt aError)
    {
    if (aError == KErrUnderflow )
        {
        // playing has ended.
        }
    else if (aError !=KErrCancel)
        {
        // error handling.
        }
    }
```

The `CMyStreamer` class is an observer for the stream in this example. In the construction phase, the audio is read first to an 8-bit buffer descriptor. The audio data, which is needed to send to the stream player plug-in, should not have any header information in the data. The `PCM` wave format audio has 44 bytes header information.

When the callback method `MaosOpenComplete()` has been called with `KErrNone` value, the audio properties and the playing volume are set. Every callback method should have appropriate error handling done by the streaming class developer to handle possible error situations correctly.

13.2 Recording

It is possible to record audio samples in different types of wave formats and codecs. Through the recorder utility it is also possible to play and edit audio samples. By using this class, it is also possible to play audio data from a certain position or play a described playing window. The recorder utility class is derived from the clip utility class that offers editing and different playing options for the recorder:

```
class CMdaAudioRecorderUtility : public CMdaAudioClipUtility
     {
public:
     enum TDeviceMode
          {
          EDefault=0,
          ETelephonyOrLocal=EDefault,
          ETelephonyMixed=1,
          ETelephonyNonMixed=2,
          ELocal=3
          };
public:

     // Static constructor
     IMPORT_C static CMdaAudioRecorderUtility* NewL(
     MMdaObjectStateChangeObserver& aObserver,
     CMdaServer* aServer=NULL,
     TInt aPriority=EMdaPriorityNormal,
     TMdaPriorityPreference
     aPref=EMdaPriorityPreferenceTimeAndQuality);

     // Open existing file to playback from / record to
     virtual void OpenFileL(const TDesC& aFilename)=0;

     // Open existing audio data in memory to playback from / record to
     virtual void OpenDesL(const TDesC8& aData)=0;

     // Open generic audio object
     virtual void OpenL(TMdaClipLocation* aLocation,
       TMdaClipFormat* aFormat,
       TMdaPackage* aArg1 = NULL, TMdaPackage* aArg2 = NULL)=0;

     // Set the audio device to use
     virtual void SetAudioDeviceMode(TDeviceMode aMode)=0;

     // Control the volume/gain on the audio device
     virtual TInt MaxVolume()=0;
```

```
     virtual TInt MaxGain()=0;
     virtual void SetVolume(TInt aVolume)=0;
     virtual void SetGain(TInt aGain)=0;
     virtual void SetVolumeRamp(const TTimeIntervalMicroSeconds&
       aRampDuration)=0;
     };
class MMdaObjectStateChangeObserver
     {
public:
     virtual void MoscoStateChangeEvent(CBase* aObject,
       TInt aPreviousState, TInt aCurrentState, TInt aErrorCode)=0;
     };
```

The enumeration `TDeviceMode` describes the behavior of the device when recording or playing. Different play modes are decided by a phone call: whether there is a phone call in progress or not. The local setting always plays the audio from the device speaker and records the audio from the device microphone.

The telephony nonmixed setting plays the audio only to a phone call; audio is not played from the device speaker. Recording is done only for audio that comes from a phone call; the device microphone is not used. For the telephony mixed setting, the recording and playing is done only when a phone call is in progress.

A default setting records the audio from a phone call and microphone and plays the audio to a phone call and speaker, if a phone call is in progress. If there is no phone call in progress, the recording is done only through the microphone and the playing is done through the device speaker.

First, the recorder utility has to be created by the `NewL()` method and after that, an appropriate open method should be called. The recorder utility has only one callback method, which has four parameters: the first represents the object that has changed the state; the previous and the current states describe what has happened, and the error code is for error handling. The different states are described in the audio clip utility class. The states are `ENotReady`, `EOpen`, `EPlaying`, and `ERecording`. To set the device mode, gain, and volume, the utility has to be in the state `EOpen`, `EPlaying`, or `ERecording`.

In the following example, the `CMyRecorder` class has been derived from the observer interface M class. In the constructing phase, the utility class is created and the recording target is chosen as a file. The sample rate and number of channels are also set. The number of channels must be one if the device does not support stereo playback; otherwise it panics. All properties are given as a parameter to the opening method, also containing the audio format and codec. In the callback method, after successful initialization of the recorder, the device mode and gain for the recording are set:

```
void CMyRecorder::ConstructL()
    {
    iMdaAudioRecorderUtility = CMdaAudioRecorderUtility::NewL(*this);

    // recording settings:
    TMdaAudioDataSettings audioSettings;
    audioSettings.iSampleRate = 8000;
    audioSettings.iChannels = 1;

    // recording file:
    TMdaFileClipLocation fileClipLocation;
    // KRecorderFile has complete path and filename of WAV file
    fileClipLocation.iName = KRecorderFile;

    TMdaWavClipFormat wavFormat;
    TMdaImaAdpcmWavCodec imaAdpcmWavCodec;
    iMdaAudioRecorderUtility->OpenL(&iFileClipLocation, &wavFormat,
      &imaAdpcmWavCodec, &audioSettings));
    }

void CMyRecorder::RecordL()
    {
    // Set the position to the beginning of audio data.
    iMdaAudioRecorderUtility->SetPosition(TTimeIntervalMicroSeconds(0));

    // delete data from playing position to the end of data.
    iMdaAudioRecorderUtility->CropL();

    // start recording.
    iMdaAudioRecorderUtility->RecordL();
    }

void CMyRecorder::Play()
    {

    // Maximum volume for playback.
    iMyRecorder->SetVolume( iMdaAudioRecorderUtility->MaxVolume() );

    // Set the position to the beginning of audio data.
    iMyRecorder->SetPosition( TTimeIntervalMicroSeconds(0) );

    // Start playing.
    iMyRecorder->Play();
    }

void CMyRecorder::MoscoStateChangeEvent(CBase* /*aObject*/,
   TInt aPreviousState, TInt aCurrentState, TInt aErrorCode)
    {
    if (aErrorCode)
        {
        // error handling.
        }
    If ( aCurrentState == EOpen && aPreviousState == ENotReady)
        {
        // set the device mode.
        iMdaAudioRecorderUtility->SetAudioDeviceMode(
          CMdaAudioRecorderUtility::ELocal);
```

```
        // Set maximum gain for recording.
        iMdaAudioRecorderUtility->SetGain(iMdaAudioRecorderUtility->
          MaxGain());

        }
    }
```

13.3 Summary

In this chapter we have described how to use sound in applications. Three audio formats are currently supported by the Series 60 Platform: sinewave tones, WAV, and MIDI. In audio programming, application priorities are important. Otherwise, it is possible that one's audio application will block the phone ringing or some other alarm tone. The preparation of playing or recording involves the use of asynchronous operations, the handling of which is done with a callback interface provided by M classes. This chapter has described how to use these classes in audio applications.

14

Customizing the Series 60 Platform

This chapter discusses issues related to customizing the Series 60 Platform for a new device as well as porting from other user interface (UI) styles to Series 60. Not everything is needed in every case, and there are different levels of customization. The Series 60 Platform is a functional software package, from the operating system to a large set of applications. Still, it is always customized before the devices are shipped. How much customization is required depends on many things; one of the most important issues to consider is self-evident: time to market.

It is quite possible that a device manufacturer can start developing multiple products at the same time. For one device, the platform is customized only as little as required to make it available as soon as possible, whereas with the other products the amount of customization and thus the difference from the original platform is greater. Figure 14.1 shows the parts that the device manufacturer must provide. One division breaks the various tasks into three tasks:

- UI customization, to achieve a certain branding of the device, through
  - modification of the application suite,
  - changing the UI appearance;
- base porting, to have the Symbian OS base work with the underlying hardware;
- telephony system integration, to enable communication of the cellular phone system and Symbian OS.

Here, the tasks are covered with some explanation of what can and cannot be done. However, this is still an introduction to the topic and the reader should not expect full details on every aspect, as the details fall beyond the scope of this book. Another thing that should be noted

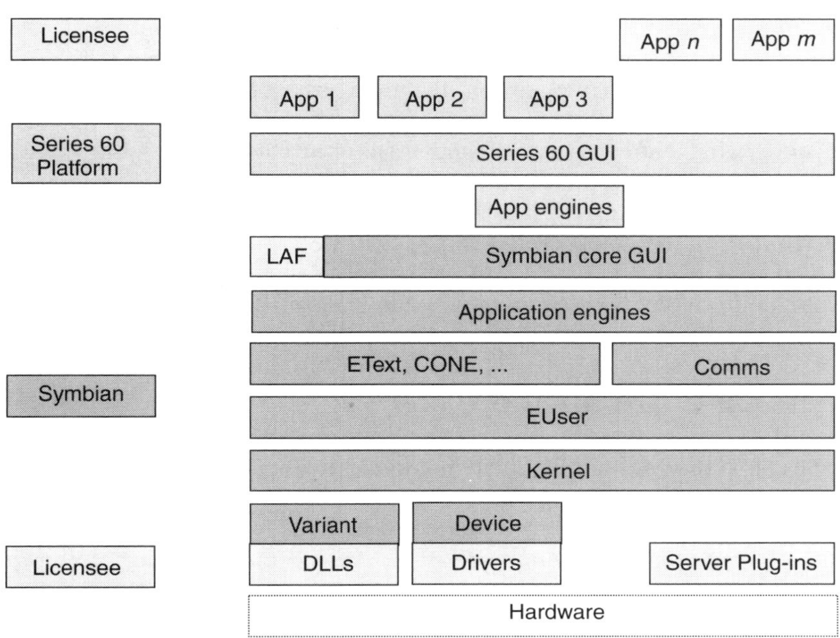

Figure 14.1 Series 60 Platform divided between the platform itself and the parts provided by Symbian and the device manufacturer. Note: App., application; GUI, graphical user interface; LAF, look and feel; CONE, control environment; comms, communications; DLL, dynamic link library

is that, when not specifically underlined, the programming interfaces for the application level developers will remain unchanged.

The term **binary compatibility** (BC) indicates the possibility of running the same program binaries on different machines. **Backward compatibility** indicates a situation where a client software built against an old version of the library can run against a new version of the library. **Forward compatibility**, in contrast, is satisfied when client built against a new version of the library can run against an old version of the library.

As an example, consider the situation where some Series 60 licensee develops a new smartphone with additional functionality and makes the new application programming interfaces available to third-party developers. The new smartphone is backwards compatible, because Series 60 applications engineered for, for example, Nokia 7650 still run on the device. However, it is not forward compatible, because the applications created by third parties that take advantage of the new application programming interfaces (APIs) will not run on the Nokia 7650 smartphone. Thus, if full BC is desired, the new interfaces may not be made available to third-party developers.

The Series 60 Platform requires customization for the devices within which it is used. This was the case even in the very first Series 60 device: the Nokia 7650. The Series 60 Platform has a certain UI style in it that the licensee will then change to suit a company's branding.

The underlying hardware can change. The computer processing unit (CPU) can at least, in theory, be different from the recommended 32-bit ARM processor (e.g. M-Core, or Intel). Peripherals can definitely change, new devices not supported by the Series 60 Platform can be introduced (e.g. a camera), and the peripherals can simply have a different chip from the those supported by the licensed code.

14.1 User Interface Customization

The Series 60 Platform is a fully fledged working software platform for smartphones, including a working user interface and a set of applications. Still, work needs to be done before it is ready for a device in terms of branding for a company's style (i.e. UI customization). Simple things that can be used for branding are changing wallpapers, startup screen, and icons for the applications. Other possibilities include status pane changes, use of company colors, and modification of the application launcher view and the set of applications that are included in the device. More difficult changes include modification of the notification system.

Most of the changes can be done either by modifying the look-and-feel module in the Uikon core or by modifying the existing applications, or creating new ones. Simple customization offers a possibility to meet tight time-to-market requirements while at the same time branding the device with the manufacturer's style. Making a few modifications to the base user interface by modifying the icons and colors and adding a few applications allows one to customize the device in a matter of a few months, including testing.

The Symbian OS UI system consists of three parts: Core, adaptable core, and device user interface, as discussed in Chapter 8. The device user interface in the Series 60 Platform is called Avkon. The Core (Uikon) and the adaptable core (standard Eikon) are implemented by Symbian and must not be modified even by UI licensing parties, except in a few exceptions.

The **core** contains a module LAF, for changing the look and feel (LAF) of the device, which can be modified to suit the needs of the device manufacturer. The possibilities the LAF module offers are changes to colors, the color scheme, and the borders, control of the behavior of a number of controls, as well as, for instance, the application UI class. Most of the changes are made to static functions, static variables, or resources. Changes to the LAF module do not affect external APIs, because only a limited number of classes link to the UikLAF.dll that contains the implementations of the various look-and-feel classes.

The **adaptable core** consists of multiple libraries. Most of these libraries contain a set of classes implemented for a limited purpose. These libraries include: EikDlg.dll, which contains the dialog framework classes; EikCtl.dll, which contains common controls; EikIr.dll, which contains infrared dialogs; and EikFile.dll, which contains the file-handling dialogs and controls. These libraries cannot be modified in any way; the adaptable nature of this module comes from the device manufacturers being allowed to remove whole libraries. For example, the Series 60 Platform does not reveal even the existence of the file system to the user and therefore the EikFile.dll can be removed to save ROM.

Application modifications are an important part of the UI customization. These do not usually require skills outside of normal application development. The most common examples of this category are the addition of new applications and modifications to existing applications.

New applications can be branding, for example Nokia could add a snakes and ladders game to the application set. Some of the applications are there to increase the fun, games, and other entertainment applications, or to provide tools such as the currency converter. Yet another reason for new applications are the addition of new peripherals such as the camera and photo album applications on the Nokia 7650.

Modification of existing applications will in most cases consist of changes to the appearance. The basic set of applications is expected to appear on each Series 60 device, and the third-party developers can, for example, expect to find a phonebook application on the device with a view to enter a new contact's information. If the phonebook application were modified incorrectly, it could result in application inoperability on that device. Other changes to views may consist of changing the normal navigation order, layout of the controls, and so on.

Two examples of modules that can be modified requiring skills not generally asked of application developers are status pane modifications and notifier framework changes. Both can be modified in a relatively short time but require the use of frameworks not accessible, except for the ROM build.

14.1.1 Customizing the Look and Feel

In theory, the look and feel of Symbian OS can be changed simply by modifying the LAF module in Uikon, as mentioned above. In practice, this is not sufficient for real customization efforts. The Series

60 graphical user interface (GUI), Avkon, has its own LAF module that needs to be modified as well.

The Series 60 LAF module is implemented in a similar way to the Uikon LAF module; however, certain additions have been made to the Uikon LAF module. These additions extend the possibilities of influencing the look and feel of the controls, not to change the core possibilities. The extensions include more ways to affect Edwins and other controls and the way in which they draw themselves in the Series 60 Platform.

Again, in theory, it would be sufficient to modify the Uikon LAF with Avkon extensions, but, unfortunately, the drawing and behavior of the controls are actually spread in more locations than only the LAF modules. The most common places to find these areas are (1) the class for the control itself and (2) one of the parent classes for the control.

Changing the implementation of a class outside the LAF modules can be tricky. Modifications to a class that is derived by multiple classes naturally affect all the deriving classes and may result in more problems than solutions. Making changes to the class itself may require removing some of the functional dependencies in the inherited classes as well as the LAF modules. After those changes, further modifications intended to modify the UI appearance are even more difficult with the decreased portability.

In practice, not just in theory, most of the modifications to the look and feel can be made by modifying the LAF modules – but not all such modifications. As mentioned above, the LAF modules offer static classes where the individual controls can get, for instance, margin width, colors, etc. In most cases, the application developers should not be concerned that there is a varying look and feel that the application can conform to each device's style. The Series 60 LAF module includes a class that should also be used by application developers: `AknLayoutUtils`. This class offers a way of laying out controls to the screen and allowing them to adapt to the current device's way of displaying each control. Addition of controls is done by using functions such as `LayoutLabel()`, `LayoutEdwin()`, and `LayoutControl()`, which are able to read the proper text margins and borders, and so on.

14.1.2 Application Shell

Application picker, application launcher, and application shell (Figure 14.2) – these are all names for the application where the user can select which application to execute. Thus, the modifications to this application have the most dramatic changes to the appearance and the overall feeling of the device. The application shell is like

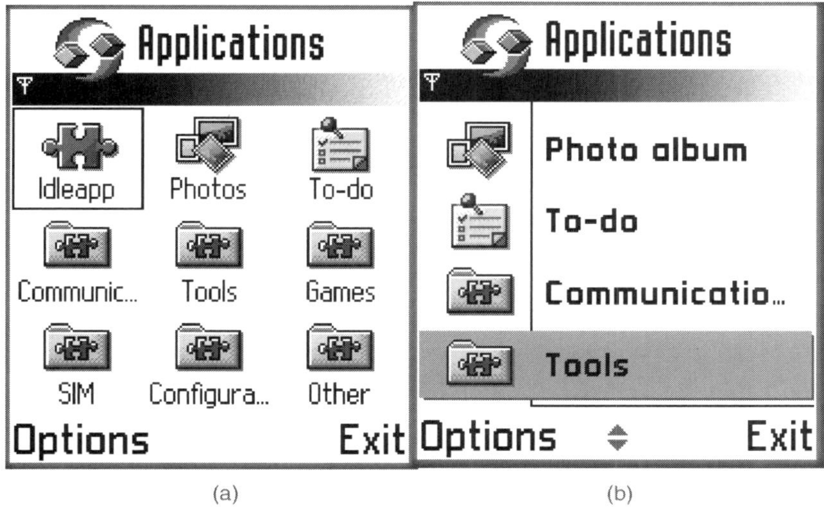

Figure 14.2 Application shell in (a) grid and (b) list format

any other application; its task is to show the user which applications are present in the device and a means of navigating and managing the applications. Managing the applications is done by organizing the applications in folders, by type. How the applications are shown to the user makes a big difference; the Series 60 Platform provides two possibilities for the appearance of the application shell to the user: a grid and a normal list, shown in Figure 14.2.

14.1.3 Status Pane

The status pane, shown in Figure 14.3, is an uncommon control in Symbian OS in the way that it is owned partially by the Eikon server and partially by applications – or other status pane clients. The status pane is divided into subpanes by defining the status pane layouts as discussed in Chapter 9.

The available layouts of the status pane and the subpanes that can be modified by the application developers are defined at ROM build. The status pane is typically used to display both system information in the

Figure 14.3 Status pane

system-owned subpanes and application-specific information in the application-owned panes. Typical system information includes icons for the signal and battery strength. Application-specific information includes the application name and the icon.

Customizing the status pane is done by modifying the available layouts, or by defining new ones, and then modifying the individual controls in the panes. It should be noted that application developers are allowed to assume a certain level of consistency across various Series 60 devices and that the possibilities are limited when making changes to application-owned panes.

The status pane layouts are defined in avkon.rss with the resource type `STATUS_PANE_LAYOUT`. The layouts in Nokia 7650 are usual layout, idle layout, and power-off recharge layout. Each layout definition is formed by defining the position of the status pane (e.g. vertical pane on top of the screen) and the panes themselves. The pane layout is organized as a tree with `SPANE_LAYOUT_TREE_NODE` resource definitions. Each definition can have the following fields:

- **size**: the width or height in pixels; which is specified depends on the tree node; the status pane is vertical then the top level size defines the height, the second level size defines the width, and so on;
- **subpanes**: each subpane consists of a number of `SPANE_LAYOUT_TREE_NODE` definitions;
- **ID**: the identifier (ID) for the subpane must be one specified in the system model;
- **flags**: these are most often `EEikStatusPaneIsStretchable` if the size is not fixed; in most cases the flags are omitted.

Laying out each individual subpane can be a tedious job, especially if some of the panes are stretchable. Each pane needs to be specified in terms of pixels, and it should be considered that, at this point, the appearance of the controls will be fixed. Most of the controls in the status pane are displayed by bitmaps, and the graphical outlook of a signal strength, for example, can be rather tightly fixed by the company brand. The controls, however, are not tied to each pane with the layout structure but with a status pane model common to all layouts – the status pane system model.

The controls that occupy each subpane are specified in the system model of the status pane. The system model is defined with the resource definition `STATUS_PANE_SYSTEM_MODEL`. Most of the definitions are formed by a list of individual pane definitions of type `SPANE_PANE`. In addition, the system model specifies the default layout and also lists all the available layouts. Each SPANE_PANE resource is defined using the following fields:

- **ID**: the id used by the layouts; if the ID is not present in the layout used, then the control is not shown;
- **flags**: mostly for specifying whether the pane is system-owned or application-owned; the default value is server-owned;
- **type**: the type of the control (e.g. EAknCtSignalPane);
- **resource**: the resource definition for the control.

The control is specified in the SPANE_PANE resource, just as it would normally be defined in the resources. It is also quite possible to define multiple panes with the same control. This will be the case where the control itself cannot adapt itself to different sizes, but will, for instance, simply draw the bitmaps associated with the control. In such cases, a simple solution is to provide multiple sets of bitmaps and define multiple sets of status pane definitions, where the initialization of the control would differ by the file from which to read the bitmaps.

14.1.4 Notifiers

The notifier framework is a means for the system to notify the user, in multiple ways, of events, regardless of the current active application or the system's state. The framework is based on events and channels. A simple reason for a need to modify the existing implementation is that the device includes a way to notify the user not present in the original Series 60 Platform implementation. An example of this type of extension would be to flash an LED (light-emitting diode) when there is an incoming phone call. Series 60 has a set of channels that can be extended with new channels, if the device manufacturer wishes to do so.

There can be multiple events at the same time, and each event can use multiple channels in informing the user about the event. The events can be specified by the device manufacturer, but they usually follow a common line. Typical events include an incoming phone call, a new message in the inbox, and battery low. Each event can be notified to the user by using multiple channels (e.g. the phone call can start a ring tone, an LED flashing, a vibrating battery, and a dialog on display).

Looking at the list of typical events, it is easy for one to conclude that certain events are more important than others. Therefore, the events can have priorities for each channel, and the order of the priorities can change between the events, depending on the channel. For example, for the loud speaker, the incoming phone call may have a higher priority than the battery-low signal, but on the display the battery-low indication may be on top of the phone call indication.

In Symbian OS v6.0 the notifier server supports plug-in notifiers. These notifiers can then be started and cancelled according to their unique identifier (UID). The notifiers are tied to a channel with a priority, and show the information passed by the client. The API for starting the notifiers is defined in the class `RNotifier` that is the client interface to the notifier server. The notification is started synchronously with `StartNotifier()` or asynchronously with `StartNotifierAndGetResponse()`.

If the notification needs to be changed (e.g. the text in the dialog is updated) then the `UpdateNotifier()` can be used. The text might be there to give some additional information about the severity of the notification (e.g. the estimated amount of battery life left in the device). Notifiers can be implemented in several ways. For example the `StartNotifier()` function, which applications will call directly without an updating possibility, can be used; alternatively an observer framework, where the application would be notifying the notifier about changes to its state, can be provided, leaving it up to the notifier to decide what to do in each case.

14.2 Base Porting

Base porting is where one gets the Series 60 Platform to run on new hardware. At this level, it is not actually getting the Series 60 to run, but Symbian OS. The Series 60 Platform is built on top of Symbian OS, but does not include Symbian OS. This means that in order to create a device running Series 60, Symbian OS must be licensed separately from Symbian.

The base-porting team can be satisfied when the text shell is running and, at this phase, it is irrelevant if the actual GUI on top will eventually be in Series 60. The amount of work naturally depends on the selected hardware. As the implementation of some parts of Symbian OS is confidential information, this section provides a rather high-level overview of the process and of Symbian OS structure, leaving out many details.

The two different ports normally developed outside Symbian are the ASSP port and Variant port. These ports are typically developed by the licensee with the help of Symbian Technical Consulting or a Symbian Competence Center, such as Digia.

The application-specific standard port (ASSP) is a piece of silicon that has the CPU and a few basic peripherals attached to it, including a serial port. The variant port is developed when one uses an existing

ASSP port as the starting point, as adding the rest of the peripherals and possibly modifying something in the existing ASSP port.

The next level in difficulty to an ASSP port is a CPU port, where CPU-level implementation is done. Development of a CPU port is, however, something that is normally left to Symbian. It is recommended that Series 60 devices have a 32-bit ARM (Advanced Risk Machines) processor and thus there is no need for a CPU port. The structure of the kernel is also supported on these various levels of porting. The Symbian OS kernel has, in fact, four layers (Figure 14.4):

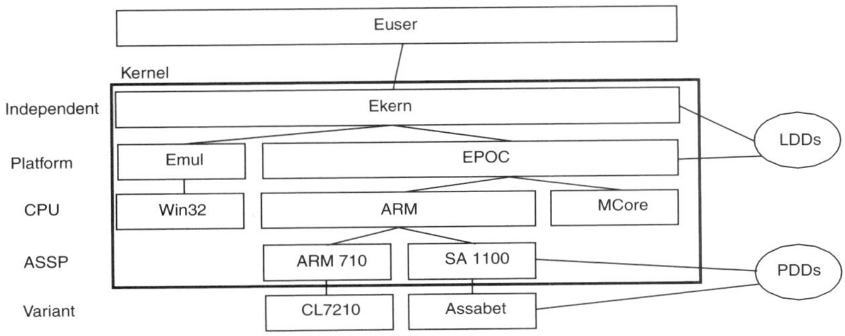

CPU = Central Processing Unit
ASSP = Application Specific Standard Port
ARM = Advanced Risk Machines (ARM is the well-known term, not the full term)
LDD = Logical Device Driver
PDD = Physical Device Driver
Other terms should not be given in full, because these are the standard used abbreviations.

Figure 14.4 Kernel modules and layering, with a few linking modules outside the kernel

- The **independent layer** is the interface to the kernel. All the modules linking to the kernel should link to the classes defined in this layer. The independent layer is common to all Symbian OS builds.
- The **platform layer** has two subpaths: emul and EPOC. The emulator uses the former, whereas all the target builds use the latter version of the platform. Most of the logical device drivers reside at the platform layer.
- At the **CPU layer**, the Symbian OS kernel is split between CPU (e.g. ARM, M-Core, and Intel). Scheduler and interrupt handler can be implemented at this layer.
- The **ASSP layer** separates, for instance, the various ARM implementations from each other. Some of the physical device drivers are implemented at this layer.

In addition to the layers, there is the variant. All the peripherals are added at the Variant level, and most of the physical device drivers are implemented at this level. Let us assume the ASSP used is already supported and that variant port is the one required. The required steps are described next.

The first thing to do is to get the code running on the device. If the already-supported methods cannot be used (e.g. JTAG) then they need to be modified or a new method needs to be implemented. Once the image is on the device, it needs to be started, and for this the bootstrap may need changing (e.g. to support a different RAM layout).

After this, it is time for the actual variant code changes (i.e. variant DLLs and device drivers). In particular, if the required screen (176 × 208 screen with recommended 0.2 mm pixel size) is not supported, the display driver should be one of the first drivers to be implemented.

The basic initialization is already done during the ASSP port, and the variant port should aim at having the text shell running at a very early phase. Once the text shell is running, the kernel tests can be run and benchmarked, the rest of the device drivers be implemented, the power model fine tuned, and so on.

After the base porting part is finalized, the text shell can be replaced by the full GUI image, and at this point the storage memory requirements come to the normal figures in a Series 60 device (i.e. approximately 16 MB ROM and 8 MB RAM). The GUI can be customized at this point, since the only thing it links to that may be changed during the base port are the new device drivers (e.g. camera drivers in the Nokia 7650).

In addition to changing the GUI and the variant, it may be necessary to modify some of the system server functionality, typically by adding or modifying the plug-in modules. One very important example would be to modify a telephony module to support underlying telephony implementation outside Symbian OS. These modules are known by their file extensions: .tsy. Implementing a proper .tsy plug-in can be considered one of the major tasks in creating a Series 60 device. Again, the details are beyond the scope of this book, but for a general explanation of .tsy modules see Chapter 15.

14.3 Porting Applications to the Series 60 Platform

Porting existing applications to the Series 60 Platform means porting a C++ Symbian OS application made for another UI style and version of the OS to the Series 60 Platform. This application can be written for Nokia 9200 Series, one of Psion's ER5 devices, or for a UIQ device. We will not be looking in detail at all the aspects of the work required but rather will try to get an overall picture and cover the main tasks. The details can be found in relevant chapters elsewhere in this book.

The application model should have been written using generic APIs of Symbian OS only, which should not have to be changed unless the changes between versions require it. Main tasks outlined here are (1) functional changes, (2) UI changes, and (3) changes to communications.

14.3.1 Functional Changes

When porting an application it should not be expected to be an exact copy of the version running on a different device type. The Series 60 Platform has a good UI specification for applications and these should be followed. These specifications obviously exist on other devices as well, and they all differ from each other and, in fact, quite a lot in some aspects. The reasons are simple:

- **Screen**: a Series 60 device with a 176 × 208 screen is not capable of holding as much information as a half VGA (Video Graphics Array) screen, for example, in a comprehensive way. The information, then, must be reduced, thus limiting the application's capabilities, or the information must be split and possibly displayed in many locations. However, if porting an ER5 application, its functionality may be increased by adding colors and multimedia.
- **Input devices**: a UIQ application and some ER5 applications rely on having a pointer as the main input device. The Series 60 Platform does have a virtual mouse, but its use should be avoided if possible. A drawing application, for example, would have to be implemented in a very different way.
- **Communications**: the first Series 60 device, Nokia 7650, supports telephony with GPRS data, Bluetooth, infrared, and messaging [including multimedia message service (MMS)] communications. Previous Symbian OS devices lack some of these technologies, but typically rely heavily on existing serial cable. The changed communications can result in changes in the application, if the communication can be changed to transparent, instant, and global.

The amount of code that requires changes varies quite dramatically. An application relying heavily on graphics in its implementation of the user interface may need as little as 10% of the code changed; in some cases even less. An application relying on the system-provided UI controls and a badly written engine with communications that must be changed might be easier to implement from scratch than to use old application as the base.

14.3.2 Changes in the User Interface

The changes in the screen size and input devices alone can make changes necessary. Typically, the whole user interaction needs to be redesigned when porting between different UI styles, as discussed in Chapter 6. However, if the application to be ported is a game with a GUI using arrow keys as the main controls, the changes can actually be very small in the view of the application.

A properly written graphical control, in this case the application view, should be implemented in a way that is easy to use at a different size. The main effort in the porting may well go into modifying the menu structure into the Series 60 Platform feel, and changing the dialog layouts, and so on.

Problems arise if the user interface utilizes ready-made controls from Uikon or the device user interface. Series 60 is not a very port-friendly environment in its implementation of ready-made controls. Many of the controls have their Series 60 version, and thus the implementation must be changed. Even if a generic control exists, it may not appear as well as a Series 60 equivalent. On many occasions, there is a new architectural layer implemented on top of an architecture found in Uikon.

Application Framework

The Series 60 Platform has implementations for all four basic classes used in the application: application (`CAknApplication`), document (`CAknDocument`), application user interface (`CAknAppUi` and `CAknViewAppUi`), and view (`CAknView`).

In the simplest form of application porting, the application UI class must be made to derive from `CAknAppUi`. Other derivations are changed if needed. The implementations of the classes may not have to be changed.

The resource file needs a small change to `EIK_APP_INFO` and the menu definitions. The menu definitions are changed to follow the specifications found in, for instance, the Nokia Series 60 UI Style Guide (Nokia, 2001g). The `EIK_APP_INFO` is typically as in the rock–paper–scissors (RPS) game (see Chapter 8), with only the menu, and the command button array (CBA) with one of the Series 60 ready-defined softkey pairs. In addition, you need to remember to react to the `EAknSoftkeyExit` command in the `HandleCommandL()` function:

```
RESOURCE EIK_APP_INFO
    {
```

```
menubar = r_rpsgame_menubar;
cba = R_AVKON_SOFTKEYS_OPTIONS_EXIT;
}
```

If the application uses dialogs, and most do, then the dialogs must in most cases be reimplemented. This applies even to simple information dialogs launched by the Eikon environment, and notes should be used instead; for example, `CEikEnv::InfoWin()` is not present in Series 60, and `CaknInformationNote`, for example, should be used.

The default implementation may lead to nonmanageable results, as shown in Figure 14.5. The dialog ported is a multipage dialog with a numeric editor as one of the controls. As Figure 14.5 shows, the tabs are hidden whenever, for instance, the numeric input icon is shown in the status pane; also, the prompt in front of the field does not show the text properly. The dialogs should be implemented so that each item is visible; but, for editing, a popup field is opened. Using the Series 60 query dialog framework produces this effect automatically.

Figure 14.5 A badly implemented multipage dialog with a numeric editor field focused

The Series 60 application view implementations fall roughly into three categories: control views, dialog views, and view architecture views. The rule of thumb in porting is: use the one used in the original implementation. In practice, this refers to the control views in most cases. The dialog views can especially be used if the original view needs to be split into multiple parts that may need editing, and the views and navigation between them are like dialogs.

The Series 60 standard applications include many examples where the main view is, in fact, a list of dialog pages, and the second view is a multipage dialog. View architecture is new to Symbian OS v6.0 and obviously is not used in ER5; Nokia 9200 series does not include many applications using view architecture, but UIQ does utilize the view architecture in many of its applications.

The bad news for those wishing to port view architecture views to Series 60 is that there is another architectural layer in Series 60 on top of the Uikon view architecture. Therefore, porting an existing view architecture view structure may not be as easy as it might be expected to be.

An application that originally used control views can, of course, be changed to use the view architecture, especially if there are many views and those views may want to pass data to each other – including between views from different applications. If the solution is made to use the view architecture, it is time for a session of very creative copy pasting.

The application user interface is changed to derive from `CAknViewAppUi`, and much of its implementation is stripped and moved to the views. It is possible that the application user interface does not own a single view or a menu – and it may handle only one `EEikCmdExit` command in its `HandleCommandL()` function.

Each view typically has its own menu and command handling, and the views are owned by the view server. In addition, the navigation must be decided on, keeping in mind that the view activation is an asynchronous call to the view server, and the application does not have any way of telling whether the activation has actually succeeded.

14.3.3 Changes to Communications

If the application has communication capabilities, the amount of work required depends on how well the implementation has been made. Communications tightly coupled with certain media and protocols, implemented in the engine and the application, is not on the list of desirables. The best option is a generic communications module implemented as a server or a separate dll. Of course, changes in the communications can force changes in the user interface if, for example, the connection type and they way it is established is hard coded in the application.

Not many things have really changed: Serial may be gone, but Bluetooth can be used as a serial device. GPRS (General Pachet Radio System) exists, but data calls were possible before. MMS is an added service that the application can utilize, if needed. In particular, if the communications module is implemented as a server with the communications protocols as plug-ins things should go smoothly.

14.4 Summary

In this chapter we have discussed application porting and platform customization of the Series 60 Platform. The customization may involve three tasks: UI customization to create a certain branch of the device, base porting to port Symbian OS base to new hardware, and telephony system integration to enable communication between the cellular phone system and Symbian OS. Application porting involves three tasks: functional changes, UI changes, and changes to communications.

Part 3

Communications and Networking

15

Communications Architecture

The microkernel structure of Symbian OS has an effect on its communications architecture. Communication services are used through **system servers**, which may be added or omitted, depending on the communication hardware supported by the smartphone. Communication software is implemented with **plug-in modules** used by the servers. New modules can be deployed and unloaded without restarting the phone. This allows an easy way to extend and configure the phone with respect to changes in the environment when, for example, the user is roaming into a new location.

This chapter contains a description of the communications architecture used in the Series 60 Platform. The architecture specifies the structure, interfaces, and behavior of communication software. It is based on Symbian OS v6.1, the Series 60 Platform consisting mainly of user interface (UI) libraries.

The architecture builds up from communications servers and plug-in modules. Servers provide basic communication services with the aid of technology and protocol-dependent communication modules. When a new service is required, a new module may be implemented and loaded to the memory without rebooting the device. Unnecessary modules may also be unloaded to save scarce memory resources.

In addition to servers and communications modules, the most commonly used communications technologies supported by the Series 60 Platform are covered in this chapter. Two following chapters provide examples on how to use communication services in applications.

15.1 The Building Blocks of the Communications Architecture

The communications (comms) architecture has a layered structure in Symbian OS, as shown in Figure 15.1. The part, which conforms with

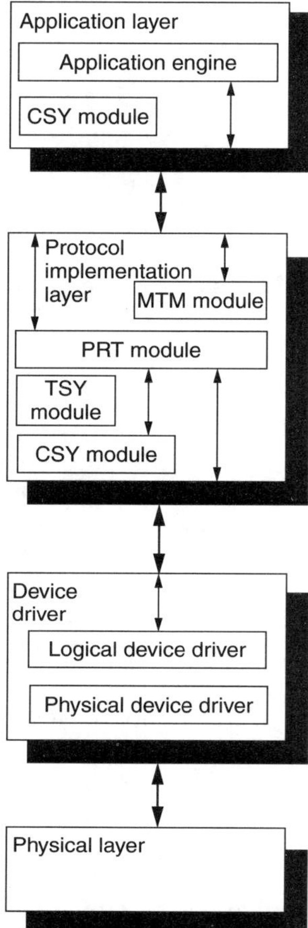

Figure 15.1 The communications architecture of Symbian OS. Note: CSY module, serial communications module; MTM, message-type module; PRT module, protocol module; TSY module, telephony control module

the well-known OSI (open systems interconnection) reference model, is the protocol implementation layer. There are four servers that can be involved in protocol implementation, depending on the characteristics of the protocol: **Serial comms**, **Socket**, **Telephony**, and **Message server**. Each server uses polymorphic interface libraries or plug-in modules that implement a specific communication service or protocol.

15.1.1 Communications Servers

As any system server, the comms servers provide a client-side application programming interface (API) for application developers. The APIs typically contain one R-type class derived from the `RSessionBase` class, which provides a handle to a server object in another process. This handle is used to create a session to the server (eventually, the `RSessionBase::CreateSession()` function is called) and query information, such as the number of supported serial ports or protocol modules, from the server.

The set of supported functions in the API is quite similar in separate comms servers, which makes it possible to define an abstract class to be used with any server. The API also contains one or more classes derived from `RSubsessionBase`, which provide functions for creating subsessions (typically, opening ports, sockets, or calls) and for transferring the actual data (taken care of by the `SendReceive()` function in `RSubsessionBase`). There are four communications servers in Symbian OS:

- the Serial communications server (C32), which provides a generic interface to serial-type communications;
- the Socket server (ESock), which provides a generic interface to communications protocols through communication endpoints known as sockets;
- the Telephony server (ETel), which provides a generic interface to initiate, control, and terminate telephone calls;
- the Message server, which provides access to message data and to the server-side message-type modules (MTMs).

In Figure 15.2, the servers and their users are depicted. In addition, some additional communications components, such as Communications database and protocols, accessed by the Socket server, are shown.

The use of communications servers depends on the application. Applications requiring simple bit transfer services need to use only simple serial-based communications, in which case the Serial comms server provides the service. The Serial comms server uses a protocol-dependent serial protocol module having a filename extension. csy. There is a separate module for RS-232 and infrared protocols. The CSY (serial communications) module uses device drivers in the kernel to access the communication hardware. Also, the device drivers are plug-in modules, which may be loaded and unloaded with respect to the hardware configuration in different locations.

Applications requiring more advanced services can use one of several available protocols for this purpose at the protocol implementation layer. The application may use the servers directly. However, it is more common that the application use a client API of one server, which may then use other servers, if required by the protocol. For example, the wireless application protocol (WAP) suite uses a WAP

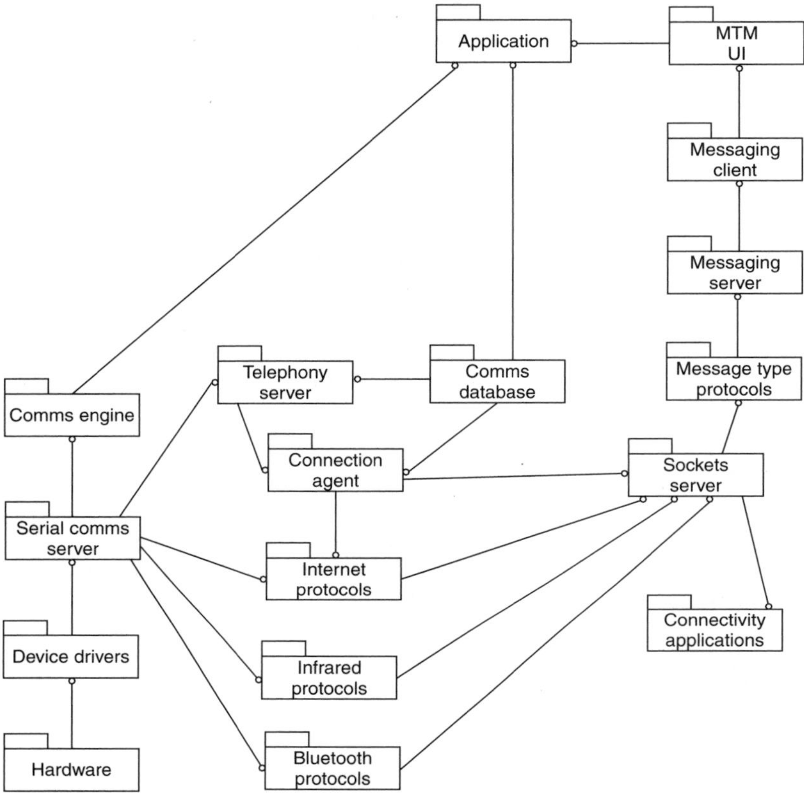

Figure 15.2 Communications (comms) servers. Note: MTM, message-type module; UI, user interface

module to access the WAP stack, which uses the short message service (SMS) protocol module to use SMS as a bearer. The SMS protocol module uses the GSM (global system for mobile) telephony control module to send the SMS over the GSM network. At the bottom, the telephony control module uses a serial protocol module to control the smartphone. All plug-in modules supply a single, unified interface in the form of a polymorphic interface dynamic link library (DLL). This enables several implementations of the same module, each capable of interfacing with different communications hardware and software.

Communications plug-in modules are used in the implementation of communication protocols. Layers of the same protocol may be implemented in separate modules. For example, physical and data link layers become part of the serial protocol module (or even device drivers), and the network and upper layers become part of the socket protocol module. The set of available plug-in module types include:

- the logical and physical device drivers used to abstract the hardware from software modules;
- the serial protocol modules used by the Serial communications server;
- the protocol modules and connection agents used by the Socket server;
- the telephony and modem control modules used by the Telephony server;
- the message-type modules, used by different message types and messaging objects.

15.1.2 Communications Modules

Device drivers come in two parts in Symbian OS. The splitting of a device driver into physical and logical drivers increases software modularity and reusability. The **physical device driver** (PDD) deals directly with the hardware, and the **logical device driver** (LDD) takes care of data buffering, flow control, **delayed function calls** (DFCs), and interrupt handling. Device interrupts are handled in two phases. When an interrupt occurs, the very short service routine is executed first. It acknowledges the device and sets a flag in the kernel to call a DFC. The DFC is called immediately in the user mode or after the execution mode has returned from kernel mode to user mode.

It is obvious that different device drivers are used in the emulator compared with those used in the target hardware. The interface to the UART (universal asynchronous receiver–transmitter) chip is implemented by the ecdrv.pdd driver in the emulator, while the euart1.pdd driver in ARM (advanced risk machine) builds. Thanks to the separation into physical and logical device drivers, it is possible to reuse the same LDD (ecomm.ldd) in both builds.

Although RS-232 is not supported in Nokia 7650 phones, the same device drivers are still used by the IrDA (infrared data association) and Bluetooth protocols. The only module that changes is the serial protocol module, which is ecuart.csy for RS-232, and ircomm.csy for infrared. For Bluetooth, the serial protocol module is named btcomm.csy. The use of device drivers and communications modules is clarified in Figure 15.3.

Serial Protocol Modules

Serial protocol modules interface directly with the hardware through a PDD portion of the device drivers. The lowest-level interface to serial

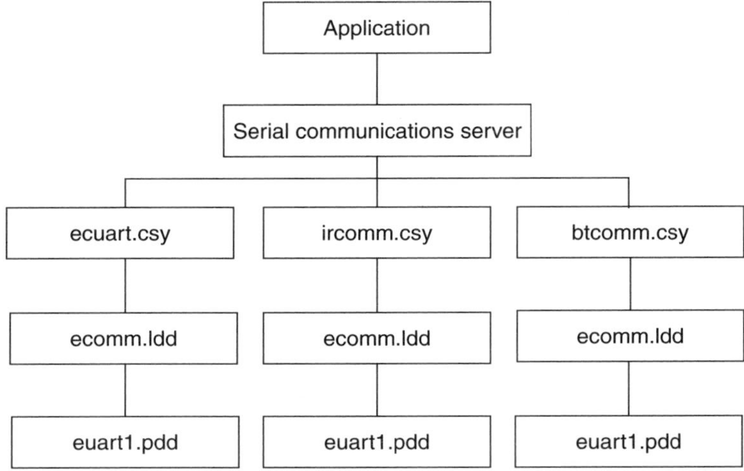

Figure 15.3 Use of device drivers and communication modules with RS-232, infrared, and Bluetooth

comms is implemented by the `RDevComm` class, which provides a handle to a channel to the serial device driver. However, instead of using `RDevComm`, the client-side should interface to the Serial comms server through the `RCommServ` class, the API of which is described in the next chapter.

Depending on the requirements of the application (and the characteristics of the smartphone) the Serial comms server uses either RS-232, infrared, or Bluetooth serial protocol module. Other modules can be implemented if new hardware interfaces are developed. Typically, clients need no knowledge of the individual CSY modules available in the phone, since they ask the Serial comms server to load whichever is defined as active in the Communications database. The Communications database will be discussed in more detail later in this chapter (Section 15.1.3).

All communication modules in Symbian OS are polymorphic interface libraries and are thus implemented according to the same principles. CSY modules are identified by the second unique identifier (UID), the value of which must be 0x10000049. A CSY module has three concepts that the developer of the module must implement: serial service provider (`CPort` class), serial protocol factory (`CSerial` class), and serial protocol information (`TSerialInfo` class). Third-party software developers do not usually implement serial protocol modules, and this is why the implementation details go beyond the scope of this book.

Protocol Modules

Protocols are used to add value to communication on top of unreliable and inefficient (in terms of channel utilization) serial-based data transfer. Added value can be, for example, error detection and correction methods, efficient and adaptive flow control, ability to avoid congestion, support of multiple simultaneous connections, or support of multipoint connections. Most protocol implementations are accessed in Symbian OS through the Socket server, ESock, which provides a similar interface to that of the well-known BSD (Berkeley Software Division) sockets. This guarantees transport independence; that is, whatever the used protocol is, it may be accessed through the similar client-side API of the Socket server.

Higher-level protocols (from network to application layer) and protocol suites consisting of several protocols are implemented in plug-in modules, called **protocol modules**, having an extension. prt. Example protocol suites supported by the Series 60 Platform include TCP/IP (transmission control protocol/Internet protocol; tcpip.prt), IrDA (irda.prt), Bluetooth (bt.prt), and WAP (wapprot.prt).

All protocols are loaded and accessed by the Socket server and accessed through a similar client-side API, regardless of their specific properties, as shown in Figure 15.4. Thus, they all have common networking operations such as sending and receiving data, establishing connections, and configuring protocols, but their implementation may be completely different. When using TCP, configuration requires binding (or connecting in the client) to a specific port in the specific IP address, whereas in IrDA and Bluetooth, configuration requires device discovery to find another device with which to communicate.

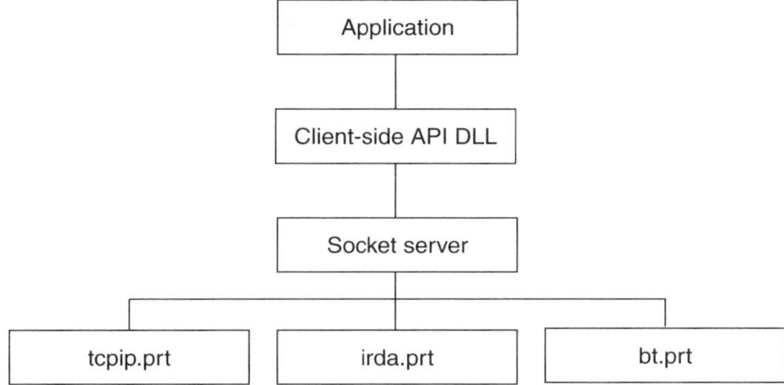

Figure 15.4 Use of protocol modules through the Socket server. Note: API, application programming interface; DLL, dynamic link library

When the Socket server is started, it scans for the esock.ini file to find out which protocols and services are defined in the system. A protocol module is loaded when a socket belonging this protocol is opened. It is also possible to load and unload protocol modules beforehand to decrease the delay during opening. In addition to connections to other devices, the Socket server provides access to other network facilities, such as host-name resolution, service resolution, and network databases, the use of which depends on the protocol. In the next chapter we will give numerous examples of the use of these services.

Since Series 60 smartphones have no permanent connection to the Internet, one must be established before transmitting or receiving any data. For this purpose **connection agents** (e.g. genconn.agt) are used. The connection agent obtains the current default **Internet access point** (IAP) settings from the Communications database and uses telephony control modules to dial up and establish a call to an IAP in case a circuit-switched data (CSD) service is used. If the General Packet Radio System (GPRS) is used instead of CSD there is no need to create a connected call.

New protocol modules may be created by system developers if new protocol standards are developed. The implementation consists of three concepts: a protocol family, a protocol, and a socket. All protocol modules are polymorphic DLLs, the second UID of which is 0x1000004A. The base classes corresponding to the three concepts are `CProtocolFamilyBase`, `CProtocolBase`, and `CServProviderBase`, defined in es_prot.h.

A protocol module can contain one or more protocols, in which case it is called a protocol family. A protocol family may build up the whole protocol suite or be a part of it, in which case functions in one protocol module may call functions in another protocol module. All protocols inside a family must be able to identify themselves and their capabilities, and at least one protocol in a family must provide socket services through the Socket server.

Individual protocols inside a protocol family are defined in an *.esk file in the system\data\ folder. The tcpip.esk file is shown below as an example. The NifMan (network interface manager) is used to start up the network interface (data link connection) and monitor the progress of connection establishment:

```
[sockman]
protocols= tcp,udp,icmp,ip
[tcp]
filename= TCPIP.PRT
index= 4
bindto= ip,icmp
[udp]
```

```
filename= TCPIP.PRT
index= 3
bindto= ip,icmp
[icmp]
filename= TCPIP.PRT
index= 2
bindto= ip
[ip]
filename= TCPIP.PRT
index= 1
loopback= loop0
[nifman]
default= genconn
```

`CProtocolFamilyBase` provides functions to list and get protocols from a protocol family. `CProtocolBase` is used to implement a socket, host-name resolution, service resolution service, or access to a network database. The class also has functions that the upper-layer protocols can use to bind to the protocol. The protocol module must contain a socket implementation derived from `CServProvider-Base` for each protocol, which supports sockets. Socket implementations use the `MSocketNotify` interface to inform the Socket server about events.

Telephony Control Modules

Basic **telephony control** (TSY) modules provide standard telephony functions, such as establishing, controlling, and terminating calls on a wide range of hardware. There are also advanced, custom TSYs, which are dependent on the specific phone hardware. Basic modules ensure that no commitment to specific telephony hardware and software configuration is required.

TSY modules are used by the Telephony server, ETel. It enables, just like the Serial comms server and Socket server, applications to be independent on specific telephony devices. The phonetsy.tsy module provides all services and is the only TSY in the device. There are also other TSY modules (hayes, gsmbsc, gprstsy), but these are written by Symbian and are intended to be used as emulators. As discussed in Chapter 14, the writing of a TSY module is part of the base porting and there is no need for application developers to do this.

Message-type Modules

Message-type modules (MTMs) implement message-handling functions. Message handling includes creation, sending, receiving, and editing messages. For each message type a separate message-type module is required. It is possible to install new MTMs dynamically to the phone. New MTMs are registered in the MTM registries. Messaging applications check from the registries the set of available MTMs in the system. Existing MTMs enable the handling of e-mail messages (SMTP, POP3, IMAP4; see the Glossary), faxes (in the Series 60 Platform, fax support is limited to fax modem functionality), short messages (short message service, SMS) or multimedia messages (multimedia message service, MMS).

There is not only one MTM per message type, but four: **client-side** MTM, **user interface** MTM, **UI data** MTM, and **server-side** MTM. In addition, there are four registries, holding records of each installed client-side MTM, server-side MTM, and so on. Every MTM is implemented as a polymorphic interface library, like any other communication module.

The client-side MTM interprets the contents of a message and provides a minimum set of generic functions. One function that is always available is `Reply()`. Using `Reply()` it is easy to implement a consistent user interface for replying to any message regardless of its type. The UI MTM provides editors and viewers. The UI data MTM provides services, which are applicable more universally. This MTM uses the message identifier to recognize the message type. Using the type the message can be shown to the user correctly without loading the UI or client-side MTMs. Thus, UI data MTMs provide universal message folders capable of storing any type of message. In addition, they are memory-efficient as the load of other MTMs may be avoided. The server-side MTM takes care of message formatting and transferring.

The role of the Message server is different from that of the other three communication servers, described above. It does not provide an API through which the services of MTMs are used. Rather, client-side MTM and UI MTM may use the server, which then uses the server-side MTM to complete the service. The Message server also has a message Index, in which each message has a UID. In addition to the identifier, the Index information includes the description of the message (header), the size of the message, and a flag indicating whether the message has attachments of binary information in file stores. The relationship of the Message server and message-type modules is shown in Figure 17.1 (page 376).

The simplest way to allow applications to create outgoing messages is to use the `CSendAs class`. It allows applications to package data into a message without using the MTM interfaces directly. An object of type `CSendAs` creates a session to the Message server, and the object can access the client-side MTM registry to get available client-side MTMs and the set of services they provide.

15.1.3 Communications Database

The communications database is the repository of communications settings. It contains information about IAPs (Internet access points), ISPs (Internet service providers), GPRS, modems, locations, charge cards, proxies, and WAP settings. Typically, the database is accessed through communications setup applications, but other software components can read from and write to the database too. The database stores the settings in a series of tables in a relational database, using the standard DBMS (database management system) architecture of Symbian OS. The DBMS enables several clients to access the database simultaneously. However, the use of the database does not require any knowledge about SQL or the proprietary C++ database interface of Symbian OS.

The contents and use of the communication database changed after Symbian OS v6.0. In v6.0 it was possible to use two database structures: IAP-type or ISP-type. The difference between the types is that the IAP type defines an IAP table, which defines sets of ISP, modem, location, and charge cards, which may be used together. The ISP type, in contrast, contains separate ISP, location modem, and charge card tables. In v6.1 (which is the basis of the Series 60 Platform) the ISP type is deprecated.

The support of GPRS has also brought changes to the comms database in Symbian OS v6.1. It is possible to define access point names, packet data protocol type (e.g. IP), packet data protocol address for the phone, and so on in incoming and outgoing GPRS tables. An other change is that there are no more separate IAP tables for outgoing and incoming connections, just a single IAP table. The global table is not used anymore to store default settings. Instead of the global table, a connection preference table is supported. This table stores connection settings, ranked by preference.

The use of the communications database has been changed in the Series 60 Platform. Applications should use the AP Engine, which is a wrapper to the database. The CApDataHandler class provides several functions for reading and updating IAP data and other settings in the database. The data item to be handled by CApDataHandler is of type CApAccessPointItem. This class provides functions for reading, comparing, and writing of individual columns in the IAP item. The columns are identified by TApMember type, which contains an enumeration to identify access points, bearers, service center addresses, GPRS parameters and so on.

15.1.4 Communications Pattern

By looking at the communications architecture and services of client-side APIs of comms servers it is easy to detect several similarities.

One of the first things to be done is to create a session to the specific comms server. After that a communications channel (port, socket, line, call) is opened, which typically means the creation of a subsession to the server. Communications settings have to be provided, after which it is possible to transfer data. Finally, subsessions and the session (connection) must be closed. More through investigation reveals the following actions must be done (here, we follow the five phases presented in Jipping, 2002):

- initialization: load required device drivers, start the communications server(s) (if not already running), and connect to it (i.e. create a session to it); load required communications modules if this is not done automatically;
- open: open the subsession (typically, open a port or a socket); depending on the server this may involve defining, for example, the access mode (exclusive or shared) or communications mode (connection-oriented or connectionless);
- configuration: set up parameters for communication, for example, baud rate or role (client or server) and undertake device and service discovery if required;
- data transfer: transmit and receive data;
- close: close handles to subsessions and session to release allocated resources.

Depending on the server and communication modules used, there may be other phases or states as well. For example, use of a connection-oriented protocol requires three phases: connection establishment, data transfer, and connection release. Data calls have possibly even more states, such as service and location selection, dialing, script scanning, login information fetching, connecting, data transfer, and hanging up.

The five states mentioned above are repeated in all comms servers except in the Message server. To make communications programming easier, we can define an abstract class having functions that relate to the identified five phases. In the next chapter we will see how this class is used in the implementation of comms classes in our application. This kind of definition is also presented in Jipping (2002).

First, we define the class to be an M class, so it provides only virtual functions that are implemented in the concrete class:

```
class MCommsBaseClass
    {
    public:
    virtual void InitializeL()=0;
```

```
virtual void OpenL()=0;
virtual void ConfigureL()=0;
virtual void Transmit( TDesC& aTransmittedData )=0;
virtual void Receive()=0;
virtual void Close()=0;
}
```

The concrete class is typically derived from CActive, so it is an active object. This is because both receiving and transmitting are asynchronous services in Symbian OS. The Initialize() function may leave if it cannot load the requested device drivers and communication module. Open() leaves if the subsession to the server cannot be created. A common reason for failure here is that the session has not been properly created earlier. The programmer has not detected the failure, because the return value of the Connect() function has not been checked.

The ConfigureL() function depends much on the server used. Sockets do not require much configuration, but there are several parameters related to serial ports. Transmit() and Receive() functions call the corresponding functions in transmitting and receiving active objects, which issue asynchronous service requests to the server (actually, it is often possible to use the synchronous versions of the functions to issue service requests, but this is not so common).

The received data are typically stored in a member variable of the receiving active object. After the receive request has been completed, the RunL() function of the receiving active object is called. The active object notifies its observer (typically, a view) to read the received data. Examples of implementing this functionality will be given in the next chapter.

The Close() function closes the handles to subsession and session. If Close() is not called, the handle remains open and there exist useless allocated objects in the server which consume memory unnecessarily. If the handle is not closed it will be closed when the thread dies and all resources allocated in the thread are freed. The Close() function is typically called in the DoCancel() function, which in addition to closing cancels all uncompleted requests.

15.2 Supported Communications Technologies

Symbian OS and the Series 60 Platform support a rich set of different communications technologies, some of which (e.g. Bluetooth) do not yet have many feasible applications and some of which are in wide use globally (TCP/IP). Personal short range communications is possible through either wired (RS-232) or wireless (IrDA, Bluetooth) protocols.

Note that a smartphone based on the Series 60 Platform does not have to provide RS-232 port, although the OS supports it. For example Nokia 7650 has no RS-232 port. Web and WAP browsing are supported on different bearers (SMS, CSD or GPRS). Data can be synchronized and converted between applications in the smartphone and desktop computer using the connectivity framework and SyncML protocol.

15.2.1 Protocols and the Open System Interconnection Model

Communications technologies are based on protocols. Formally defined, a protocol is a communicating state machine. To enable the communication between the state machines (also called **peer entities**) we need a formal language, which is used to define the set of protocol messages (**protocol data units**) between the state machines. This language may be based on binary frames or higher-level application messages, which are defined with **ASN.1**, for example (Larmouth, 1999).

A task of a protocol is to provide a service to other protocols or protocol users. Thus, a set of protocols build up a layered structure, where protocols at higher layers use services provided by protocols at lower layers. ISO (International Organization for Standardization) has defined a seven-layer reference model in which protocols at a specific layer have well-defined functions. This helps protocol implementation and interconnection of separate systems, as the name of the reference model, open systems interconnection (OSI), states (Rose, 1990).

The OSI model defines interconnections between protocol layers and between protocols at the same layer, as shown in Figure 15.5. Protocol services are accessed by means of service primitives, which usually are mapped to function calls in the implementation. Service

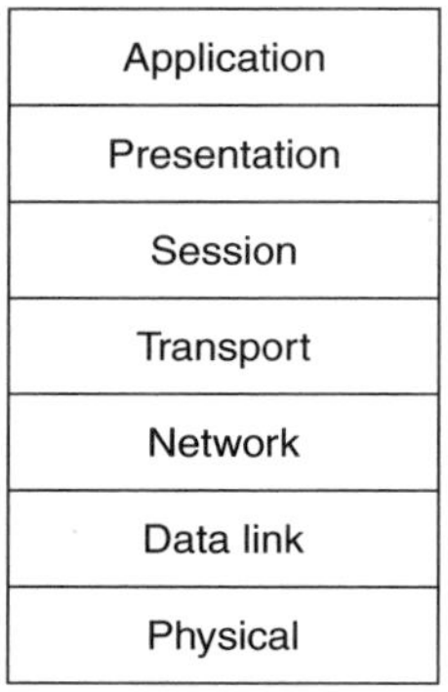

Application
Presentation
Session
Transport
Network
Data link
Physical

Figure 15.5 Protocol layering according to the open systems interconnection (OSI) model

primitives are available in the SAP (service access point), which is an identifier (address or port number) of the service provider protocol. In Symbian OS, higher protocol layers are implemented into protocol modules; lower layers may be part of the serial protocol module. The implementation details of a protocol are hidden from the user, who sees only the client-side API of the Serial comms or Socket server.

Although, a protocol may provide an arbitrary number of services, there are only four service primitive types. Unreliable service requires only two primitive types: request and indication. The former is used to request the service, and the latter indicates that a peer entity has requested a service. The other case is reliable, where the service is confirmed with two primitive types: response and confirmation. The response is issued by the entity that received the indication to give information about the success of the service request. Finally, this information is passed to the original service requester by the confirmation primitive.

All protocol definitions include the definition of messages and service primitives. However, the user of a service API of Symbian OS has no idea about the underlying service implementation or service interface because she or he always accesses the service through a certain communications server.

15.2.2 RS-232

RS-232 is a simple point-to-point protocol supporting both synchronous and asynchronous communication, where data are transmitted serially one bit at a time. Synchronous transfer is faster but requires clock synchronization; asynchronous transfer requires only that the same baud rates are used in the transmitter and the receiver. Although synchronous transfer is faster, it is used very seldom. Data are transmitted inside frames that consist of 5–8 data bits. Other bits are used for indicating the start and stop of transmission (in asynchronous communication), and error detection (parity bit). The frame structure is shown in Figure 15.6.

Start bit	Data	Parity bit
1	101010	1

Figure 15.6 RS-232 frame (asynchronous communication)

A single parity bit indicates whether the number of 1 bits is even or odd in the code word (data frame). If one bit is changed in the code word, the parity bit changes as well, which increases the minimum distance between code words (the **hamming distance**) from one to two.

If the Hamming distance is n, it is possible to detect all $n-1$ bit errors (Stallings, 1999). Thus, a single parity bit is capable of detecting one-bit errors in data frames. Two-bit errors cancel out and cannot be detected at all. Three-bit errors are interpreted as one-bit errors, and so on.

For flow control there are two options in RS-232: **software handshaking**, based on XON and XOFF signals, and the more complicated **hardware-based handshaking**, based on clear to send (CTS) and ready to send (RTS) signals. XON simply indicates that the receiver is ready to receive data, and XOFF indicates the transmitter is to seize up until XON is received again.

Hardware handshaking is much more complicated. CTS is used by the receiver to indicate that it is ready to receive data. The transmitter uses RTS to request a data transmission, but before any data are transferred data set ready (DSR) and data carrier detected (DCD) signals must be set too. DSR indicates that the carrier (e.g. a modem) is ready, and DCD that the modem has a good carrier signal. Thus, all signals (DSR, DCD, and CTS) must be set before the transmission takes place.

15.2.3 IrDA

IrDA (infrared data association) is not a single protocol but a protocol stack that hides the physical characteristics of infrared (IR) communication. So, IrDA is not implemented in the serial protocol module but as a Socket server protocol module in Symbian OS. The stack is depicted in Figure 15.7. Framing, taking place at the physical layer and controlled by the IrLAP (infrared link access protocol), is more complicated than in RS-232. Seven-bit start and stop sequences are used, and multibit errors in one frame can be detected with the cyclic redundancy check (CRC) method. Using the method, a transmitted frame is divided with a

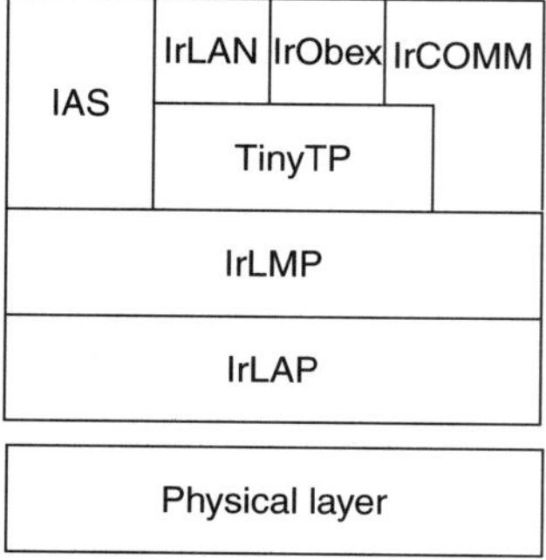

Figure 15.7 The IrDA (infrared data association) stack. Note: IAS, information access services; IrLAN, infrared local area networks; IrOBex, infrared object exchange; IrCOMM, serial communication server; TinyTP, tiny transport protocol, IrLMP, infrared link management protocol; IrLAP, infrared link access protocol

generating polynome (bit sequence) and the remainder of the division is added at the end of the frame. The receiver divides the received frame together with the received remainder with the same generating polynomial used by the transmitter; if the remainder is zero, the frame is received correctly with high probability.

IrLAP (infrared link access protocol) provides the reliable transfer of bits over physical media. Reliability is achieved with the CRC method, simple **media arbitration**, acknowledgements, and retransmissions. The arbitration method is very simple. A device that is not transmitting must first listen to the media for 500 ms and establish a connection after that, provided there are no other devices transmitting during the listening period. After the connection has been established, it is active until disconnected.

Connection is always established by using the same configuration:baud rate of 9600 bps (bits per second), eight-bit frames, and no parity. After the connection establishment, the devices negotiate the configuration parameters. IrLAP also addresses each datastream, so there may be several logical connections over one physical IR link. There are three different framing methods defined:

- asynchronous SIR (slow IR), from 9.6 to 115.2 kbps;
- fast IR (FIR) with synchronous SDLC-like (Synchronous Data Link Control) framing from 57.6 kbps to 1.152 Mbps;
- FIR with synchronous pulse position modulation (PPM) up to 4 Mbps.

IrLMP (infrared link management protocol) is responsible for connections between applications (link multiplexing, device discovery). Each datastream is assigned with a one-byte number, which is a logical service action point (SAP) selector. The communication process has the following phases:

- *Device discovery*: exchange of information with IrLMP peer entity to discover information about the device;
- connection: data connection with service IDs;
- data delivery: exchange of application data;
- disconnection: close of IrLMP connection and resetting the service IDs.

Service IDs are recorded into an information access service (IAS), which is a network database accessible through the Socket server. This enables several services to share the same IR link.

TinyTP (tiny transport protocol) is an optional protocol in the IrDA stack but is included in Symbian OS. It is responsible for flow control and data segmentation and reassembly. Each IrLMP connection may be controlled individually, which enables efficient flow control and link utilization. The size of the service data unit (SDU; i.e. the data frame) is negotiated with the IrLMP. Datastream is segmented into SDUs, and each SDU is numbered. This enables efficient interleaving of the IR link. In addition, errors affect smaller data units, which improves the efficiency of data transfer.

IR programming APIs are shown in Figure 15.8. It is possible to use the client-side API of the Serial comms server (IrCOMM), in which case an IR link is emulated to be an RS-232 link. The IrDA protocol is accessed through the Socket server. There are two services: the IrMUX and TinyTP protocols. In addition, it is possible to use the IAS. IrTinyTP provides a reliable transport layer, which corresponds to a sequenced packet service, while the IrMUX protocol corresponds to unreliable datagram service. Note that IrCOMM provides only a fixed data transfer of 9600 bps. Other services, such as IrObex (infrared object exchange) and IrLAN (infrared local area network), uses IrMUX or TinyTP. For example, IrObex uses IrMUX in port numbers, defined in the IAS.

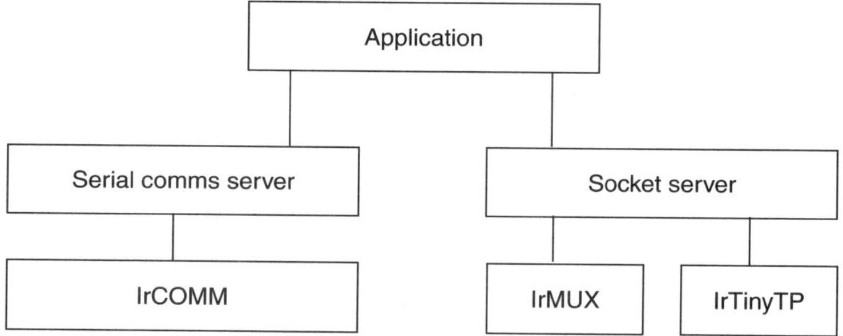

Figure 15.8 Infrared programming application programming interfaces (APIs). Note: IrCOMM, Serial communications server; IrMUX, protocol corresponding to the unreliable datagram service; Ir Tiny TP, infrared tiny transport protocol

15.2.4 mRouter

The mRouter protocol has replaced the psion link protocol (PLP), which was used to synchronize and convert data between the smartphone and desktop computer. The protocol may use either TCP, UDP(user datagram protocol), or IP on top of the point-to-point protocol (PPP) link to enable data transfer between the PC and the smartphone. In addition, the protocol may be used, as the name indicates, as a gateway allowing the smartphone to access Internet through the PC. The mRouter protocol is described in more detail in Chapter 18.

15.2.5 SyncML

SyncML is a standard for universal device synchronization and device management. It enables the use of one program to synchronize several devices and use networks as well as serial cables on multiple platforms for synchronization. The SyncML server is a device containing a sync agent and sync engine. The SyncML client first sends its modifications to the server. After receiving the modification, the server does sync analysis to check which information has changed. After client modification, the server may send server modifications to the client to complete the synchronize procedure. In addition to synchronization, SyncML can be used for device management, although the current implementation in the Series 60 Platform does not support this yet. For more detailed description of the protocol, refer to Chapter 18.

15.2.6 TCP/IP

TCP/IP (transmission control protocol/Internet protocol) is a well-known protocol suite consisting of several protocols. At the transport layer there are the unreliable UDP and the reliable TCP protocols. UDP is a packet-oriented protocol, which means that the transmission units are packets, whereas TCP is stream-oriented or byte-oriented in which individual bytes may be sent instead of whole packets, for example in the case of retransmissions.

At the network layer, there are two protocols: IP, which is responsible for addressing and routing and ICMP (Internet control messages protocol), which is responsible, for example, for checking the status of different routes in the network. Currently IPv4 is supported, but in v7.0 onwards both IPv4 and IPv6 are supported simultaneously. Compared with IPv4, IPv6 has greater address space (128-bit addresses instead of 32-bit addresses) and better support for mobile routing, authentication, and real-time streaming. Between the data link layer and network layer there are two protocols: the address resolution protocol (ARP) and reverse address resolution protocol (RARP) for address resolution from IP addresses to physical addresses, and vice versa.

The TCP/IP stack has been implemented into a protocol module, which provides two services. The protocol is not selected explicitly as in IrDA, but only the protocol family is selected to be TCP/IP. After that, if a programmer chooses a reliable socket type, the used protocol is automatically TCP. On the other hand, if an unreliable socket type is chosen, the used protocol is UDP.

15.2.7 Dial-up Networking

Connections from a smartphone to the Internet are established through dial-up networking (or by using the mRouter protocol, mentioned above, in Section 15.2.4). This includes the data link layer protocol support for IP protocol and the use of ETel services to establish a physical call to the IAP.

In Symbian OS the data link layer connection is provided by the network interface (NIF), such as PPP.nif. The network interface is started by the networks interface manager (NifMan), which is also responsible for monitoring the progress of the connection.

The route to the network address is set up by a NIF agent. If there is no route available, the system uses the NifMan to start up an appropriate NIF. NifMan, together with its connection agent, enables the use of dial-up networking via an IAP according to the following phases:

- A component opens a session with the Socket server and tries to connect to a specific IP address.
- ESock dynamically loads the TCP/IP protocol family when the socket is opened.
- IP asks NifMan for a route to the IP address.
- As there is no connection, NifMan loads its connection agent to set one up.
- Optionally, the connection agent opens a dialog with which the user may accept connection settings in the Communications database.
- The connection agent opens a session with the ETel server and asks it to make a call and connect.
- ETel looks for relevant details for the current port, location, IAP, and modem settings in the Communications database. Note that the Series 60 Platform is used only in smartphones with a built-in GSM modem, so there cannot be external modems.
- ETel loads the appropriate TSY module (phonetsy.tsy).
- The TSY module opens a session with C32 (Serial comms server).
- The TSY tells C32 and Symbian OS kernel to load the correct CSY module and appropriate device drivers.
- At last, a hardware connection exists. ETel can talk to the phone or modem and tell it to dial up the IAP.
- On connection, ETel loans the port back to the connection agent, which can now run any scripts necessary to log on to the Internet (password checking).
- The connection agent hands control back to NifMan, which now activates, for example, the PPP link for any authentication, and to control the remainder of the call.

When the PPP asks for a line to be dropped, NifMan returns control to the connection agent, which returns the port to ETel for termination of a call, and the whole stack now unwinds.

PPP

PPP is a simple data link layer protocol enabling access to the Internet through a serial link. Although it is a protocol, it is not accessed through the Socket server, but rather the NifMan uses it for data transfer. The PPP implementation in Symbian OS contains several protocols for providing advanced features of PPP. The link control protocol (LCP) uses echo–reply packets to test and ensure link quality. Authentication of the user may be performed with the PPP authentication protocol (PAP), which uses plaintext passwords, or the challenge handshake authentication protocol (CHAP), which uses MD5 and DES (data encryption standard) encryption algorithms. Further information about encryption in communications will be provided at the end of this chapter (Section 15.4.3). PPP supports also several compression methods to improve the link efficiency: PPP compression control protocol (CCP), Microsoft PPP compression, PPP predictor compression, and Van Jacobsen header compression.

15.2.8 Bluetooth

Bluetooth is an open data communication standard developed by the Bluetooth special interest group (www.bluetooth.com). It is a low-power, full duplex, digital radio technology with a gross data rate of 1 Mbps. The standard covers both hardware and software layers, as shown in Figure 15.9. The radio uses ISM band (industrial, scientific, and medical) at 2.4 GHz spread-spectrum, fast acknowledgement, frequency hopping, and forward error correction. **Spread spectrum** spreads the transmission signal into a broad frequency range, which makes it less vulnerable to interference. In addition, there are 79 frequencies at which the signal hops 1600 times per second. This also increases immunity from interference.

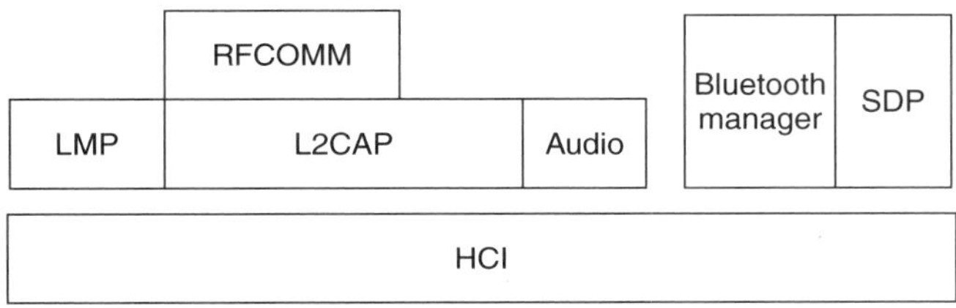

Figure 15.9 Bluetooth stack. Note: RFCOMM, see text; CMP, link management protocol; L2CAP logical link control and adaptation protocol; SDP, service discovery protocol; HCI, host controller interface

Forward error correction means that there is an attempt to correct erroneous received frames without retransmission, by using the redundant bits in the frames. This decreases the need for retransmissions, which again improves the performance. If the link quality is good, there is no reason to use redundant bits for error correction and this function may be turned off. Errors are still detected with **fast acknowledgement**, which means that received packets are acknowledged immediately in return packets. The acknowledgements do not contain any number information, only either ACK (acknowledgement) or NACK (Negative ACK). If the transmitter receives the NACK, it immediately retransmits the previous packet.

The Bluetooth HCI (host controller interface) interfaces with the baseband protocol, which is responsible for data encoding, transmitter management, frequency hopping, security, error detection, and correction and framing. The protocol supports both circuit [synchronous connection-oriented (SCO)] and packet switching [asynchronous connectionless (ACL)]. SCO and ACL channels can be used simultaneously for asynchronous data and synchronous voice transfer. There may be up to three simultaneous SCO channels open. Each voice channel supports a 64 kbps voice link. The asynchronous channel can support an asymmetric link of 712 kbps in either direction, while permitting 57.6 kbps in the return direction. If a symmetric link (i.e. both directions simultaneously, with equal throughput) is used, the maximum transfer speed is 432.6 kbps.

The link manager protocol (LMP) establishes, secures, and controls links between devices. It is not used by upper layers, but gives authenticated, established, links to the logical link control and adaptation protocol (L2CAP). L2CAP provides connectionless and connection-oriented data services, protocol multiplexing, and datastream segmentation and reassembly.

The service discovery protocol (SDP) allows Bluetooth devices to discover, search, and browse services. It is a distributed protocol implemented on top of L2CAP. RFCOMM is not a core Bluetooth layer, but allows an interface between existing services and the Bluetooth stack. RFCOMM emulates the RS-232 protocol and, for example, to IrOBEX it presents an interface fully similar to RS-232.

The Bluetooth device may register services, which it provides to other devices, in the Bluetooth Security Manager and SDP database. The Bluetooth Security Manager enables services to set appropriate security requirements that incoming connections to that service must meet. Security settings define simply whether authentication, authorization, or encryption is required or not. The SDP database contains the attributes of the service. One attribute may specify, for example, the port number in which the service is available. If a device knows that a service is available in a specific port in the remote device and no authentication or encryption is required, it may start using the service directly.

15.2.9 GPRS

GPRS (General Packet Radio Service) is a radio-based wide-area packet data service extending the CSD service of the GSM network. Compared with CSD, data transfer is packet-switched rather than circuit-switched. Packet switching suits well the transfer of IP packets, for example. Billing can be based on the number of transferred packets, not on connection time. However, operators may choose whatever billing method they wish. There is no need to maintain the connection between the terminal and the network if packets are not transferred. This enables several users to share the same radio interface channels, which in turn improves the utilization of the network.

Channel allocation is very flexible in GPRS. In principle, it is possible to allocate 1–8 time slots for each user independently in the down and up link. However, the number of supported time slots depends on the capabilities of the terminal. In theory, with use of the maximum number of time slots and lightest coding scheme it is possible to achieve a transfer rate of 171 kbps. There are four coding schemes, which differ from each other with respect to the sensibility to transfer errors. If the link quality is low, a heavier coding scheme is used, which reduces the transfer rate (Cai and Goodman, 1997).

There are three types of GPRS terminals: class A, class B, and class C. Classes differ from each other in terms of complexity and the support of simultaneous circuit-switched and packet-switched connections. Class A terminals allow the use of simultaneous circuit-switched and packet-switched calls. It is possible, for example, to browse web pages at the same time a speech call is active. A class B terminal allows the creation of both circuit-switched and packet-switched connections, but they cannot be active simultaneously. Class C terminals do not allow the establishment of circuit-switched and packet-switched connections simultaneously. There may be class C terminals, that do not support circuit-switched connections at all. For UMTS (Universal Mobile Telecommunications System), there are three other GPRS terminal types. The interested reader should read GSM specification 02.60 for further information.

In Symbian OS, applications may choose GPRS as one bearer of application data, but it does not provide any programming interface.

15.2.10 HTTP

HTTP (hypertext transfer protocol) is a simple text-based protocol for content transfer on the Internet. It was originally a stateless protocol, where only one request–response pair was served per connection; in contrast, HTTP1.1 supports leaving the connection open after transmission and using it for sending further data. Basic client server communication over TCP/IP is possible, but there are often intermediaries, such as proxies and gateways, between the browser and server. A **proxy** is a forwarding agent, that rewrites all or part of the request before forwarding it to the server. A **gateway** receives requests and translates them to the underlying servers. In addition to proxies and gateways, there are **tunnels**, which are relay points between two endpoints. By using a tunnel, a message can be passed unchanged, for example, across the firewall.

There is no open programming API available for application developers in the Series 60 Platform yet. HTTP1.1 is the only supported technology for the transfer of midlets, because it is the only protocol required by the current MIDP (mobile information device profile) specification. More information about MIDP and HTTP1.1 is provided in Chapter 20.

15.2.11 WAP

WAP (wireless application protocol) is a protocol stack for content browsing on low-bandwidth, high-latency networks. It is suitable for many applications and it is used for delivering MMS messages. The protocols of the WAP stack are shown in Figure 15.10. WAP is bearer-independent. Any network capable of conveying WAP datagrams may be used.

WDP (wireless datagram protocol) provides network and transport layer functions, such as port level addressing, datagram segmentation and reassembly, selective retransmission, and error recovery.

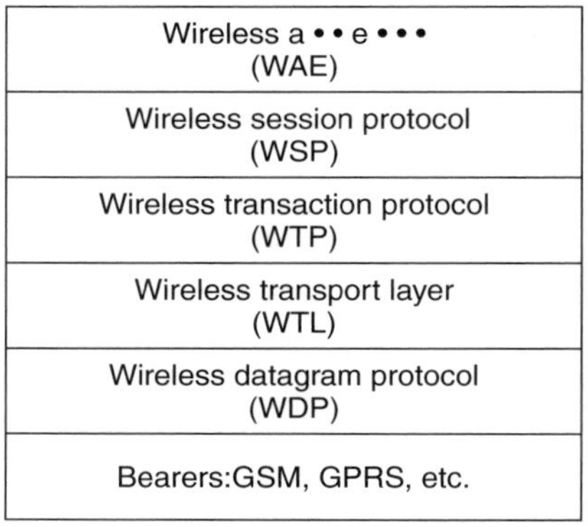

Wireless a • • e • • • (WAE)
Wireless session protocol (WSP)
Wireless transaction protocol (WTP)
Wireless transport layer (WTL)
Wireless datagram protocol (WDP)
Bearers:GSM, GPRS, etc.

Figure 15.10 WAP (wireless application protocol) stack. Note: GSM, global system for mobile; GPRS, General Pachet Radio System

WTP (wireless transaction protocol) supports connectionless and connection-oriented transactions. It also handles the retransmission of datagrams. WTLS (wireless transport layer security) provides data encryption between client and server. It can also ensure data integrity and can detect whether data have been repeated. WSP (wireless session protocol) emulates HTTP (is modeled after HTTP) in a restricted way. Using connectionless service, the WSP communicates directly to WDP without WTP. Connection-oriented service takes place through WTP. This type of service enables capability negotiations and header caching.

15.2.12 Summary of Series 60 Platform Communication Technologies

All transport protocols supported by Symbian OS are positioned into the OSI model listed in Table 15.1. Note that we have omitted messaging protocols, because these will be discussed in more detail in Chapter 17. As can be seen, none of the protocols fully covers all OSI layers. Usually, one layer in the real protocol covers several OSI layers. In many cases, functions of the higher layers are left for applications and thus are not needed in protocols. This can be clearly seen in serial and telephony protocols.

Table 15.1 Communications technologies and protocols supported by Symbian OS

Layer	Serial	IrDA	TCP/IP	WAP	Bluetooth	Telephony
Application				WAE		
Presentation						
Session				WSP		
Transport		Tiny TP/ IrMUX	UDP, TCP	WTP WTLS		
Network		IrLMP	IP, ICMP	WDP	L2CAP/ RFCOMM	
Data link	RS-232	IrLAP	Ethernet, PPP	Bearers	Base band	Telephony
Physical	UART chip and physical link	Infrared			BT radio	Wired, mobile telephony

Note: for abbreviations, see the Glossary.

15.2.13 Using Real Devices with the Software Development Kit

Communications software is difficult to test in the emulator without real communicating hardware. Serial ports of the PC running the emulator may be used directly, but there are two differences. First, in the emulator, the name of the port is COMM::, whereas in the PC it is

COM. Second, the numbering starts from 0 in the emulator, but from 1 in the PC. Thus, the emulator port COMM::0 is mapped to PC port COM1, and so on.

IR ports can be used in a PC which does not have the ports by attaching a mobile phone, for example, to the PC. This allows the use of the IR port of the mobile phone. When the emulator is started, it looks for IR devices and will use one if found.

For Bluetooth, the emulator looks for Nokia's Bluetooth device (i.e. Nokia Connectivity Card DTL-4; go to www.forum.nokia.com). If one is found, it is possible to send and receive Bluetooth packets. The kit has to be connected to the computer via both the USB (universal serial bus) and the serial cables, because it draws power from USB, but the emulator supports only the serial cable for data transfer. Note that for a Quartz 6.1 emulator one must use Ericsson's Bluetooth Application Tool Kit (www.tality.com).

A GSM phone may be connected through an IR port to the emulator. In addition, network connections can be established by using a null modem cable.

15.3 Communications Security

Threats to communications security may be divided into three categories: confidentiality, integrity, and availability. **Confidentiality** ensures that no piece of datum is exposed to nontrusted persons. Data **integrity** means that data cannot be modified by unauthorized persons. **Availability** means that data and services remain available for intended users.

15.3.1 Confidentiality

Local communications, such as RS-232, infrared, or Bluetooth are difficult to 'snoop' on without getting caught. The eavesdropper needs proper receivers and must be in the intermediate space between communicating parties. Global or wide-area communications, such as the Internet or GSM, are more easily 'snooped' with proper devices. To gain access to application-level information, protocol headers and other framing information may be extracted with a protocol analyzer tool. The protocol analyzer knows the encoding and decoding functions (and protocol message structure) of each protocol in the suite and may be used to find the relevant information from a large amount of redundant protocol control information. If this extracted information contains, for example, passwords in plain text, the eavesdropper has free heir to utilize the user's account in any way.

15.3.2 Integrity

Confidentiality does not imply data integrity. In fact, data may be highly integrity critical but still not require any confidentiality. Public keys are good examples of such data. Although it is quite easy to protect against eavesdropping by encrypting the transmitted data, it is impossible to prevent any modifications, either sensible or absurd. Cryptography provides a means of detecting if integrity of the data is lost, but there are no means to guarantee that no modifications are made.

One common type of attack against integrity is forgery of critical data. E-mail messages are easily forged by using directly the SMTP protocol to upload forged messages to the mail server. Also, IP addresses may be forged. If we trust some party and download software from a forged IP address, we may, in the worst case, cause a lot of damage to the existing software.

15.3.3 Availability

Availability ensures that any server or service is always available. Attacks against the availability aim at authentication, protocol, or service. By getting the right password, the eavesdropper may do anything with the credentials of the owner of the password. Attacks on the service and protocol try to run the service into a state in which it accepts anything, even commands from the snooper. The attacher attempts to create such a State by causing exceptions in the attacked software. If exceptions are not handled properly, the server may enter an undefined state, accepting any input from the attacker. One common way is to use buffer overflow. If there is no checking of boundaries of the buffer in software, external users may write data into areas not intended for users, which may break down the state of the server. The attacker may close down the server or read confidential data from it.

15.4 Protection Mechanisms

Very little can be done to prevent the physical 'snooping' on a communications medium, especially if data are transferred through several networks around the world. Encryption ensures the confidentiality of data with a high probability, although there is no encryption algorithm that cannot be broken given enough computing and memory resources. Although encryption makes it more difficult to read the data, it does not prevent any modifications to the data. To ensure integrity in addition to confidentiality, a message certificate, such as a digital signature, may be used. This reveals whether the message has been modified, but it cannot prevent the modification in any way. Availability may be ensured by authentication and by preventing the distribution of the authentication information, such as passwords, to nontrusted persons.

15.4.1 Protocol Security

Protocols provide either built-in security features or use security services of other protocols. The former approach is used for example in SSH (secure shell). It is the de facto standard for remote logins, which include secure use of networked applications, remote system administration, and file transfer. SSH encrypts both the passwords and the transferred data, which makes it very suitable. It also notifies host identification changes if someone has forged an IP address and pretends to be the remote host.

The latter approach is used, for example, in WAP, in which the security is provided by a separate protocol WAP transfer layer security (WTLS), which encodes the data after they have been sent from the application. The latter approach enables several applications to use the same security service, which makes the implementation of applications easier. When using sockets, security is provided by TLS (transfer layer security) and SSL (secure socket layer). The TLS module is, effectively, an enhancement of the SSL protocol, which means that clients can interoperate with both SSL and TLS servers.

Instead of using insecure IP protocol, an optional IP layer protocol, IPSec, may be used to secure host-to-host or firewall-to-firewall communications (Doraswamy and Harkins, 1999). IPSec is actually a set of protocols that seamlessly integrate security into IP by providing data source authentication, data integrity, confidentiality, and protection against reply attacks. The encapsulating security payload (ESP) provides data confidentiality and limited traffic load confidentiality in addition to all features provided by the authentication header (AH), which provides proof of data origin on received packets, data integrity, and antireply protection.

AH and ESP can be used in two modes to protect either an entire IP payload or the upper-layer protocols of an IP payload. The transport mode may be used to protect upper-layer protocols, whereas the tunnel mode may be used to protect entire IP datagrams, as shown in Figure 15.11.

In transport mode, an IPSec header is inserted between the IP header and the upper-layer protocol header. In the tunnel mode, the entire IP packet to be protected is encapsulated in another IP datagram, and an IPSec header is inserted between the outer and inner IP headers.

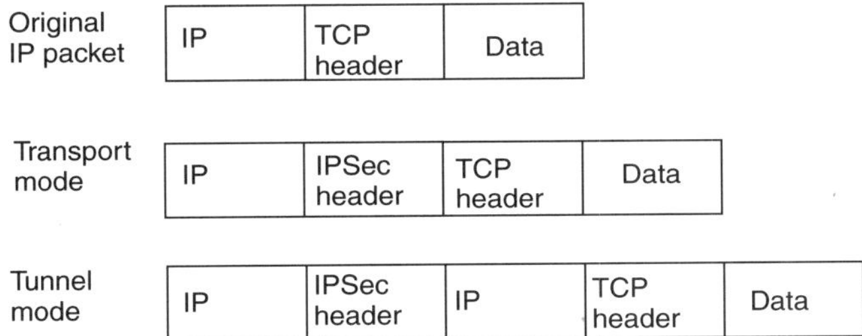

Figure 15.11 IPSec in transport and tunnel mode. Note: IP, Internet protocol; TCP, transmission control protocol

The IPSec architecture allows a user to specify his or her own security policy for different traffic flows. Some flows may require light security whereas others may require completely different security levels. For example, the security policy may determine that in the network gateway all traffic between local and remote subnets be encrypted with DES and authenticated with MD5. In addition, all Telnet traffic to a mail server from the remote subnet require encryption with Triple-DES (3 DES) and authentication with SHA-1. These algorithms will be described in more detail later in Section 15.4.3.

15.4.2 Authentication

Authentication may be based on information, biometrics, or physical properties. The security is lost if someone can 'snoop' the information or forge or steal the property. When the piece of information is used in an open network, it should be encrypted each time of use. Some protocols for downloading mail (POP3 and IMAP4) use authentication, but they do not encrypt the passwords, making it easy to 'snoop' the passwords and read others' mail messages. In PPP, which also uses authentication with plain text passwords, the risk of revealing the password is small, because a point-to-point connection to an IAP is used.

In addition to a password, a biometric property, such as blood vessel organization in the retina, or a fingerprint, may be used. The third option is to use some physical, owned property, such as a SIM (subscriber identity module) card, for authentication. Because physical properties may be stolen, some application in which money transactions are transferred over a Bluetooth from a smartphone to a cashier server should not rely only on this means of authentication. Thus, the future smartphones should have a built-in fingerprint identification device to authenticate the user in a sufficiently reliable way.

15.4.3 Data Encryption

There are several encryption algorithms supported in Symbian OS: RSA, DES, RC2, RC4, SHA-1, MD5, Diffie Hellman, and Triple-DES (3DES). Algorithms use either symmetric or asymmetric ciphering. The term 'symmetric' means that the ability to encrypt implies the ability to decrypt, whereas with asymmetric algorithms it is impossible to decrypt what has been encrypted.

Symmetric ciphers operate on input, either in blocks or as a stream. For example, DES has a block size of 64 bits, whereas stream ciphers operate on one bit or byte. In block ciphers, the block of plain text is encrypted into a cipher text by using a key, usually in several sequential steps, and the result will be fed into the next operation, ciphering the next block of code. Since the use of the same key will encrypt the same block of plain text into the same block of cipher text, it is possible to generate a code book of all possible cipher texts for a known plain text. If we know something about the data, for example that its first bytes represent an IP protocol header, we may determine the key by using the code book.

Asymmetric ciphers or public key ciphers are based on one-way functions. Thus, knowledge of the public key does not reveal the other key, which is the secret key. The other key is used for encryption and the other for decryption, depending on how the keys are used. If a person encrypts the data with someone's public key, only the holder of the secret key may decrypt that data. However, only the holder of the secret key may digitally sign the data, but anyone may verify the signature using the sender's public key. Further information on cryptography can be found in (Schneier, 1995).

RSA

The first public key algorithm was invented by Whitfield Diffie and Martin Hellman; the most popular algorithm is RSA, named after its inventors Ron Rivest, Adi Shamir, and Leonard Adleman. The security of the RSA algorithm is based on the difficulty in factoring the product of two very large prime numbers. A drawback of the algorithm is that it is quite slow and can operate only on data up to the size of the modulus of its key. Fortunately, several blocks up to the size of the modulus of the key can be encrypted in sequence. A 1024-bit RSA public key can only decrypt data that is less than or equal to that size. Actually, the size is little bit smaller (1013 bits), because the encoding requires 11 additional bits. The RSA method is the de facto standard for key exchange and digital signatures.

DES and 3DES

DES (data encryption standard) is a symmetric block cipher that transforms a 64-bit binary value into a unique 64-bit binary value using a 64-bit key. In fact, only 56 bits of the key are used and the rest is used for error detection. Thus, there are 7×10^{16} possible key combinations. By changing the key frequently, the risk of data to be revealed becomes smaller. 3DES applies the DES cipher three times and uses a key that is three times longer than that of DES (156 bits). Thus, it is also three times slower compared with DES. It is planned that 3DES replace DES.

RC2 and RC4

RC2 and RC4 are variable-key-size cipher functions. They may be used as drop-in replacements for DES if a more secure (longer key) or less secure (shorter key) is required. RC2 also supports triple encryption (i.e. it may be used instead of 3DES). Software implementation of RC2 is approximately twice as fast as that of DES. RC2 is a symmetric block cipher like DES, but RC4 is a symmetric stream cipher. The software implementation of RC4 is 10 times as fast as the implementation of DES.

MD5 and SHA-1

MD5 is a one-way hash algorithm that generates strings, the decryption of which requires a lot of computing power. Thus, it is used to generate digital signatures. MD5 is used, for example, in RSA as a hashing function to sign a message. MD5 is designed so that finding two messages hashing to the same value is infeasible. The hashing generates a digest value of the message. This value is decrypted with the sender's secret key. The receiver decrypts the digest with sender's public key and compares the value with the one calculated from the original message. If these match, the receiver may be sure of the authenticity of the sender. SHA-1 is another hash algorithm in addition to MD5. It produces a message digest of 160 bits. Three key values may be used: 128 bits, 192 bits, or 256 bits.

Software Certificates

Although encryption ensures the confidentiality of data, it cannot guarantee integrity. A digital signature uses one-way hash functions to reduce a document down to a digest, which is then encrypted by using the private key and, for example, the RSA algorithm. The digest is appended to an original document. Verification of the signature entails running the original document through the identical hash function to produce a new digest value, which is compared with the decrypted value in the received document. If these digests do not match, the document has lost its integrity.

In the same way, it is possible to use digests as software certificates to ensure that the integrity of downloadable software has not been lost. In Symbian OS there is a PC command line tool – The Certification Generator – that creates a private–public key pair and issues certificate requests. The resultant private key is used digitally to sign installation files, enabling the install system to authenticate them.

The secure installation process first checks the signature on the installation file by using the public key in the developer's certificate to ensure that the install package is signed by the developer's private key. It then checks the signature on the developer's certificate by using the organization's public key from the organization's certificate to ensure that the key pair really does belong to the person named on the certificate.

15.5 Other Security Issues in Symbian OS

In addition to security support, built into the protocols supported by Symbian OS, there are other security issues to be taken into account. No security method is useful if one's software does not apply it. For example, use of a Bluetooth connection to a service provider entails a significant risk if we cannot trust the provider. An untrustworthy party may upload harmful software modules into our smartphone when we think that we are downloading the newest and 'coolest' game. Before accepting any data, we should check whether the origin may be trusted. By default, we should not trust anyone.

Communications software should be tolerant of the most common types of attacks against security. Encryption provides good protection against confidentiality attacks, and digital signatures may be used to detect whether the integrity of the message is broken; but how do we detect in software its unauthorized use? There are two ways to try to break in through the authentication: either by guessing or knowing the right password or by generating an absurd password containing meaningless or too much data. If the authenticated service is unable to deal with this absurd data, it may be possible to write data into parts of software not intended for external users. This may cause the breakdown of authentication. Thus, the validity of passwords and any external input data should be checked in terms of content and size before used by the software. In Symbian OS, the right way to implement this is to use `ASSERT` macros.

15.6 Summary

In this chapter we have described the communications architecture of Symbian OS. The architecture enables dynamic extension of communications features, without rebooting the device, with the help of loadable plug-in modules. The architecture enables dynamic extension of communications features via plug-in modules without rebooting the device. These modules are used by communications servers in the realization of different communications technologies and protocols. There are four communication-related servers in Symbian OS and the Series 60 Platform: the Serial communications server, the Socket server, the Telephony server, and the Message server, each providing an own client-side API through which the services, provided by communications modules, are accessed.

There are several techniques supported by Symbian OS for data transfer, content browsing, and messaging. Data may be transferred locally through RS-232, infrared, or Bluetooth or globally by using TCP/IP, GSM, or GPRS. Both Web and WAP browsing are supported, and messages may be sent as e-mail, short messages, multimedia messages, or fax.

There are three kinds of threats to communications security: confidentiality, integrity, and availability. Confidentiality may be ensured to a great extent by encryption algorithms. Although integrity cannot in general be ensured, it is still possible to detect when the integrity of data is lost. The main method for implementing this is to use digital signatures with the help of asymmetric ciphers. Availability can be divided into two issues: availability for intended users by an access mechanism based on authentication, and denial of service, where the service is blocked from intended users where there is an enormous number of forged service requests. A user may be authenticated based on information, biometrics, or physical properties, but there are no general protection mechanisms against denial-of-service attacks.

16

Communications Application Programming Interface

As described in the previous chapter, use of any communications service is based on one of the communications servers: the Serial communications server, the Socket server, the Telephony server, and the Message server. Each server has a client-side application programming interface (API), through which the communications server and the underlying plug-in communication module are accessed. Thus, the service API provided to the client is similar regardless of the specific communications module.

The use of each communications server is similar. The client-side API always includes a class that provides a handle to the server. The handle is used to create a session between the client and server processes. The actual communication takes place in subsessions provided by one or more classes derived from the `RSubsessionBase` class.

This chapter contains a description of how to use different communications APIs in the Series 60 Platform. As a concrete example, the development of a generic data-transfer engine is shown. The engine is an active object derived from the `CActive` class. In addition to the engine itself, there are usually two other active objects used: one for transmitting data and one for receiving data. The engine is an active object because it takes care of the connection establishment, which is an asynchronous service. Three classes instantiated from `CActive` are needed, because one active object may issue only one request at a time so that a read request in the receiving object does not block the send request in the transmitting object. The communications engine of the rock–paper–scissors (RPS) game implements the virtual functions of the `MCommsBaseClass` interface, presented in Section 15.1.4, to ensure it conforms to the generic communications pattern.

16.1 Communications Infrastructure

In addition to client-side APIs of communications services, there are a few other concepts that are frequently used by the developer and user of communications software. The communications database is a record of common communications settings used by several communications services. It was described in Section 15.1.3. NifMan (network interface manager) is responsible for monitoring network connections and is required to start up an appropriate network interface (NIF) established by network agents. The interface provides a connection established by the Telephony server (ETel), through which a route is created to a destination station or network. NifMan is described in detail in the section on the socket server (Section 16.3).

16.2 Serial Communications Server

It may seem useless to discuss serial communications, because the Nokia 7650 smartphone does not have an RS-232 port, and, if it had, the use of the port would be very simple to learn. However, there are several reasons why we have included this section in the book. First, it is really easy to learn how serial communication works in Symbian OS. Because other communications servers are based on the same principles, it makes their use easier. Second, almost all communication eventually uses serial communication for data transfer in the Series 60 Platform. Telephony services are used through an internal modem, the communication of which is based on serial communication. Bluetooth is used through serial communication, and so on. Thus, it is obvious we should start studying the communications APIs from the Serial communications server.

The client-side API of the Serial communications server, C32, is provided by two classes, `RCommServ` and `RComm`, which are defined in the c32comm.h header file. `RCommServ` provides a client session to the server, and `RComm` provides an API to the serial port. The use of the Serial communications server requires the following actions in different phases of the communications pattern:

- initialization: the physical and logical device drivers are loaded, the server (C32) is started, and the appropriate serial communications (CSY) module is loaded, depending on the target build [Windows single process (WINS) multiprocess advanced risk or machine (MARM)] and transfer technology (RS-232, infrared, or Bluetooth) used;

- opening: the port is opened and for instance, the access mode to the port is determined: exclusive, shared, or pre-emptive;
- configuration: serial communications have a rich set of communications parameters, with the use of start and stop bits, number of data bits, use of parity bit based error detection, baud rate, and handshaking options;
- data transfer: reading and writing of a serial port, takes place;
- Closing: the session and subsessions are closed to release the port for other applications, if opened in the exclusive mode.

The declaration of the serial communications engine class is as follows:

```
class CSerialEngine : public CActive, public MCommsBaseClass
    {
    public:
        enum TCommType      // These may be used to choose the type of
            {               // communication
            ERS232, EIRComm
            };
        static CSerialEngine* NewL(void);
        ~CSerialEngine();
    public:      // New functions
    private:     // From MCommsBaseClass
         void Initialize();
         void OpenL();
         void ConfigureL();
         void Transmit( TDesC& aTransmittedData );
         void Receive();
         void Close();
    private:      // From CActive
        void RunL();
        void DoCancel();
    private:
        CSerialEngine();
        void ConstructL(void);
    private:      // Data
        RCommServ iServer;
        RComm iCommPort;
        TCommType iCommPortType;
        CTxAO *iTransmittingActiveObject;
        CRxAO *iReceivingActiveObject;
     };
```

`TCommType` may be used to identify the type of communications: RS-232 or infrared. It results in only a few changes to the source code. The `CSerialEngine` class is designed to take care of connection establishment. Thus, `RunL()` is called by the active scheduler when the connection is ready. However, there is no connection in serial communication and so the implementation of `RunL()` is empty. It may be implemented, for example, in the sockets engine. `CSerialEngine` creates the active objects, taking care of data transfer. Note that the engine also has handles to services providers as members. This is a common way to implement any active object.

The first phase (initialization) in the communications pattern is quite straightforward to implement. Static functions `LoadPhysicalDevice()` and `LoadLogicalDevice()` from the `User` class are used to load the physical and logical device drivers, respectively. In WINS builds, the file server must be started before the device drivers can be loaded. It is a common error not to ignore the `KErrAlreadyExists` value when loading the drivers. This value should not cause a leave but should be handled in the same way as the `KErrNone` value.

The names of physical device drivers are different for WINS and MARM builds. The driver is ecdrv.pdd in WINS builds and is euart1.pdd in MARM builds. In both builds, the same logical device driver (ecomm.ldd) is used. Device drivers can be defined as literal constants in the header file. It is then easy to change the device driver names:

```
// Device driver names
_LIT(KLddName,"ECOMM");

#if defined (__WINS__)
_LIT(KPddName,"ECDRV");
#else
_LIT(KPddName,"EUART1");
#endif

// Communication modules
_LIT(KRS232,"ECUART");
_LIT(KIRCOMM, "IRCOMM");
```

When both device drivers are in place, the Serial communications server may be started. This requires only one function call: `StartC32()`. The name of the CSY module depends on whether RS-232-based or infrared-based communications are required. In the RS-232 case, ecuart.csy is used, whereas infrared communications require the ircomm.csy module. It is not necessary to load the communication module in Bluetooth. The only thing needed to be done with respect to serial communications is to load the appropriate device drivers. Other initialization procedures will be performed by the Bluetooth protocol module.

The `InitializeL()` function of the serial communication will look like the following piece of code:

```
void CSerialEngine::InitializeL()
    {
    TInt err;
```

```
#if defined (__WINS__) // Required in WINS to enable loading of
     RFs fileServer;    // device drivers
     User::LeaveIfError( fileServer.Connect() );
     fileServer.Close();
#endif
     // Load device drivers
     err=User::LoadPhysicalDevice( KPddName );
     if ( err != KErrNone && err != KErrAlreadyExists )
          User::Leave( err );
     err = User::LoadLogicalDevice( KLddName );
     if ( err != KErrNone && err != KErrAlreadyExists )
          User::Leave( err );

     // Start the server and create a session
     StartC32();
     User::LeaveIfError( iServer.Connect() );

     // Ask the server to load the CSY module
     if ( iCommPortType == ERS232 )
          User::LeaveIfError( iServer.LoadCommModule(KRS232) );
     else
          User::LeaveIfError(iServer.LoadCommModule(KIrComm));
     // Create receiving and transmitting active objects
     // using normal two-phase construction.
     // Code omitted.
     }
```

There are several functions in the `RComm` class that may be used to open the port. The arguments of the functions contain the session handle, the port to be opened, the access mode, and, optionally, the role. The access mode may be exclusive, shared, or preemptive. The exclusive mode prevents the use of the port from other processes. The processes may share the port if both open the port in the shared mode. The preemptive mode allows another client to open the port in the two other modes. Note that when using ircomm.csy, only the exclusive mode should be used.

The role defines whether the port is opened in the data terminal equipment (DTE) or the data circuit terminating equipment (DCE) role. DCE is usually a device, such as a modem, used by the DTE for data transfer. In the DCE mode, the smartphone's serial port is configured as a port of a modem.

The serial ports are numbered from zero, so the port corresponding to the COM1 port in Windows is port `COMM::0` in Symbian OS. Another thing affecting the port name is the type of serial technology used. RS-232 ports are named `COMM::0`, but infrared ports `IRCOMM::0`. The port name is the only difference in addition to the CSY module, which is different in infrared communication compared with communication based on RS-232:

```
void CSerialEngine::OpenL()
	{
	User::LeaveIfError( iCommPort.Open(iServer, KPortName,
	  ECommExclusive) );
	}
```

Serial communications have several configuration parameters. Before changing the configuration it is advisable to save the existing configuration, because the port may be shared with another process. The existing configuration may be read with the `Config()` function. It may also be used in the initialization of a `TCommConfig` structure, which makes it unnecessary explicitly to set every member of the structure. It is good programming practice to check whether the required configuration is supported by the port before changing it. This is done with the `TCommCaps` structure, which is used in a way similar to the `TCommConfig` structure, by calling the `Caps()` function:

```
void CSerialEngine::ConfigureL()
	{
	TCommCaps currentCaps;
	iCommPort.Caps( currentCaps );
	TCommConfig portConfiguration;
	iCommPort.Config( portConfiguration );

	// Check port capabilities
	if ( (currentCaps().iRate & KCapsBps57600) == 0 )
		User::Leave( KErrNotSupported );
	// Change the configuration
	portConfiguration().iRate = EBps57600;
	User::LeaveIfError( iCommPort.SetConfig
	  (portConfiguration) );
	}
```

Before configuration, all pending transmitting and receiving requests should be cancelled; otherwise a panic will result. Requests are cancelled simply by calling the `Cancel()` function of the `RComm` class.

The `RComm` class provides several functions for transmitting (`Write()`) and receiving (`Read()`) data. Before any data transmission the serial communications chip must be powered up. This is done with the `Read()` function, in which the data length argument is zero. The Serial server always handles 8-bit data, even in the Unicode system, which may require that the data be cast into the correct type before transmission and after receiving.

What happens when data is received by the engine? When reading, the receiving active object issues a read request with an empty buffer. Application continues with its other duties and the Serial server monitors the incoming data. When the data are received, the server fills the buffer, and the receiving active object signals the application. The engine may then read the data from the buffer.

Active objects can have only a single request active at any time, so wrapping read and write into the same class would mean that these requests cannot be active at the same time. Either two objects of the same class or instances of two different classes are needed. The latter approach is the method used in our examples:

```
void CSerialEngine::Receive()
    {
    if ( !iReceivingActiveObject->IsActive() )
        {
        // There is another function to read the received
        // data from a member variable
        iReceivingActiveObject->Recv();
        }
    }
void CSerialEngine::Transmit( TDesC& aTransmittedData )
    {
    if (!iTransmittingActiveObject->IsActive())
        {
        iTransmittingActiveObject->Send( aTransmittedData );
        }
    }
```

The following two functions issue the read and write requests. The received data are read into an 8-bit buffer, from which it is read and cast to Unicode, if necessary, by the functions of the `CSerialEngine` class. There are several `Read()` functions to be used in the `RComm` class. If we do not know the amount of received data beforehand, the version used in the next example is very useful. It returns after a time interval without waiting for the whole buffer in the descriptor to be filled. When a timeout occurs, the received data buffer may be empty, although data have been received. It depends on the serial protocol module implementation whether the data are stored in the buffer or must be read using, for example, `ReadOneOrMore()`. The following example assumes that `iDataBuf` contains any received data after the timeout:

```
void CRxAO::Recv()
    {
    // Change from ASCII to Unicode in RunL() method
    // Data should be read by the CSerialEngine object
    // using a separate function
    iCommPort.Read( iStatus, KMaxTimeInterval, iDataBuf );
    SetActive();
    }
void CTxAO::Send( TDesC& aData )
    {
```

```
    if (!IsActive())
        {
        TPtrC8 outputBufPtr.Copy(aData);
        iDataBuf = outputBufPtr;
        iCommPort->Write(iStatus, iDataBuf);
        SetActive();
        }
    }
```

Finally, the port is closed. It is not necessary to cancel the requests before closing the port because the `Cancel()` function is called automatically. The session is closed in a similar way:

```
void CSerialEngine::Close()
    {
    iCommPort.Close():
    iServer.Close();
    }
```

In addition to the session to the Serial communications server, the `RCommServ` class provides information about the number of CSY modules loaded by calling the `NumPorts()` function. The `GetPort-Info()` function is used to obtain a `TSerialInfo` class of a named port. `TSerialInfo` contains information about the name of the port and the lowest and highest unit numbers.

16.3 Socket Server

The Socket server is used according to the same principles as the Serial communications server. What is different from serial communications is that there are typically several subsessions instead of one. There may be several sockets open for data transfer between applications. In addition, some sockets may be used for special services such as host-name resolution or device discovery in case of IrDA (infrared data association) and Bluetooth sockets.

The six protocol stacks implemented in the Series 60 Platform are TCP/IP (transmission code protocol/Internet protocol), IrDA, Bluetooth, short message service (SMS), Psion Tink protocol (PLP), and wireless application protocol (WAP). Although there is the PLP protocol module, it has been replaced by mRouter (see Section 15.2.4), which should be used. In addition, SMS should be used through the messaging and not the protocol interface, as will be shown in the next chapter.

The client-side API of the Socket server is provided by two classes: `RSocketServ`, which provides a handle to the server, and `RSocket`, which implements the communications socket. Both classes are defined in es_sock.h. In addition to the RSocket class, RHostResolver may be used for the host-name and device-name resolution services, and RServiceResolver may be used for service-lookup facilities. RNetDatabase is the fourth class that may be used for accessing a network database. Note that the network database is not the communications database. The latter is used for storing general settings of the connections, whereas the network database contains protocol-specific information, such as service identifiers (IDs) stored into the IrDA information access service (IAS).

The RSocketServ class provides the same kind of services to the Socket server as the RCommServ provides to the Serial server. The NumProtocols() function may be used for getting the number of protocol modules known by the Socket server, in the same way that NumPorts() returns the number of CSY modules in the case of serial communications. GetProtocolInfo() returns information on all protocols, as GetPortInfo() returns information about a serial port. Another way to get information about the protocol is to use the FindProtocol() function. It returns similar information to that provided by GetProtocolInfo() about a named protocol.

16.3.1 Support Classes

TSockAddr is a base class for socket addresses. It has a family field and a port field. The maximum length of the address is 32 bytes. TInetAddr for TCP/IP addresses, TIrdaSockAddr for IrDA addresses, and TBTSockAddr for Bluetooth addresses are all derived from TSockAddr. The address type of the TInetAddr class is TUint32, into which an ordinary IP address, such as 127.0.0.1, may be converted by means of the INET_ADDR macro. TIrdaSockAddr stores 32-bit IrDA device addresses, and TBTSockAddr stores 48-bit Bluetooth device addresses.

TNameEntry is a support class used in name resolution. It packages the TNameRecord class so that it can be passed between a client and a server. TNameRecord has three members to store information about the name: address of type TSockAddr, name, and flags indicating, for instance, whether the corresponding name is an alias.

TProtocolDesc has information about a specific socket. It includes, for example, address family, name, protocol, and socket-type information. The information may be read with the GetProtocolInfo() and FindProtocol() functions of the RSocketServ class.

16.3.2 Communications Pattern for Sockets

Initialization of socket connections is simple compared with serial communication, because the Socket server loads the required protocol automatically when a socket is opened. In some cases, it may be desirable to preload the protocol to reduce the delay of opening a socket. In that case, the loading of a certain protocol can be done asynchronously with the `StartProtocol()` function of the `RSocketServ` class. There is also a corresponding function to unload the protocol: `StopProtocol()`. In addition to the protocol name, `StartProtocol()` and `StopProtocol()` functions require the address family and socket type as parameters. Address families are, of course, protocol-dependent and are as follows:

- KAfInet = 0x0800 for TCP/IP,
- KIrdaAddrFamily = 0x100 for IrDA,
- KBTAddrFamily = 0x101 for Bluetooth,
- KFamilyPlp = 273 for PLP,
- KSMSAddrFamily = 0x010 for SMS,
- KWAPSMSAddrFamily = 0x011 for WAP.

There are four different socket types that may be used: `KSockStream` is used for reliable stream-based communication; `KSockDatagram` is used for unreliable datagram-based communication; `KSockSeqPacket` is used in packet-oriented data transfer; and `KSockRaw` allows the user to define the transfer mode. The selection is made on the basis of the data-transfer requirements between applications. Some protocols support only one type of socket. For example, Bluetooth RFCOMM uses sockets of the stream type, and L2CAP uses sequenced packets.

16.3.3 NifMan

The network interface manager, NifMan, is responsible for route setup for sockets protocols. In addition, it selects the correct network interface for the route. NifMan provides progress information on connection establishment through the `RNif` class, which provides both asynchronous `ProgressNotification()` and synchronous `Progress()` functions for getting the current state of a dial-up connection.

16.3.4 TCP/IP Sockets

Before a TCP/IP socket can be opened, a session to the Socket server must be created. This is the only action required in the initialization phase. One argument, the number of message slots, can be passed to the function. The number indicates the number of asynchronous operations that are allowed to be in progress at any time. The default value is `KESockDefaultMessageSlots`, which is 8. Too small a value may result in the `KErrServerBusy` return value.

The suitable value for message slots is $N + 2$, where N is the number of asynchronous services that the client may request (connection establishment, data receive, data transmit, etc.). There should always be one slot available to cancel the service request. If the client exits but it cannot cancel the requests, it panics. The last slot is reserved for a synchronous service request. One is enough, because there can be only one synchronous request active at one time. An initialization function creating a session to the socket server is as follows:

```
void CTcpIpSocketServerEngine::InitializeL()
      {
      User::LeaveIfError( iServerSocket.Connect() );
      }
```

After initialization, a socket is opened. A socket may be opened in two ways: either as a blank socket, or as a socket associated with a specified protocol. Blank sockets cannot be used for data transfer until they have been paired with a client by the `Accept()` method. When opening a socket, its type and role must be decided. The role is either a server or a client. In the server role, two sockets must be opened. One for listening to incoming connection requests and one blank socket for data transfer to the client after the acceptance of the connection request:

```
void CTcpIpSocketServerEngine::OpenL()
      {
      TInt result;
      if ( iSocketRole == EClient )
            {
            result = iSocket.Open( iServerSocket, KAfInet,
              KSockStream, KUndefinedProtocol);
            User::LeaveIfError( result );
            }
      else
            {
            result = iListernerSocket.Open( iServerSocket,
              KAfInet, KSockStream, KUndefinedProtocol );
            User::LeaveIfError( result );
            result = iSocket.Open( iServerSocket );
            User::LeaveIfError( result );
            }
      }
```

Configuration depends on the socket type and its role. A connectionless socket is easy to configure. A local address is simply assigned to it, using the `Bind()` function. Clients of the socket stream type must assign the address of the remote socket so that the sockets remain together during the connection. This is done by using the `Connect()` method. In the server side of streaming sockets, the listener socket listens to incoming connection requests and the other socket serves them:

```
void CTcpIpSocketServerEngine::ConfigureL()
    {
    if ( iSocketRole == EClient )
        {
        TInetAddr serverAddr( INET_ADDR(127,0,0,1), KServerPort );
        iSocket.Connect( serverAddr, iStatus );
        }
    else
        {
        TInetAddr anyAddrOnServerPort( KInetAddrAny, KServerPort );
        iListenerSocket.Bind( anyAddrOnServerPort );
        // Set up a queue for incoming connections
        iListenerSocket.Listen( KQueueSize );
        // Accept incoming connections
        iListenerSocket.Accept( iSocket, iStatus );
        }
    SetActive();
    }
```

There are different functions for reading and writing data, depending on the socket type and role. For datagram sockets `SendTo()` and `RecvFrom()` are used. For stream sockets, several versions of `Send()` and `Recv()` may be used. All functions take a buffer to be sent or into which the received data are stored, as an argument.

Note that, as in the case of serial communications, data are always transferred in 8-bit bytes, even on a Unicode system. An optional parameter may be used to set the transmission flags, such as urgent data. Another optional argument is the length, which indicates the actual data received or sent. The write object should in most cases use a timer to avoid infinite waiting in case the write cannot be done:

```
void CTcpIpSocketServerEngine::Receive()
    {
    if ( !iSocketRxAO->IsActive() )
        {
        iSocketRxAO->Recv();
        }
    }
void CRecSocketAO::Recv()
    {
    if (!IsActive())
```

```
        {
        iSocket.RecvOneOrMore( iBuffer, 0, iStatus, iRecvLen );
        SetActive();
        }
    }
```

After use, the socket is closed by calling `Close()`. Before closing, all pending asynchronous operations should be cancelled with `CancelAll()`. This function call cancels all operations (read, write, input–output control, connect, and accept), except shutdown, which cannot be cancelled. Although `Close()` is a synchronous function, it calls the asynchronous `Shutdown()` function, which closes the socket in the normal way (i.e. completes) after both socket input and output have been stopped.

LIBC

Because of the large amount of communications software written in pure C, there is also a C language implementation of the BSD 4.3 sockets in Symbian OS called LIBC. This helps the porting of communications software, because it enables the use of communications engines implemented in C language. The LIBC library contains a standard TCP/IP library including the address resolution protocol (ARP), reverse address resolution protocol (RARP), Internet protocol (IP), Internet control messages protocol (ICMP), transmission control protocol (TCP), and user datagram protocol (UDP).

Host Resolution

In addition to data transfer, sockets may be used for other services, such as host-name resolution and device discovery. The `RHostResolver` class provides host-resolution services. It is used in exactly the same way as the objects of the type `RSocket`. The host resolution is applied in four services: To obtain names from addresses or vice versa, or to get and set local host names.

The actual service is protocol-dependent. For TCP/IP sockets the host resolution accesses the domain name service (DNS) to convert between textual and numeric IP addresses. For IrDA, the host resolution resolves all devices in the neighborhood of the requesting devices. Examples will be provided in the following sections.

16.3.5 IrDA Sockets

The IrDA protocol stack is used through the socket interface to access the infrared communications facilities. Two protocols may be used through a client-side API: IrMUX, which provides unreliable service, and IrTinyTP, which provides reliable data-transfer service. `RHostResolver` is used for device discovery rather than for name resolution, but otherwise IrDA sockets are used in the same way as TCP/IP sockets.

IAS (information access service) is a database for saving information about IrDA connections and their properties. The IAS database may be accessed after a successful device discovery. The database may be queried, for example, for remote port numbers of IrTinyTP, IrCOMM, or IrOBEX services (see Glossary) so that several services may share the same link by using different port numbers. The client-side API of the IrDA sockets has five concepts:

- IrTinyTP socket protocol for reliable transport, used through `RSocket` handle;
- IrMUX socket protocol for unreliable transport, used through `RSocket`;
- IrDA discovery for host resolution, used through `RHostResolver`;
- IAS database queries, used through `RNetDatabase`;
- IAS database registration, used through `RNetDatabase`.

The initialization and opening phases of IrDA sockets are similar to those of TCP/IP sockets. However, configuration is different, because an IrDA device must discover another device with which to communicate. Device discovery is implemented through the name resolution interface by using an instance of the `RHostResolver` class and its `GetName()` function. The server does not perform device discovery, rather, it must be discovered:

```
void CIRSocketEngine::ConfigureL()
    {
    TInt result;
    TPckgBuf<TUint> buf(1);
    // IrLAP does the device discovery, when device discovery
    // socket option is set
    sock.SetOpt( KDiscoverySlotsOpt,KLevelIrlap,buf );
    result = iHostResolver.GetByName( KHostname,iHostEnt );
    resolver.Close();
    User::LeaveIfError( result );

    addr = TIrdaSockAddr( iHostEnt().iAddr );
    }
```

The following piece of code performs the IAS query:

```
TIASResponse CIRSocketEngine::IASQueryL(TDesC8& aClassName,
  TDesC8& aAttributeName)
```

```
{
TInt result;
RNetDatabase rnd;
TIASQuery query;
TIASResponse response;
result = rnd.Open( iSocksvr, iProtocolInfo.iAddrFamily,
  iProtocolInfo.iProtocol );
User::LeaveIfError( result );
query.Set( aClassName, aAttributeName,
  addr.GetRemoteDevAddr() );
rnd.Query( query, response, status );
User::WaitForRequest( status );  // Freezes the thread
return response;
}
```

The class names and their attributes are given in Table 16.1. The `TIASResponse` return value may have four types, from 0 to 3. Type 0 results are missing results (i.e. results are not given owing to errors), type 1 results are integers; type 2 results are a sequence of bytes; and type 3 results are strings.

Table 16.1 Information access service (IAS) class and attribute names; for more details, see the Glossary

Class name	Attribute name	Response
Device	Device name	Remote device hostname
	IrLMPSupport	Level of support by remote device for IrMUX
IrDA:IrCOMM	Parameters	Collection of parameters detailing IrCOMM support
	IrDA:IrLMP:LsapSel	Range of remote port numbers for IrCOMM service
	IrDA:TinyTP:LsapSel	Range of remote port numbers for TinyTP service
	IrDA:InstanceName	Name of IAS instances on remote device
IrDA:IrOBEX	IrDA:TinyTP:LsapSel	Range of remote port numbers for IrOBEX service

16.3.6 Bluetooth Sockets

Bluetooth sockets are based on a Bluetooth host controller interface (HCI), which specifies a command interface to a Bluetooth device. There are two client APIs: a socket API for new applications, and a serial API for legacy applications (Nokia, 2001a).

Through the client-side API of the Socket server, it is possible to access L2CAP (logical link control and adaptation protocol) and the RFCOMM protocol (see the Glossary) as well as the service discover protocol (SDP) and Bluetooth Security Managers, as shown in Figure 16.1. The Bluetooth manager is executed in its own thread, created by the Bluetooth protocol module. It interacts with the user through the configuration manager. Three distinct service security levels are defined: high, for services that require authorization and authentication (automatic access is granted to trusted devices only);

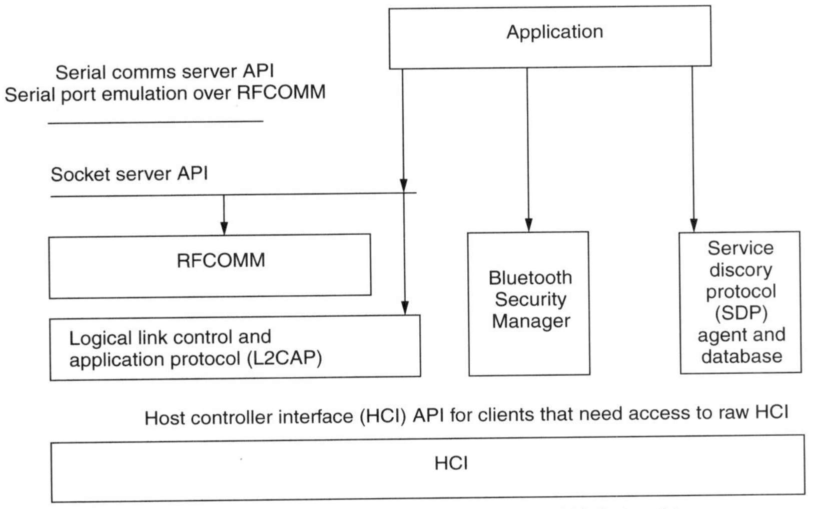

Figure 16.1 Bluetooth client-side application programming interfaces (APIs) in the Series 60 Platform

medium, for services that require authentication only; and low, for services open to all devices.

Before providing any service, it must be registered with both the Bluetooth security manager (if security is required) and the service database. In the same way, the use of any service requires not only device discovery but also service discovery. If no special service parameters are required (a client already knows in which port the service is available) a data connection may be established between devices without service discovery.

The pattern of using Bluetooth services depends on the role of the communicating software module. It may be a client or a server. In each case the initialization starts as in the case of the Serial communications server. Thus, first, the required device drivers and communications modules are loaded. Before a connection can be established, a client:

- finds a device, by inquiry or registry lookup;
- raises a SDP query for services and their attributes; if the client already knows the port of a specific service, it is not necessary to execute service discovery.

After device and service discovery, the client creates a connection to the Bluetooth server host. On the server side, the service might have been registered to the SDP and Bluetooth security manager databases. Thus, in the server, the service initialization may require:

- local channel allocation for the service,
- service registration in the SDP (e.g. reserved channel stored),
- security registration in the Bluetooth security manager.

After the connection has been established, the generic data-transfer and connection-closing phases are executed both in the client and server side. The following header file gives the server-side functions:

```
//  INCLUDES
#include <e32base.h>
#include <btmanclient.h>
#include <bt_sock.h>
#include <btextnotifiers.h>

// CONSTANTS
const TInt KRPSServiceId = 0x5021;

//##ModelId=3C67AF470095
class CBTServer : public CActive
{
    public:     // // Constructors and destructor
        static CBTServer* NewL();      // Static constructor
        static CBTServer* NewLC();      // Static constructor
        virtual ~CBTServer();      // Destructor (virtual)

    protected:
        CBTServer();      // Default constructor, protected to
                          // allow derivation
        void ConstructL();      // Second phase construct

    public:     // Methods derived from CActive
        void RunL();
        void DoCancel();

    public:     // New methods
        void CheckBluetoothL();
        void CheckedBluetoothL();

        void LoadDriversL();

        void ReserveLocalChannelL();

        void RegisterBTMan();
        void UnregisterBTMan();
        void RegisteredBTManL();

        void ConnectSdpL();

        void ConnectSocketServerL();
        void WaitForConnectionL();
```

```
    void CloseRemoteL();
    void ReadSizeL();

    void ReadDataL();
    void ProcessDataL();

    CArrayFix<TInt16>* ReadTableL( TPtrC8& aData );
    TInt16 ReadShortL( TPtrC8& aData );

protected:  // Data
    enum
        {
        EStateNotConnected,
        EStateCheckingBluetooth,
        EStateRegisteringSecurity,
        EStateAcceptingConnections,
        EStateReadingSize,
        EStateReadingData
        };

    TInt iStateMachine;

    CCxContextBoard* iBoard;
    TUint8 iLocalChannel;

    RSocketServ iSocketServ;
    RSocket iListenSocket;
    RSocket iRemoteSocket;

    RBTMan iBTMan;

    RBTSecuritySettings iSecurity;

    RSdp iSdp;
    RSdpDatabase iSdpDb;
    TSdpServRecordHandle iSdpHandle;

    RNotifier iNotifier;
 TPckgBuf<TBool> iNotifierResult;

    HBufC8* iData;
    TPtr8 iDataPtr;
};
```

The client has functions for device and service discovery. Most of the other functions are used for browsing the service attributes when the service record has been received from the remote device; we will explain these functions in detail later. The following is a declaration of the client class:

```
#include <e32base.h>
#include <e32std.h>
```

```
#include <es_sock.h>
#include <bt_sock.h>
#include <btsdp.h>
#include <btextnotifiers.h>

class CSdpAgent;

class CBTClientEngine : public CActive,
    public MSdpAgentNotifier, public MSdpAttributeValueVisitor
    {

    public:
        IMPORT_C static CBTClientEngine* NewL();
        ~CBTClientEngine();

    protected:
        CBTClientEngine();
        void ConstructL();

    public:     // from CActive
        void RunL();
        void DoCancel();

    public:     // from MSdpAttributeValueVisitor
        void VisitAttributeValueL( CSdpAttrValue &aValue,
          TSdpElementType aType);
        void StartListL(CSdpAttrValueList &aList);
        void EndListL();

    protected:     // from MSdpAttributeValueVisitor
        void NextRecordRequestCompleteL(TInt aError,
          TSdpServRecordHandle aHandle,
          TInt aTotalRecordsCount);
        void AttributeRequestResult(
          TSdpServRecordHandle aHandle,
          TSdpAttributeID aAttrID,
          CSdpAttrValue* aAttrValue);
        void AttributeRequestComplete(
          TSdpServRecordHandle aHandle, TInt aError);

    protected:
        void DiscoverRemoteL();
        void DiscoveredRemoteL();
        void QueryRemoteL();
        void QueriedRemote();
        void ConnectRemoteL();
        void CloseRemoteL();
        void WaitForDataL();
        void WriteDataL();

    private:
        enum TState
            {
```

```
            EStateNotConnected,
            EStateDiscovering,
            EStateQuerying,
            EStateQueried,
            EStateConnecting
            };

        RFs iFs;
        RNotifier iNotifier;
        TBTDeviceResponseParamsPckg iDiscoveryResponse;
        TBTSockAddr iRemoteAddr;
        RSocketServ iSocketServ;
        RSocket iSocket;
        RTimer iTimer;
        CSdpAgent* iSdpAgent;
        TInt iStateMachine;
        TInt iRemoteChannel;
        TUUID iPreviousUUID;

    };
```

Bluetooth Server

The server component creates a session to the Bluetooth security manager and SDP database servers to register the service. A session to the Socket server is required to enable the use of the Bluetooth protocol module. Note that the server is an active object, that uses asynchronous service requests to the server. The first function to be called is `CheckBluetoothL()`, which is used to check that the Bluetooth module is powered on:

```
void CBTServer::ConstructL()
    {
    User::LeaveIfError( iBTMan.Connect() );
    User::LeaveIfError( iSecurity.Open( iBTMan ) );
    User::LeaveIfError( iSdp.Connect() );
    User::LeaveIfError( iSdpDb.Open( iSdp ) );
    User::LeaveIfError( iSocketServ.Connect() );

    CActiveScheduler::Add( this );
    CheckBluetoothL();
    }
```

The `CheckBluetoothL()` function uses an instance of the `RNotifier` class to check whether the power is switched on. Some session handles provide an extended notifier server that provides support for plug-in notifiers. The notifiers are useful for checking the status of a device or a connection, for example. The used plug-in notifier is identified by a unique identifier (UID), given as a parameter to the asynchronous StartNotifierAndGetResponse() function. It is also possible to send and receive data from the notifier by using the same function. In the example below, we send a boolean value and store the result in the iNotifierResult member variable:

```
void CBTServer::CheckBluetoothL()
    {
    // Make sure the Bluetooth module is powered on
    TPckgBuf<TBool> pckg;

    User::LeaveIfError( iNotifier.Connect() );

    iNotifier.StartNotifierAndGetResponse(
        iStatus,
        KPowerModeSettingNotifierUid,
        pckg,
        iNotifierResult );
    iStateMachine = EStateCheckingBluetooth;
    SetActive();
    }
```

After an asynchronous service request and SetActive() call, it is up to the active scheduler to decide which RunL() function is called. The function shows us the order of actions in the Bluetooth server. First, the result of the Bluetooth device status is checked in the Checked-BluetoothL() function, as shown in the next example. After that, the device drivers are loaded. The interface to the Bluetooth device uses the serial communications. Thus, first physical and logical device drivers for the serial port are loaded and the Serial communications server is started, if it is not already running. The function is omitted in the following, because it is similar to that in previous examples:

```
void CBTServer::RunL()
    {
    switch( iStateMachine )
        {
        case EStateCheckingBluetooth:
            CheckedBluetoothL();
            LoadDriversL();
            ReserveLocalChannelL();
            RegisterBTMan();
            break;

        case EStateRegisteringSecurity:
            RegisteredBTManL();
            ConnectSdpL();
            ConnectSocketServerL();
            WaitForConnectionL();
            break;
```

```
        case EStateAcceptingConnections:
             ReadSizeL();
             break;

        case EStateReadingSize:
             ReadDataL();
             break;

        case EStateReadingData:
             ProcessDataL();
             break;
        }
    }
void CBTServer::CheckedBluetoothL()
    {
    // Bluetooth module is now checked on
    iNotifier.CancelNotifier( KPowerModeSettingNotifierUid );
    iNotifier.Close();

    User::LeaveIfError( iNotifierResult[0] );
    }
```

The `ReserveLocalChannelL()` function is used to allocate an available port to a new service. The reservation can be made with socket options. The used option gets the next available RFCOMM server channel, as seen below:

```
void CBTServer::ReserveLocalChannelL()
    {
    // Reserve a local channel for us to use
    iLocalChannel = 1;     // To init channel
    TInt port = 0;
    RSocket socket;

    TProtocolDesc pInfo;

    User::LeaveIfError( iSocketServ.FindProtocol(  KRFCap(),
      pInfo ) );

    User::LeaveIfError( socket.Open( iSocketServ, KRFCap ) );

    User::LeaveIfError( socket.GetOpt(
      KRFCOMMGetAvailable ServerChannel,
      KSolBtRFCOMM, port ) );
    iLocalChannel = STATIC_CAST( TUint8, port );
    socket.Close();
    }
```

Next, the service is registered to the Bluetooth security manager. For each service in the manager, service UID, protocol, and channel IDs are stored. The service UID is used to identify the service. The protocol

ID identifies the protocol layer below the service. For most Bluetooth services, this is RFCOMM. The protocol ID is specified as a Bluetooth socket level options constant, for example `KSolBtRFCOMM`. The channel ID identifies the port (channel), on which the service is running.

There are three possible security settings, depending on whether an incoming service requires authentication, authorization, or encryption:

```
void CBTServer::RegisterBTMan()
    {
    TBTServiceSecurity serviceSecurity(
    TUid::Uid (KRPSServiceId),
    KSolBtRFCOMM, iLocalChannel );

    serviceSecurity.SetAuthentication(EFalse);
    serviceSecurity.SetEncryption(EFalse);
    serviceSecurity.SetAuthorisation(EFalse);
    serviceSecurity.SetDenied(EFalse);

    iSecurity.RegisterService( serviceSecurity, iStatus );
    iStateMachine = EStateRegisteringSecurity;
    SetActive();
    }
```

In addition to the Bluetooth security manager, a new service is also registered to the SDP database. For each service there is a service record with service-specific attributes in the database. Each attribute has an ID, a type, and a value. The types and some attribute IDs are predefined. Some universal attributes are defined in the btsdp.h: service record handle, service class ID list, service record state, service ID, protocol descriptor list, and so on. Complex attributes can be formed through lists of attributes. The lists can themselves contain lists of attributes. Attribute types may be boolean, data element alternative (DEA), data element sequence (DES)[1], integral, nil, string, an unsigned integer, or an URL(universal resource locator).

A key attribute is the service class. Each service is an instance of a service class, which provides an initial indicator of the capabilities of the server. In addition, it defines other attributes which may be or must be in the service record. One application may also have several services, in which case they are stored in separate service records. Each service record may have many service classes. The service classes must be specializations or generalizations of each other and be ordered from most derived to least derived.

The `CreateServiceRecordL()` function creates a new service record with a single service class. On return, a handle to the created

[1] Not to be confused here with data encryption standard.

record is provided. An unsigned integral value for the attribute is provided by the `CSdpAttrValueUint` class. Its `NewUintL` allocates and constructs a new instance of `CSdpAttrValueUint`. The `UpdateAttributeL()` function sets a new value to an attribute. In the example below, the service record state (ID = 0x02) is set to 0.

The next attribute is updated in the same way. The protocol list has an ID value of 4 and its type is a data element sequence. The start of the list is indicated with the `MSdpElementBuilder::StartListL()` function, showing that the subsequent elements belong to the DES (or DEA).

`BuildDESL()` adds a DES into the DES, and `BuildUUIDL()` adds the identifier of the L2CAP protocol to the list. The next integer identifies the used channel, which in this case is KRFCOMM. For RFCOMM we use the previously reserved channel:

```
void CBTServer::ConnectSdpL()
    {
    // Create an SDP record
    TBuf8<4> value1;
    TBuf8<4> value2;

    CSdpAttrValue* attrVal;
    CSdpAttrValueDES* attrValDES;

    // Set Attr 1 (service class list)
    iSdpDb.CreateServiceRecordL( KRPSServiceId, iSdpHandle );

    // Set Attr 2 (service record state) to 0
    value1.FillZ( 4 );
    attrVal = CSdpAttrValueUint::NewUintL( value1 );
    CleanupStack::PushL( attrVal );
    iSdpDb.UpdateAttributeL( iSdpHandle, 0x02, *attrVal );
    CleanupStack::PopAndDestroy(); // attrVal

    // Set attr 4 (protocol list) to L2CAP, and RFCOMM
    value1.FillZ( 4 );
    value1[3] = KRFCOMM;
    value2.FillZ( 4 );
    value2[3] = STATIC_CAST(TUint8, iLocalChannel );
    attrValDES = CSdpAttrValueDES::NewDESL( 0 );
    CleanupStack::PushL( attrValDES );

    attrValDES
         ->StartListL()
               ->BuildDESL()
               ->StartListL()
                     ->BuildUUIDL( TUUID( TUint16( KL2CAP ) ) )
                     ->BuildUintL( value1 )
               ->EndListL()
               ->BuildDESL()
```

```
            ->StartListL()
                ->BuildUUIDL( TUUID( TUint16( KRFCOMM ) ) )
                ->BuildUintL( value2 )
            ->EndListL()
        ->EndListL();
    iSdpDb.UpdateAttributeL( iSdpHandle, 0x04, *attrValDES );

    CleanupStack::PopAndDestroy(); // attrValDES
    }
```

The server starts listening to the channel as in the case of any other application using the Socket server. After the connection has been established, data may be read and written using the `Read()` and `Write()` functions of the `RSocket` class:

```
void CBTServer::ConnectSocketServerL()
    {
    // Connect to the socket server
    TProtocolDesc pInfo;

    // Open a socket
    iListenSocket.Open( iSocketServ, KRFCap );

    // Set up address object
    TBTSockAddr addr;
    addr.SetPort( iLocalChannel );
    User::LeaveIfError( iListenSocket.Bind( addr ) );

    // Begin to listen
    User::LeaveIfError( iListenSocket.Listen( 2 ) );
    }
```

Bluetooth Client

If the client knows that there is a Bluetooth device in its neighborhood and the remote device has a service in a specific port, the client can connect directly to the device. In the normal case, device discovery must be performed to find any Bluetooth devices in the neighborhood.

The `ConstructL()` function is similar to the corresponding function of the server, although the client does not need to connect to the Bluetooth security manager and SDP database. By looking at `RunL()` we see previously mentioned phases in the use of Bluetooth in the client side. We start by device and service discovery, after which the client connects to the server and starts transferring data. The last function call in the `ConstructL()` starts the device discovery:

```
void CBTClientEngine::RunL()
    {
    switch( iStateMachine )
```

```
    {
    case EStateDiscovering:
        DiscoveredRemoteL();
        QueryRemoteL();
        break;

    case EStateQueried:
        WaitForDataL();
        break;

    case EStateConnecting:
        WriteDataL();
        CloseRemoteL();
        WaitForDataL();
        break;

    }
}
```

The `DiscoverRemoteL()` function tries to find devices, that support the service that is requested. Device information is stored in an instance of the `TBTDeviceSelectionParams` class. The search is based on the device class or service disarray protocol (SDP) universally unique identifier (UUID), whichever is used:

```
void CBTClientEngine::DiscoverRemoteL()
    {
    // Connect to the RNotifier
    User::LeaveIfError( iNotifier.Connect() );

    // Look for devices with services with our UUID
    TBTDeviceSelectionParams filterParams;
    TBTDeviceSelectionParamsPckg fpPckg( filterParams );
    TUUID targetServiceClass( KUUID );
    filterParams.SetUUID( targetServiceClass );

    // Start the notifier. The response will be in
    // iDiscoveryResponse
    iNotifier.StartNotifierAndGetResponse(
        iStatus,
        KDeviceSelectionNotifierUid,
        fpPckg,
        iDiscoveryResponse );

    // Change the state
    iStateMachine = EStateDiscovering;
    SetActive();
    }
```

The `DiscoveredRemoteL()` function simply cancels the notifier and sets the remote address, in case it is valid:

```
void CBTClientEngine::DiscoveredRemoteL()
    {
    // Close the notifier
    iNotifier.CancelNotifier( KDeviceSelectionNotifierUid );
    iNotifier.Close();

    // If the result is not a valid Bluetooth address,
    // leave with KErrGeneral. Otherwise copy the address
    // of the remote device.
    if( iDiscoveryResponse().IsValidBDAddr() )
        {
        iRemoteAddr.SetBTAddr( iDiscoveryResponse().BDAddr() );
        }
    else
        {
        User::Leave( KErrGeneral );
        }
    }
```

The final state before connection establishment and data transfer is to query the service in the server's SDP database. An instance of `CSdpAgent` is used to make SDP requests to a remote device. `SetRecordFilterL()` sets the classes of service to query for. The result will contain records for services that belong to the classes listed in the filter. The filter is a UUID, which represents service classes. In the example below, the KUUID (constant UUID) value is 0x5021.

```
void CBTClientEngine::QueryRemoteL()
    {
    // Create agent.
    iSdpAgent = CSdpAgent::NewL( *this, iRemoteAddr.BTAddr() );

    // Create a search pattern and add a service classes to it
    CSdpSearchPattern* list = CSdpSearchPattern::NewL();
    CleanupStack::PushL(list);
    list->AddL( KUUID );

    // Set the search pattern on the agent
    iSdpAgent->SetRecordFilterL(*list);

    // Get first search result: results in call back to this
    iSdpAgent->NextRecordRequestL();

    // Advance the state machine
    iStateMachine = EStateQuerying;

    CleanupStack::PopAndDestroy();     // list
    }
```

The actual handling of the service record in the remote SDP database is taken care of by the `NextRecordRequestCompleteL()` function.

We stored values in the protocol list attribute in the server, so they should be readable in the client. An instance of `CSdpAttrIdMatchList` holds a list of attribute IDs that may be retrieved from a remote device:

```
void CBTClientEngine::NextRecordRequestCompleteL( TInt aError,
  TSdpServRecordHandle aHandle, TInt /*aTotalRecordsCount*/)
    {
    // We got a service record.
    if( aError != KErrNone )
        {
        // Some error
        QueriedRemote();
        }
    else
        {
        // Ask for the contents of the record

        // Create a match list
        CSdpAttrIdMatchList* matchList =
          CSdpAttrIdMatchList::NewL();
        CleanupStack::PushL(matchList);

        // Add an attribute ID to get
        matchList->AddL( TAttrRange( KSdpAttrIdProtocol-
          DescriptorList ) );

        // Set the match list on the agent
        iSdpAgent->AttributeRequestL( aHandle,
          KSdpAttrIdProtocolDescriptorList );

        CleanupStack::PopAndDestroy(); //matchList
        }
    }
```

`AcceptVisitorL()` is a virtual function that requests a callback to pass the attribute value. If there are attributes that are not lists (the type of an attribute is different from DEA or DES), the `VisitAttributeValueL()` function in `MSdpAttributeValueVisitor` is called. As a parameter, the attribute value object and type are passed. This is a useful way to enumerate and go through each attribute in a list:

```
void CBTClientEngine::AttributeRequestResult( TSdpServRecord-
  Handle /*aHandle*/, TSdpAttributeID aAttrID,
  CSdpAttrValue* aAttrValue)
    {
    // We got the contents. Use an attribute
    // visitor to parse the results.
    if( aAttrID == KSdpAttrIdProtocolDescriptorList )
```

```
        {
        aAttrValue->AcceptVisitorL( *this );
        }
    }
```

The implementation of the `VisitAttributeValueL()` function is given below; it is used to read the channel port number in which the service is available:

```
void CBTClientEngine::VisitAttributeValueL(
  CSdpAttrValue& aValue, TSdpElementType aType)
    {
    // This will be called for each attribute
    // received. It just waits for attributes
    // of certain type and parses them.
    // All other attributes are ignored.

    TUUID rfcomm( TUint16( KRFCOMM ) );
    switch( aType )
        {
        case ETypeUUID:
            iPreviousUUID = aValue.UUID();
            break;

        case ETypeUint:
            if( iPreviousUUID == KRFCOMM )
                {
                iRemoteChannel = aValue.Uint();
                }
            break;

        default:
            break;
        }
    }
```

Finally, the attribute request has been completed. We simply call the `QueriedRemote()` function, which updates the state of the state machine:

```
void CBTClientEngine::AttributeRequestComplete(
    TSdpServRecordHandle /*aHandle*/, TInt aError)
    {
    // Request complete. Announce that we're done
    // to the state machine.
    QueriedRemote();
    }
void CBTClientEngine::QueriedRemote()
    {
    // Just advance the state machine
```

```
iStateMachine = EStateQueried;
// Continue immediately
SetActive();
TRequestStatus* status = &iStatus;
User::RequestComplete( status, KErrNone );
}
```

16.3.7 WAP

The Series 60 Platform provides access to a WAP 1.2.1 stack. Through API, it is possible to access wireless session, transaction, or datagram protocols (WSP, WTP, and WDP, respectively). The WAP stack is used in the same way as the other communications servers in Symbian OS; first, a session to a WAP server is created. Note that this means a session to the WAP server in Symbian OS, not the WAP server serving WAP requests in the network.

The session is created using the `RWAPServ` class. The naming analogy to Serial and Socket servers is obvious. The `Connect()` method creates the session. Unlike the corresponding methods of the Serial and Socket servers, this method also starts the server, if it is not already running.

There is a common class (`RWAPConn`) from which all subsession classes are derived. This base class supports operations that may be used for connections in all WAP stack layers. WDP access is provided by the `RWDPConn`, and WSP access is provided by either `RWSPCOConn` (connection-oriented session) or `RWSPCLConn` (connectionless session).

The bearer used by the WAP stack is selected by using the communications database. The protocol stack of the bearer, such as SMS or IP, are plug-in protocol modules. The WAP API is used by application engines using or providing WAP services.

Nokia (2001h) has published a guide to WAP service development. This publication is very useful not only to application developers but also to WAP service developers.

16.4 ETel Server

The telephony server of Symbian OS is called ETel. It provides an interface and client-side API to telephony hardware. The user is not dependent on the vendor, because all applications use the standard API. The client-side session to the server is provided by the `RTelServer` class, which supports analogous functions to those of the Serial and Socket servers, in order to load telephony modules (`LoadPhoneModule()`), enumerate phones (`EnumeratePhones()`), and to get information on the specified phone (`GetPhoneInfo()`). In addition to

RTelServer, the client-side API provides three classes: RPhone, RLine, and RCall.

RPhone provides functions to answer and initiate lines, query phone and line status and capabilities, access phone-based information such as GSM phonebooks, pass active calls between one another, support messaging such as SMS, and register to receive phone event notifications.

RLine provides access to the functionality of lines. These include line capability and status queries and event notifications associated with the line.

Calls are established and disconnected using the RCall class. In addition to call control functionality, this class enables queries on call capabilities, call status, different bearers, and call progress.

An important function is LoanDataPort(). It allows a client to gain control of the port used to communicate with the modem. The client can send data directly to the port. While the client is using the port, the ETel server queues any commands which it has to the modem. The control shifts back to the ETel server when the RecoverDataPort() function is called:

```
CCallEngine::CreateCallL()
   {
   _LIT(KTsyName, "PhoneTsy.tsy");
   _LIT(KDataLineName, "Data");
   _LIT(KTelNumber, "+350401234567890");
   TInt ret;
   RPhone phone;
   RLine line
   RCall call;

   ret = iServer.Connect();
   User::LeaveIfError( ret );
   ret=server.LoadPhoneModule( KTsyName );
   User::LeaveIfError( ret );
   ret = iServer.GetPhoneInfo( 0, phoneInfo );      // Number
   // of phone
   User::LeaveIfError( ret );

   ret = phone.Open( iServer, phoneInfo.iName );
   User::LeaveIfError( ret );

   ret = line.Open( phone, KDataLineName );
   User::LeaveIfError( ret );

   ret = call.OpenNewCall( line );
   User::LeaveIfError( ret );

   // Dial out to remote number entered by user
   ret = datacall.Dial( iStatus, KTelNumber );
```

```
User::LeaveIfError( ret );

call.LoadDataPort();
HangUpDataCall();
call.Close();
line.Close()
phone.Close();
iServer.Close()
}
```

16.4.1 Fax

ETel supports fax services through three APIs: fax client, fax store, and header line API. The fax client API is used to send and receive fax messages, the fax store API includes services to store and retrieve fax information, and the header line API supports generation of the header line on the top of fax page. The architecture of the ETel faxing is shown in Figure 16.2.

The fax printer driver creates the required fax pages, which are stored as a fax store file. This fax information is then taken from the file and passed by the fax client to ETel, which sends the fax. The fax message-type module (MTM) creates a header line data file, which is used by ETel to create a header line to add to the top of every page sent. The persistent settings are stored in the communications database component.

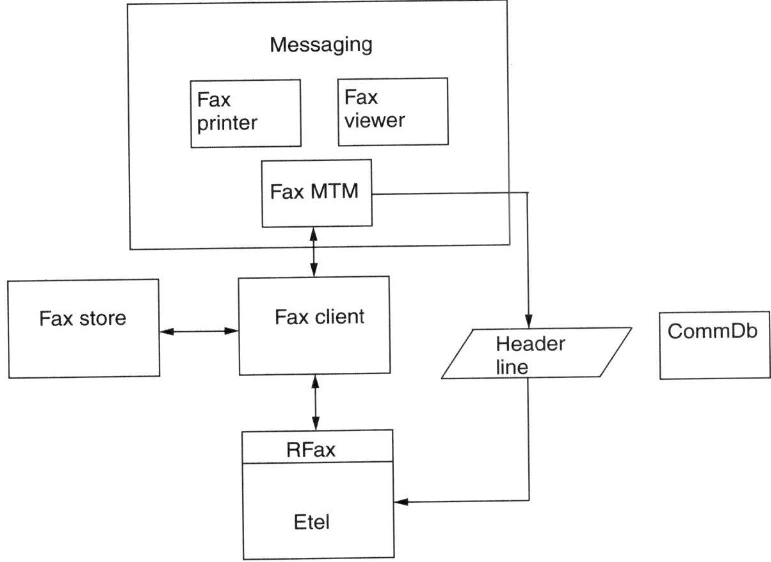

Figure 16.2 ETel faxing architecture; Note: MTM, message-type module

16.5 Summary

The communications architecture of Symbian OS and the Series 60 Platform enables not only easy expandability of the communication system but also an easy means of programming. Each server provides a client-side API, which is used almost in an identical way to the server. In addition, if a protocol implementation changes, but the used server is the same, there are only small changes to the use of the protocol.

In this chapter we have provided examples of how to use the client-side APIs of the Serial communication, Socket, and Telephony servers. We have emphasized the first two servers because application developers do not use the Telephony server directly. The interested reader may find out more on the Telephony server in Jipping (2002), which contains a very through description of any communications programming in Symbian OS. We have not yet discussed the Message server yet. The message server is the subject of the next chapter.

17

Messaging

Messages are self-contained data objects that are transferred between messaging applications. There are numerous types of messages, but most of them share some common features. Each message has a **header**, **body**, and zero or more **attachments**.

The message header contains the source and the destination of the message and perhaps a time stamp showing when it was transmitted. The content of the message is included in the message body, and it is meaningful only to those programs (or to a user) that may use the message. Attachments can be arbitrary binary or textual data.

Messages are delivered according to the **pull** or **push** model. In the pull model, messages are stored in the server, from which the messaging client downloads the received messages. In the push model, the server contacts the receiver directly and pushes the message to it.

In this chapter, the messaging architecture of Symbian OS and the Series 60 Platform is described. In general, all message types are supported with a set of plug-in modules. The supported types include e-mail, short and multimedia messages, fax, and BIO (bearer-independent object) messages. All these message types and the use through the application programming interfaces (APIs) of the plug-in modules are described.

17.1 Messaging Architecture

Symbian OS implements a powerful, extensible framework for multiprotocol messaging. The messaging architecture consists of the message server and layered implementations of different messaging protocols that are implemented as message-type modules (MTMs).

Series 60 has support for a short message service (SMS), a multimedia message service (MMS), e-mail, BIO messages (smart messaging), infrared, Bluetooth, and fax, although fax is typically used through the Telephony server (ETel). New protocols can be added to the framework, and it is possible to find out at runtime which message types are supported in the device. The design also allows generic message access, the execution of certain operations, and the extraction of the same information from all message entries without knowing their type.

For sending messages with use of different protocols, Symbian OS has an easy-to-use Send As API. A user interface (UI) is also available for the Send As in Series 60 (SendUI); SendUI encapsulates the whole message-sending process, including the required UI components, such as progress dialogs and menu items. SendUI and its usage will be introduced later in this chapter, in Section 17.6.

The messaging framework is presented in Figure 17.1. The abstract MTM base classes are supplied by Symbian, as are the message server and the session object. Other modules visible in the figure can be implemented by third parties. Symbian delivers concrete MTMs for the message types mentioned earlier. Application means, for example, a message editor implementing UI functionality or any other application that uses the messaging subsystem. The purpose of each component is presented in this chapter.

The basic type in messaging is an **entry**. An entry is an abstraction of the message or part of the message that is stored in the message

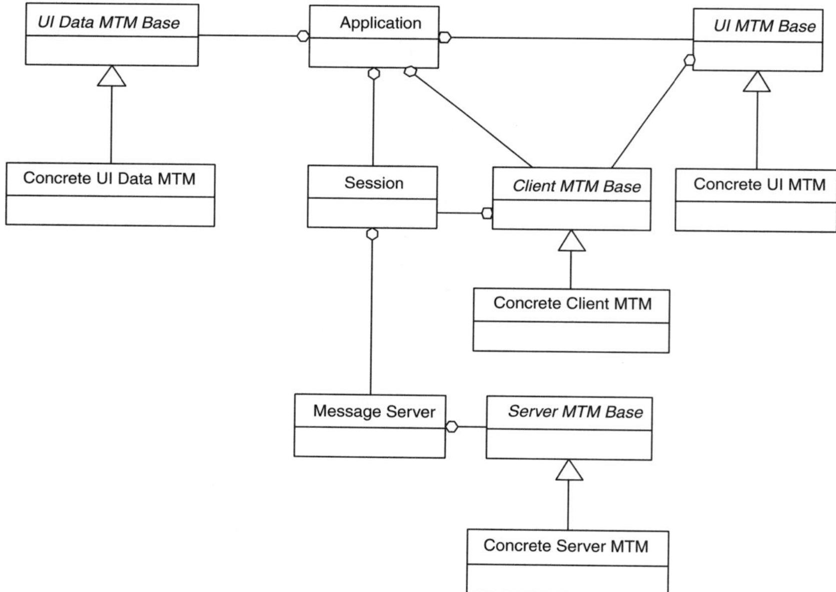

Figure 17.1 Messaging framework. Note: MTM, message-type module; UI, user interface

server. For example, one short message is always one message entry, but multimedia messages and e-mails can have multiple entries, since attached objects are entries as well.

Entries are created by the messaging engine when a message is received or created. Also, folders and services (e.g. mailbox, SMS service-centre address) are stored as entries. All entries are identified by a unique identifier (UID) and they can be retrieved by using this UID.

The header file msvids.h defines the identifiers for common folders and services such as the local Inbox and Outbox. Entries can be retrieved by opening a session to the message server and using a constructor of `CMsvEntry` or the services of a session object. The classes that are used to access the entries are as follows:

- `CMsvEntry`, which is defined in msvapi.h, encapsulates the actions to create, manipulate, move, delete, and access the entries. It is of rather a heavyweight class, since it caches information about its children entries as well.
- `CMsvServerEntry`, defined in msventry.h, is similar to `CMsvEntry`, but it is used by server MTMs only.
- `TMsvEntry` from msvstd.h encapsulates the index entry and can be used to obtain information, such as size and type of the message. There are protocol-specific derivations of this class such as `TMsvEmailEntry` and `TMmsMsvEntry`. Changes to `TMsvEntry` are not written directly into the message server; they must be committed by using the `CMsvEntry::ChangeL(const TMsvEntry&aEntry)` method.

The message server is a standard Symbian OS server that accepts asynchronous requests from clients. Its main responsibilities are controlling access to message data, delegating operations to modules implementing messaging protocols, and assuring that message data are consistent.

Temporary, exclusive access to message data can be given to a requester. Access can consist of either reading or writing; multiple readers are allowed access to one message entry at a time. The server stores messages as dictionary file stores in the file system of the device. Every entry has at least two streams, which store body text and a backup copy of the entry's index data. Usually, the MTM creates additional streams to store information about recipients, settings, and other protocol-specific data.

Generally, there is no need to use services provided by the message server directly; instead the MTMs may be used. Clients communicate with the message server using session objects that are instances of the

CMsvSession class. Every client owns exactly one session instance. As mentioned earlier, entries can be retrieved by using the services of CMsvSession. The following example show how to open a session to the message server, access entries, and count the total size of the messages in Inbox:

```
#include <msvapi.h>  // CMsvEntry, MMsvSessionObserver
#include <msvstd.h>  // CMsvEntrySelection
#include <msvids.h>  // KMsvGlobalInBoxIndexEntryId
#include <msvuids.h> // KUidMsvMessageEntry
#include <e32base.h> // CBase

// Observer needed to connect server, derived from MMsvSessionObserver
class CMySessionObserver : public MMsvSessionObserver, public CBase
    {
    public:
    void HandleSessionEventL(TMsvSessionEvent aEvent, TAny* aArg1,
      TAny* aArg2, TAny* aArg3);
    };

void CMySessionObserver::HandleSessionEventL(TMsvSessionEvent aEvent,
  TAny* aArg1, TAny* aArg2, TAny* aArg3)
    {
    switch (aEvent)
        {
        // close session if something wrong with server
        case EMsvGeneralError:
        case EMsvCloseSession:
        case EMsvServerFailedToStart:
        case EMsvServerTerminated:
            User::Exit(KErrGeneral);
            break;
        default:
            // Do nothing
            break;
        }
    }

TInt CountSizeOfMessagesInInboxL()
    {
    TInt size(0); // total size of the messages

    // Create session observer
    CMySessionObserver* observer = new(ELeave) CMySessionObserver;
    CleanupStack::PushL(observer);

    // Open session to message server
    CMsvSession* session = CMsvSession::OpenSyncL(*observer);
    CleanupStack::PushL(session);

    // Create object defining grouping, ordering and if invisible
    // entries are included into folder entry to be retrieved
    TMsvSelectionOrdering order(KMsvNoGrouping, EMsvSortByNone,ETrue);
    // Retrieve entry representing Inbox
    CMsvEntry* inboxEntry = CMsvEntry::NewL(*session,
      KMsvGlobalInBoxIndexEntryId, order);
    CleanupStack::PushL(inboxEntry);
```

```
// Get list of message entries in Inbox, folders are not included
CMsvEntrySelection* selection =
  inboxEntry>ChildrenWithTypeL(KUidMsvMessageEntry);
CleanupStack::PushL(selection);

// Count size of the messages
const TInt count(selection->Count());
for (TInt i = 0; i < count; i++)
     {
     // access message entry
     TMsvEntry messageEntry;
     TMsvId owningServiceId;
     User::LeaveIfError(session->GetEntry((*selection)[i],
       owningServiceId, messageEntry));
     // size of the entry is in TMsvEntry::iSize
     size += messageEntry.iSize;
     }

// Delete selection, inboxEntry, session and observer
CleanupStack::PopAndDestroy(4);

return size;
}
```

Protocols are implemented as MTMs that implement general and protocol-specific functionality. General actions include creating, editing, and deleting messages. Usually, editing and viewing are implemented by separate dynamic link libraries (DLLs) to reduce the memory footprint and complexity of the MTM. An example of protocol-specific functionality is the ability to fetch one attachment from a mail server to the device by using the Internet Mail Access Protocol version 4 (IMAP4) MTM. This functionality is not supported by the Post Office Protocol version 3 (POP3) mail protocol or by its MTM.

There are four types of MTMs:

- UI MTM handles message editing, viewing, folder manipulation, and connection handling. It also displays progress information and error messages.
- UI data MTM provides icons and menu items for the message type. These are not included in UI MTM, to make actions lighter.
- The client-side MTM handles address lists and subject, when appropriate, creates reply and forward messages, validates messages, and manages the MTM-specific data in the message store.
- The server-side MTM copies messages between local and remote folders, handles messages in remote servers, interprets entry data to and from the protocol presentation, and manages interaction with transport components, such as sockets and ETel.

Messaging applications use mainly UI MTMs, UI data MTMs, and client-side MTMs to handle entries. If the user interface is not needed, all functionality is available in the client-side MTM. Owing to UI dependencies, UI MTMs and UI data MTMs must be adapted to the UI style used; client-side and server-side MTMs are generic technology used in all UI styles.

The MTMs are loaded into the memory when needed by using services of the MTM Registry. The loading of an MTM is shown later in the SMS example (Section 17.3). The registry keeps count of instantiated MTMs and automatically unloads modules that are not needed again immediately in order to save memory. Unloading can be delayed to prevent sequential loading and unloading; appropriate timeout depends on the use of the protocol. For example, instant messaging MTMs are needed regularly, and messages need a fast reaction time, so they can always be kept loaded, whereas fax modules can be unloaded immediately, because they are probably rarely used.

New message types can be added to the device by implementing and installing the relevant MTMs. If the protocol does not need user interaction, only the client and server modules are needed. Installation to the device is possible, for example, by using an SIS (symbian installation system) file.

The new message type must be registered with the message server by storing some basic data about the MTMs, such as UIDs and a human-readable name to be shown in Send via menus of the applications using SendUI. Also, the ordinals of the factory methods in the DLLs are registered so that the MTM classes can be instantiated.

Prior to Symbian OS v 6.1, a file containing the required registration information was generated either by using the mtmgen utility and setting a file or by externalizing the CMtmGroupData object. In Symbian OS v 6.1 and subsequent versions, the MTMs can also be registered by writing a resource file that is compiled by using a resource compiler. Benefits achieved by using the resource file are localization, version control, and easier registration file generation.

The resource definition is in the mtmconfig.rh file. The registration file syntax for messaging types requiring client-side and server-side MTMs is shown in the following example:

```
#include <mtmconfig.rh>       // Resource definition
#include "mymtmuids.h"       // UIDs for MTM
RESOURCE MTM_INFO_FILE
     {
     mtm_type_uid = KMyProtocolUid;      // this differentiates
                                        // protocols
     technology_type_uid = KMyMessageTypeUid;
     components =
          {
          MTM_COMPONENT
               {
               human_readable_name = "My server MTM";
```

```
            component_uid = KUidMtmServerComponentVal;
            specific_uid = KMyServerMTMUid;
            entry_point = 1;      // factory method ordinal
            version = VERSION { major = 1; minor = 0; build =
              100; };
            },
        MTM_COMPONENT
            {
            human_readable_name = "My client MTM";
            component_uid = KUidMtmClientComponentVal;
            specific_uid = KMyClientMTMUid;
            entry_point = 3;      // factory method ordinal
            version = VERSION { major = 1; minor = 0; build =
              100; };
            }
        };
    }
```

The technology-type UID given in the definition tells the messaging type to which the MTM belongs. For example, all e-mail MTMs [IMAP4, POP3, and simple mail transfer protocol (SMTP)] have the same technology type, but different MTM-type UIDs. The BIO messaging types have a very similar resource for their registration. When the registration file is generated either by compiling the resource, using mtmgen, or externalizing a CMtmGroupData object, it is copied to the \system\mtm directory. An installation program that calls `CMsvSession::InstallMTMGroup` with a path to the created registration file as the argument must be run after the files are copied. This adds the MTM to the registry file maintained by the message server. Finally, the message server must be restarted before the installed module can be used. All steps can be done in the installation program, so the process is not visible to the user. The MTMs are uninstalled in a similar way, by first calling `CMsvSession::DeInstallMTMGroup()` and then removing the files copied into the device during installation.

The user-defined observers can be registered to be notified about events relating to entries of the message server. `MMsvSessionObserver` provides an interface to react to events sent by the message server. The example shown above had a simple implementation of this interface. All messaging applications implement this interface, since it is needed when a session to the message server is established. The interface consists of one method, the argument of which indicates an event. Notified events include message server shutdown an indication to close the session, and entry creations and deletions. For a full list of notified events, see the documentation of `MMsvSessionObserver`.

`MMsvEntryObserver` provides an observer interface to entries. Notification is sent, for example, on entry creation, deletion, and movement. The observed entry may be a folder, and, in this case, notifications are received from all events having that folder as the context. Thus, it is possible, for example, to observe all incoming short messages by defining the observer for the Inbox folder and reacting to messages that are short messages.

17.2 E-mail

Series 60 messaging supports all common e-mail protocols: POP3, IMAP4, and SMTP. Secure connections to servers implementing the feature are available for all protocols. The sending of e-mail is possible from any application by using the SendUI interface. Symbian OS supports MIME (multipurpose Internet mail extension) encoding for created and received messages.

What differentiates e-mail from SMS and MMS is the concept of remote mail servers. SMS and MMS messages are received directly into the device, and there are no copies of the message elsewhere. E-mail messages are received into the remote e-mail server, and the device has only a snapshot of a certain situation in the mail server. Thus, it is possible that the device has messages that are deleted from the server and that the server has new messages that are not synchronized with the device. It is necessary to implement e-mail clients in order to handle such situations.

Before the e-mails can be created or read, one or more e-mail accounts must be created. This is done by creating two services into the message server – one for sending mail via an SMTP account and another for receiving messages via a POP3 or IMAP4 account. The account entries have their own stores into which the settings are saved.

The SMTP account is always invisible; only the receiving service is shown to the user. Sending and receiving services are linked to each other by storing the identifier (ID) of the opposite service into the `iRelated` member variable of the `TMsvEntry` of the service. Thus, the sending service can be restored from `iRelated` of the POP3 or IMAP4 service entry, and the receiving service is found in `iRelated` of the SMTP service entry.

E-mail has its own entry type defined, `TMsvEmailEntry`, which is derived from `TMsvEntry`. This class provides access to such features as 'Will new messages include a signature or Vcard?' and 'Is message formatted as plain text or HTML?' IMAP4 flags are accessed and modified using this class. E-mail messages can be modified by using the `CImHeader` and `CImEmailMessage` classes. `CImHeader` encapsulates the header information, such as recipient, sender, subject, and date information about the message. `CImEmailMessage` is used to access and modify the body text and attachments of the message.

17.2.1 Sending E-mail

New e-mail messages can be created by using the `CImEmailOperation` class. It provides static methods to create new messages, replies, and forwarded messages. In the following example, a new e-mail message is created into the local drafts folder and filled with the sender, recipient, subject, and body text:

```
void CExampleAppUi::CreateAndSendMailMessageL()
    {
    // message parts that are created
    TMsvPartList partList(KMsvMessagePartBody |
        KMsvMessagePartAttachments);

    // message attributes, define that message is in preparation
    TMsvEmailTypeList typeList(KMsvEmailTypeListMessageInPreparation);

    // create wrapper object to synchronize operation
    CMuiuOperationWait* wait = CMuiuOperationWait::NewLC();

    // create mail message
    CImEmailOperation* operation = CImEmailOperation::CreateNewL(
        wait->iStatus,      // reports progress of the operation
        iSession,      // pointer to CMsvSession object
        KMsvDraftEntryId,      // folder into which message is created
        KMsvNullIndexEntryId,      // SMTP service, using
                                   // default service
        partList,      // defines message parts to be created
        typeList,      // the kind of message created
        KUidMsgTypeSMTP);      // the protocol the message is using
    CleanupStack::PushL(operation);
    wait->Start();      // this waits for completion of
                        // create operation

    // next find out id of the new message
    TMsvId id;
    TPckgC < TMsvId > paramPack(id);
    const TDesC8& progress = aOperation.FinalProgress();
    paramPack.Set(progress);
    CleanupStack::PopAndDestroy(2);      // wait, operation

    // get created message
    CMsvEntry* entry = CMsvEntry::NewL(iSession, id,
      TMsvSelectionOrdering());
    CleanupStack::PushL(entry);

    // restore message header and set recipient and subject
    CMsvStore* store = entry->ReadStoreL();
    CleanupStack::PushL(store);
    CImHeader* header = CImHeader::NewLC();
    // set own information as sender
    header->SetFromL(_L("Tester <tester@domain.does.not.exists>"));
    // add recipient to recipient array
    header->ToRecipients().AppendL(
      _L("tester2@domain.does.not.exists"));
    // set subject
    header->SetSubjectL(_L("Test mail message"));
```

```
// commit changes
header.StoreL(store);
CleanupStack::PopAndDestroy();      // header
// create body text
CImEmailMessage* message = CImEmailMessage::NewL(id);
CleanupStack::PushL(message);
// create richtext object using default formatting
CRichText* richText = CRichText::NewL(EikonEnv->
  SystemParaFormatLayerL(), iEikonEnv->
  SystemCharFormatLayerL());
CleanupStack::PushL(richText);
_LIT(KTestMailMsg, "Hello world!");
richText->InsertL(0, KTestMailMsg);      // Insert body text
wait = CMuiuOperationWait::NewLC();
message->StoreBodyTextL(id, richText, wait->iStatus);
wait->Start();      // wait for completion
CleanupStack::PopAndDestroy(5);      // wait, richText,
                                     // message, store,
                                     // entry
}
```

The above example execution is stopped in order to wait for asynchronous message creation and body text insertion. This will affect the responsiveness of the applications, and this technique should be used with care.

An easier way to do the same thing is to use the Send As interface, which is described later in this chapter (Section 17.6). Like all e-mail interfaces, `CImEmailMessage` provides an asynchronous interface. Before an e-mail can be sent from the device, an SMTP service must be defined. The service stores all default settings for the created messages in the `CMsvStore` stream, and these settings are represented by the class `CImSmtpSettings`. This class encapsulates the information about user identification, secure connections, e-mail address, scheduling, and signature attachment. The settings can be stored and restored by using the `StoreL()` and `RestoreL()` methods of `CImSmtpSettings`.

Some preferences can be changed on a per message basis, so a message can be sent using different settings from those defined for the used service. Such settings include, for example, scheduling, receipt request, and signature attachment. It must be noted, though, that not all e-mail clients support changes to all attributes. Scheduling settings enable the message to be sent immediately, opening a new network connection if needed, during the next connection or at the user's request.

The created messages can have any supported character set used in the header and body texts. One can define whether the message use MIME encoding or not. The body text can be in plain text only, MHTML only, or both. Encoding and the body text character set can be defined for a single message in the mailbox. `CImSmtpSettings` provides access to set the default values for the service, and, by using the methods of `CimEmailMessage`, it is possible to alter the settings for the message. `CImHeader` has methods for setting the header character set.

Symbian OS v 6.1 also has a new class for the settings, `TImEmail-TransformingInfo`. A `TImEmailTransformingInfo` object can be associated with a service when it is used as a default value for all new messages, or it can be associated with a single message. The class has 'getters' and 'setters' for all mentioned attributes concerning character sets and message encoding. The advantages of using `TImEmailTransformingInfo` is that all conversion and encoding information is stored in one place in which they can be easily accessed and modified.

17.2.2 Receiving E-mail

E-mail messages can be received from the mail server into the device by using POP3 and IMAP4. Messages are never retrieved to the device automatically; the operation must be initiated by user action or by other means, such as a timer. The fetching of messages is possible with use of `CImPop3GetMail` or `CImImap4GetMail` APIs. These classes are not meant to be used directly; rather, they are used through the client MTMs, which use the above-mentioned APIs.

Usually, when a new message is received into the e-mail server, the header information is fetched during the next connection and the whole message is fetched on the user's request. It is also possible to restrict the number of headers to be fetched to reduce network traffic, when an IMAP4 mailbox is used. This is done by using the `SetInboxSynchronizationLimit()` method of the `CImImap4Settings` object related to the mailbox. `SetMail-boxSynchronizationLimit()` is used to control how the other folders compared with Inbox are synchronized.

The header information includes sender and receiver, subject, size, character set, encoding, and information about the attachments. Attachments can be fetched from the server on demand, one by one if IMAP4 is used (POP3 does not support this). The messages fetched can be deleted from the device and left on the server, or deleted from the device and the server.

Settings for receiving e-mail are defined by using the `CImPop3Set-tings` and `CImImap4Settings` classes. The settings are used and stored as SMTP settings, discussed in Section 17.2.1. The classes encapsulate settings, such as 'is secure connection used?', 'What is the login name?', 'What is the password?', and 'What are the synchronization preferences?'. There are no settings per message; there is no need for them as there were with SMTP messages.

Both protocols implement the offline queue to which operations made offline are appended. The most common operation queue is the deletion of a message. When a connection to the server is opened, all operations in the queues are executed.

17.3 SMS

SMS enables the transfer of short messages (160 characters, at the most, of 7-bit text, or 140 bytes of 8-bit binary data) to and from a mobile phone, even while a call is in progress. The 8-bit messages are often used for smart messaging in addition to 8-bit text. The 16-bit messages of 70 characters or less are used for Unicode (UCS2) text messages for Chinese and Arabic characters, for example. Unlike e-mail, SMS uses a push model to deliver the messages. The messages are stored in the SMSC (short message service center) and immediately pushed to the receiver, if it is attached to the network. Otherwise, the messages are stored in the SMSC for a specified time (e.g. 24 hours), to wait for delivery to the receiver's phone.

SMS can be used in either a text or PDU (protocol data unit, or protocol description unit) mode. When the text mode is used, a 160-character text message is encoded in binary format. There are several encoding alternatives, the most common of which are 8859-1, PCDN, PCCP437, IRA, and GSM (global system for mobile). In the PDU mode, binary data can be transmitted as such. The PDU mode (binary mode) enables the use of custom character coding. For example, with 5-bit coding one can send a message containing 224 characters at maximum. There are four PDU types defined in the SMS relay protocol. The relay protocol is a part of the SMS service of the GSM network:

- SMS-SUBMIT is used for sending short messages.
- SMS-DELIVER is used for receiving short messages. The PDU type changes from SMS-SUBMIT to SMS-DELIVER after transmission.
- SMS-STATUS-REPORT is used for sending queries about submitted but not yet delivered short messages, which are stored in the SMSC.
- SMS-COMMAND PDU type is used for indicating the status of a previously submitted short message to the sender (e.g. confirmation of a successfully delivered short message).

In addition to the textual or binary contents, the SMS PDUs have a user data header. The header contains the sender and the receiver addresses (phone numbers), as well as the address of the SMSC through which the message was delivered. Other fields are used to store the time stamp, message class, concatenation information, and application port addressing. The header takes up the first 7 bytes of user data, which leaves 133 bytes for the content, or 128 bytes in case of a concatenated SMS message (www.forum.nokia.com).

In the header, the time stamp field indicates when the message was submitted from the sender's device. The class field of the header is used to indicate one of the four classes to which the message may belong. Class 0 or a flash message is used to activate the SMS indicator, but the message is not stored in the receiver's phone. Class 1 is used for standard text messages, stored in the phone's memory. Class 2 messages are stored in the memory of the SIM card. Class 4 messages are stored in an external device (e.g. in the PC, if it is connected to the phoneline).

Several messages may be concatenated together, in which case the concatenation information denotes, for example, the number of message PDUs concatenated together. The application port addressing is used to deliver the message to the correct application. For example, smart messages contain no sensible information to be shown in the receiver's SMS viewer.

17.3.1 Creating and Sending Short Messages

There are several ways of creating and handling short messages in Symbian OS. An easy way to create short messages is to use the `CSendAs` class described in Section 17.6. Another way is to use the SMS MTMs directly, as shown in this section. As usual, a session to the message server is first required:

```
TInt CSMSAppUi::CreateAndSendNewSMSL()
    {
    // ... session to the Message server created
    TMsvEntry newEntry;      // Entry in the Message server index
    newEntry.iMtm = KUidMsgTypeSMS;      // Message type is SMS
    newEntry.iType = KUidMsvMessageEntry;      // Entry type is message
    // Local service ID
    newEntry.iServiceId = KMsvLocalServiceIndexEntryId;
    newEntry.iDate.HomeTime();      // Date set to home time
    newEntry.SetInPreparation( ETrue );      // Message is in
                                             // preparation

    // Create an object to access an entry. The entry is a draft, no
    // grouping and sorting of entries
    CMsvEntry* entry = CMsvEntry::NewL( *session,
      KMsvDraftEntryIdValue, TMsvSelectionOrdering() );
    CleanupStack::PushL( entry );
    // Create new entry
    entry->CreateL( newEntry );
    // Set the context to the created entry
    entry->SetEntryL( newEntry.Id() );
    // MTM client registry for creating new messages
```

```
CClientMtmRegistry* mtmReg;
mtmReg = CClientMtmRegistry::NewL( *session );
CleanupStack::PushL( mtmReg );
// Create a client-side SMS MTM object
CBaseMtm* mtm = mtmReg->NewMtmL( entry->Entry().iMtm );
CleanupStack::PushL( mtm );
mtm->SetCurrentEntryL( entry );
CleanupStack::PopAndDestroy(3);     // entry, mtmReg, mtm
```

The `CMsvEntry` class accesses the particular entry in the Message server. A `TMsvEntry` object is used to represent an index of the message in the server. The SMS-specific parts of the message are handled by means of the SMS MTM. Generic fields, such as the destination address, may be set by using the virtual functions of `CBaseMtm`, but SMS-specific fields are accessed through the `CSMSHeader` class. In addition to the header fields described above, other settings include delivery methods, handling of special messages, and SMS bearer options.

The SMS MTM is used to access the standard message parts (i.e. the header and the body) in an SMS-specific way:

```
// Insert the contents of the message
CRichText& mtmBody = mtm->Body();
mtmBody.Reset();
_LIT(KTestSmsMsg, "Hello world!");
mtmBody.InsertL( 0, KTestSmsMsg );      // Insert body text
newEntry.SetInPreparation( EFalse );      // Preparation done
newEntry.SetSendingState( KMsvSendStateWaiting );
newEntry.iDate.HomeTime();    // Set time to home time
entry->ChangeL(newEntry);

// To handle the sms specifics SmsMtm used
CSmsClientMtm* smsMtm = static_cast< CSmsClientMtm* > ( mtm );
CleanupStack::PushL( smsMtm );

// Cache changes of header fields and header options
smsMtm->RestoreServiceAndSettingsL();

// CSmsHeader provides access to SMS header fields
CSmsHeader& header = smsMtm->SmsHeader();
CSmsSettings* sendOptions = CSmsSettings::NewL();
CleanupStack::PushL( sendOptions );

// Restore existing settings
sendOptions->CopyL( smsMtm->ServiceSettings() );

// Set send options
sendOptions->SetDelivery( ESmsDeliveryImmediately );
header.SetSmsSettingsL( *sendOptions );
CleanupStack::PopAndDestroy(2);     // sendOptions and smsMtm

// Set destination address
_LIT(KDestinationAdd, "+35899876543210")
smsMtm->AddAddresseeL( KDestinationAdd );
```

```
// Commit the cached changes
smsMtm->SaveMessageL();
```

Now, the message is ready for sending. First, it is stored in the Outbox from which it is sent to its destination:

```
// Move message to the Outbox
// Handle to a parent entry
CMsvEntry* parentEntry = CMsvEntry::NewL(*session,
  newEntry.Parent(), TMsvSelectionOrdering());
CleanupStack::PushL(parentEntry);
// Move original message from the parent to the Outbox
// create wrapper object to synchronize operation
CMuiuOperationWait* wait = CMuiuOperationWait::NewLC();

// create SMS message
CMsvOperation* op = parentEntry->MoveL( newEntry.Id(),
                   KMsvGlobalOutBoxIndexEntryId, wait->iStatus );
CleanupStack::PushL( op );
wait->Start(); // this waits completion of create operation

TMsvLocalOperationProgress prog=
  McliUtils::GetLocalProgressL( *op );
User::LeaveIfError( prog.iError );

id = prog.iId; // Id of the moved entry
CleanupStack::PopAndDestroy(4);      // op, wait, parentEntry,
                                     // CCommandAbsorbingControl
```

Send the message:

```
// We must create an entry selection for message copies (although
// now we only have one message in selection)
CMsvEntrySelection* selection = new (ELeave) CMsvEntrySelection;
CleanupStack::PushL( selection );

selection->AppendL( movedId );      // Add created message
                                    // to the selection
SetScheduledSendingStateL( selection );      // Schedule the
                                             // sending with
                                             // the active
                                             // scheduler
CleanupStack::PopAndDestroy( selection );

delete mtm;
delete mtmReg;
delete session;
return KErrNone;      // At this point the message has been sent
}
```

Sending of the message uses the `SetScheduledSendingStateL()` function given below:

```
void CSMSAppUi::SetScheduledSendingStateL(
  CMsvEntrySelection* aSelection )
   {
```

```
CBaseMtm* smsMtm = iMtm;
// Add entry to task scheduler
TBuf8<1> dummyParams;
CMuiuOperationWait* wait = CMuiuOperationWait::NewLC();
CMsvOperation* op= smsMtm->InvokeAsyncFunctionL(
  ESmsMtmCommandScheduleCopy, *aSelection,
  dummyParams, wait->iStatus);
CleanupStack::PushL( op );
wait->Start();
CleanupStack::PopAndDestroy(2); // wait, op
}
```

Again, an asynchronous function is used. The function `ScheduleCopy` schedules the message to be sent through the Telephony server.

17.3.2 Receiving Short Messages

Each creation of a new message generates an event handled by the `MMsvSessionObserver` class. When receiving messages, we are interested in the new created messages in the Inbox (i.e. the session events), the entry ID of which is `KMsvGlobalInBoxIndexEntryId`:

```
TInt CSMSAppUi::ReceiveSMSL (TMsvId aEntryId
/* From session observer */)
     {

     TMsvEntry msvEntry = (iSession->GetEntryL(aEntryId))->Entry();

     // First create a new mtm to handle this message
     CBaseMtm* smsMtm = iMtmReg->NewMtmL( msvEntry.iMtm );
     CleanupStack::PushL( smsMtm );
     smsMtm->SwitchCurrentEntryL( aEntryId );
     smsMtm->LoadMessageL();      // Load the message

     // Check, if the message is targeted to us
     if (smsMtm->Body().Read(0,4).Compare( KGDSMSTag )==0)
          {
          // Process the message
          // Flash the message in the upper right corner of the screen
          iEikonEnv->InfoMsg( smsMtm->Body().Read(0,4) );

          // Delete message from inbox, first take a handle to Inbox...
          TMsvSelectionOrdering sort;
          // Also invisible messages handled
          sort.SetShowInvisibleEntries( ETrue );

        // Take a handle to the parent entry
        CMsvEntry* parentEntry = CMsvEntry::NewL( *iSession,
          msvEntry.Parent(), sort );
        CleanupStack::PushL( parentEntry );

        // Delete message from Inbox
        CMuiuOperationWait* wait = CMuiuOperationWait::NewLC();
       // ParentEntry is the Inbox (must be so that DeleteL can
```

```
        // be called)
        CMsvOperation* op = parentEntry->DeleteL( msvEntry.Id(),
        wait->iStatus );
        CleanupStack::PushL( op );
        wait->Start();
        CleanupStack::PopAndDestroy(4); // smsMtm, parentEntry,
                                        // op, wait,
        }
    }
```

17.4 MMS

Using the SMS it is possible to deliver tones and simple black-and-white pictures up to 72 × 28 pixels in addition to 160-character text greetings. MMS, the multimedia message service, is a new technology to deliver multimedia content between terminals (Nokia, 2001d). The content may be a picture, data, text, audio, or video.

MMS is implemented on top of WAP 1.2, as shown in Figure 17.2, which describes the MMS architecture (WAP Forum, 2001). Instead of WAP, HTTP may also be used, allowing the use of MMS also in Java applications in the Series 60 Platform. The MMS Proxy relay delivers multimedia messages to the destination MMS server, which stores the message for later forwarding. In a typical network architecture, the Proxy relay and MMS server are implemented in the same network component, the multimedia message service center (MMSC).

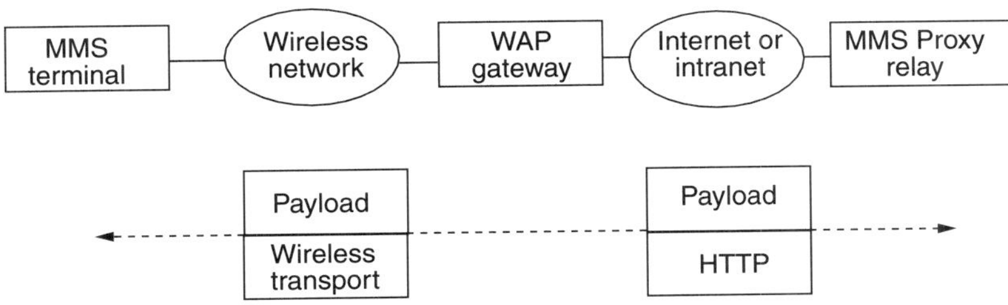

Figure 17.2 Multimedia message server (MMS) architecture. Note: HTTP, hypertext transfer protocol; WAP, windows application protocol

Compared with SMS, the MMS delivery model is more complicated. There are six MMS PDUs: M-Send, M-Notification, M-Retrieve, M-NotifyResp, M-Acknowledge, and M-Delivery. In addition, WSP (Wireless Session Protocol) messages are used. PDUs also have different types (request, indication, response, and confirmation), which were described in Chapter 15. The M-Send.req and M-Retrieve.conf PDUs have a normal message structure, which consists of a header and a body. All other PDUs contain only the header part.

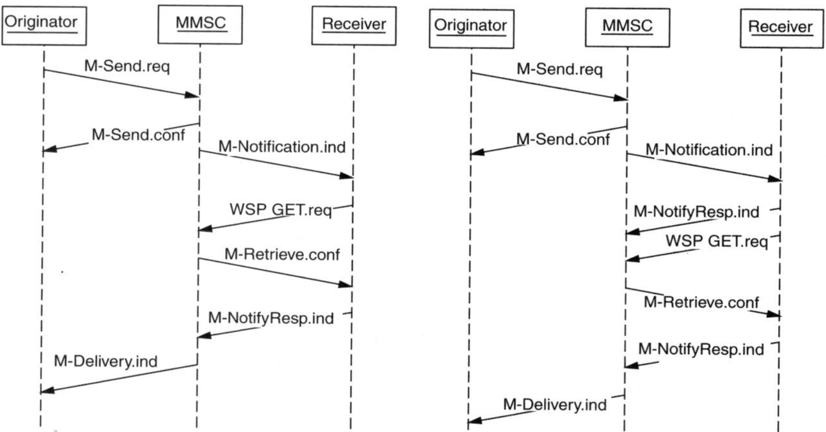

Figure 17.3 Immediate and postponed delivery of an multimedia message. Note: MMSC, multimedia message service center; WAP, window application protocol; WSP, wireless session protocol

The MMS PDUs are transmitted in the content section of WSP or HTTP messages over the circuit-switched data (CSD) or General Pachet Radio System (GPRS) connection. The use of messages in two different scenarios is shown in the sequence chart in Figure 17.3. First, the sender uses a WSP POST request to send M-Send.req, which contains also the content or the body of the message. If the MMSC receives the message properly, it replies with a WSP POST response, containing an M-Send.conf PDU, to the sender. This is, in principle, similar to sending a short message. However, the message is not delivered immediately to the receiver; first, a notification is sent using WAP PUSH (sent as an SMS) containing an M-Notification.ind message. The receiver may accept the message, postpone it, or discard it. The receiver uses a WSP GET request and response having the M-Retrieve.conf message to get the sent message from the MMSC. The received message is acknowledged with a WSP POST request containing either the M-NotifyResp.ind or M-Acknowledge.req PDU, depending on whether the message was delivered immediately or after a time delay. The MMSC confirms message delivery to the sender by using WAP PUSH, including an M-Delivery.ind message (WAP Forum, 2002a).

17.4.1 MMS Protocol Data Unit Structure

MMS PDUs are delivered inside WSP or HTTP messages. In the WSP header the content type is defined to be an application/nvd.wap.mms-message. The WSP body contains the MMS PDU (WAP Forum, 2002b).

The header of each MMS PDU starts with the same parameters, which should be given in this order:

- X-Mms-Message-Type, which defines the PDU, (e.g. M-Send.req);
- X-Mms-Transaction-ID, which identifies the M-Send.conf PDU and the corresponding M-Send.req;
- X-Mms-MMS-Version, which is currently 1.0.

Other parameters in the header depend on the PDU type. A complete description is given in WAP forum (2002b).

As mentioned, there are only two PDUs that have the body in addition to the header: M-Send.req and M-Retrive.conf. The content-type definition is the last part of the header and is either application/vnd.wap.multipart.related and application/vnd.multipart.mixed. The former content type is used if the content is going to be presented to the user. Thus, there must be presentation instructions included in the message. The instructions are typically defined in SMIL (synchronized multimedia integration language), which is based on XML (extensible markup language). The latter content type is used in messages, the content of which is not shown to the user, in which case no SMIL part is needed. An example MMS message is as follows:

```
Content-type: application/vnd.wap.mms-message
...
X-Mms-Message-Type: m-send-req
X-Mms-Transaction-ID: 0
X-Mms-Version: 1.0
Date: 1026453567
From: 10.10.0.0/TYPE=IPv4
To: +35840123456789/TYPE=PLMN
Subject: Test message
X-Mms-Expiry: 1026539967
X-Mms-Priority: High
Content-type: application/vnd.wap.multipart.related;
type="application/smil";
start="<0000>"
...
Content-type: application/smil
```

Content Presentation

The content is presented with the SMIL language, which supports the presentation of timed multimedia objects and animation. There are several SMIL profiles, which define a collection of modules to some application domain. The modules, in turn, define the instructions – the way in which multimedia objects are presented – or, more formally, they define the semantics and syntax for certain areas of functionality. Some defined modules concern layout, timing, synchronization, and animation. The presentation instructions and presentable objects are packaged in the same MMS PDU. The message content contains pointers [e.g. URLs (universal resource locators)] to the presented objects (Nokia, 2002).

SMIL support is still quite limited in the Series 60 Platform. Currently, it is possible to present slide shows using SMIL, but, for example, a video clips lasting several minutes are not yet supported. Each slide may have at most one image and one piece of text shown on the screen. Additionally, the message may contain played or spoken sound. The layout of the slide is presented using SMIL.

In SMIL, <par>and</par> tags are used to indicate objects that should be presented at the same time. Similarly, <seq>and</seq> tags are used to indicate the sequential order of presentation. The image and the text in one slide may be shown in parallel, but it is impossible to present two slides at the same time. The <body> tag is the sequence time container. Each slide is a parallel time container, which enables the parallel presentation of slide content.

The example below shows an example of the MMS PDU body. The layout specifies that the upper part of the screen is reserved for the image and the lower part for the text. There are two slides, shown in sequence: one from Rome and another from San Francisco. Both slides contain image and text. Additionally, a sound clip is played during both slides:

```
<smil>
  <head>
    <layout>
      <root-layout width="160" height="140"/>
      <region id="Image" width="160" height="100" left="0" top"0"/>
      <region if="Text" width="160" height"40" left="0" top="100"/>
    </layout>
  </head>

  <body>
    <par dur="60s">
      <img src="rome.gif" region="Image"/>
      <text src="rome_greetings.txt" region="Text"/>
      <audio src="rome.amr"/>
    </par>
    <par dur="60s">
      <img src="san_francisco.gif" region="Image"/>
      <text src="san_fancisco.txt" region="Text" />
      <audio src="san_francisco.amr" />
    </par>
  </body>
</smil>
```

Supported Contents

The smallest maximum size of the multimedia message supported by several device manufacturers is 30 kB (CMG *et al.*, 2002). The MMS Conformance document (CMG *et al.*, 2002) specifies the minimum set of content formats that devices must support. The Series 60 Platform supports additional formats too. In principle, the content may be of any MIME type. Thus, in addition to actual multimedia messages, it is possible to send Nokia ring tones, vCard and vCalendar objects, Symbian OS application and Java application installation packages, WML (wireless markup language) content, or third-party file formats.

Supported picture formats are JPEG with JFIF exchange format, GIF87a, GIF89a, and Windows BMP. In addition to these, Nokia 7650 supports animated GIF89a and png image formats, sent in JPEG. For audio, AMR format is supported. Nokia 7650 supports also WAV, with ADPCM, PCM, aLaw, μLaw, and GSM 6.10 encodings. For text us-ascii, utf-8, and utf-16 is supported. A complete list of supported contents can be found in (CMG *et al.*, 2002).

17.5 Smart Messaging

Smart Messaging is a Nokia proprietary concept for sending and receiving special messages, (e.g. ring tones, picture messages, operator logos, and business cards) over SMS. Infrared and Bluetooth are also supported as bearers. As the messaging is bearer-independent, applications can communicate with a wide variety of handsets. The smart messaging architecture is extensible and enables application developers to design and implement new kinds of message types. The most common smart message types are documented in the Smart Messaging Specification, is available at www.forum.nokia.com.

In order to support a new kind of smart message, the following things need to be done:

- implement a BIO information file;
- define the message format;
- implement the plug-in viewer component that plugs into the SMS Viewer; the same viewer is used for infrared and Bluetooth smart messages;
- implement functionality into the viewer for decoding and validating the message content;
- implement functionality for a special command, (e.g. 'Save to MyApp' or 'Play').

In the following, we use the same rock–paper–scissors (RPS) game example as in earlier chapters to demonstrate the implementation of a new smart message type. The smart message type is used for playing the game over SMS.

17.5.1 Bio Information File

A BIO (bearer-independent object) information file (BIF) is needed for every smart message type. A BIF ties together all the information that is needed for displaying, receiving, and sending a smart message. BIFs are resource files. They are located in the folder c:\system\bif\.

Below is the BIF file from our example; the meaning and use of the fields are explained further after the example:

```
NAME    RPSG

#include <biftool.rh>
#include "rpsviewer.loc"

RESOURCE BIO_INFO_FILE
    {
    message_type_uid     = 0x0FFFFFFB;
    message_parser_uid   = 0x0FFFFFFD;
    message_app_uid      = KUidUseDefaultApp;
    message_appctrl_uid = 0x0FFFFFFC;
    file_extension       = ".rps";
    description          = "Rock, paper & scissors";
    icons_filename       = "none";
    icon_zoom_levels     = {1};
    ids=
         {
         ID
            {
            type = EWap;
            confidence = ECertain;
            port = 9500;
            },
         ID
            {
            type = ENbs;
            confidence = ECertain;
            text = "RPS msg";
            },
         ID
            {
            type = EIana;
            confidence = ECertain;
            text = "text/rps"
            }
         };
    }
```

- The field `message_type_uid` is the UID for the smart message type. Received smart messages are stamped with the type when they are created in the Inbox by the framework. The UID should be obtained in the same way as normal application UIDs.
- The field `message_parser_uid` is used for some built-in smart message types. You should set it to NULL.
- The `message_app_uid` should have the value `KuidUseDefaultApp`. This means that the SMS Viewer application with a plug-in viewer is used. Another application cannot be used because of architecture reasons.
- The UID of the plug-in viewer DLL is specified in `message_appctrl_uid`.
- The `file_extension` field is optional. The smart messaging framework itself does not use it.
- The `description` field defines the Inbox header. This description can be localized in the standard way because the BIF is a resource file.

The values of the `icons_filename` and `icon_zoom_levels` fields do not have any effect in the Series 60 Platform.

The supported sending and receiving ways are defined by an array of ID blocks. Each ID block defines a possible way of receiving or sending the message. When sending, the first suitable block for the chosen bearer is used. This means that you can control the sending method for a bearer by placing the appropriate block first in the array.

The most common way of sending and receiving smart messages is to use the narrow band socket (NBS) port number in the user data header of the short message. This is done by using the type EWap and a port number. The EWap also makes possible the receiving of messages with an old legacy-style NBS text header. If the received message is to be recognized using textual tags, the ENbs type (Symbian OS enumeration for NBS) should be used. In this case, the text field is used instead of the port number. The NBS port number space is controlled by Nokia.

For IR and Bluetooth smart messaging, the IANA type is used. The MIME type must be defined in the text field. An Apparc recognizer with the same MIME type must exist so that the message will be sent correctly.

17.5.2 Defining the Message Data Format

The message data format needs to be defined. The protocol must allow us to transmit moves from the challenger to the opponent, and the opponent's response back to the challenger. We call the first message sent by the challenger a **challenge message**, and a response to that from the opponent a **response message**.

The challenge message contains the message identifier, challenger name, and his or her moves. The identifier is used to distinguish between different challenges. The player name is an optional field. The moves are encoded as integers so that 'rock' is represented as 0, 'paper' as 1, and 'scissors' as 2. When the opponent has received the message and played the game, the results will be sent back to the challenger. The response message contains the original ID, challenger name, and his or her moves and, in addition, to these it will contain the responder name and his or her moves. The message format is as shown in Figure 17.4.

Msg-id		Chal-name		Chal-moves		Resp-name		Resp-moves

Figure 17.4 The message format for the rock–paper–scissors game

Suppose we have two players, Joe and Tom. At first, Joe challenges Tom, and the message might have the following content: 1307 Joe 002. Tom accepts the challenge and plays the game. When the game is finished the response is sent back to Joe and the message content might now be: 2005 Joe 002 Tom 102.

17.5.3 Implementing the SMS Viewer Plug-in Component

The smart message is displayed using a viewer component, which plugs into the SMS Viewer application. The viewer control should be derived from `CMsgBioControl`, which is the base class for the plug-in viewers (see MsgBioControl.h). The DLL location is should be located in \system\libs\. The UID2 value for the DLL is 0x10005F5F. The UID3 you need to allocate yourself. The only exported function from the DLL is the `NewL()` function, which must have a certain signature required by the smart messaging framework:

```
/**
 * Two-phased constructor
 * @param aObserver Reference to the Bio control observer.
 * @param aSession Reference to Message Server session.
 * @param aId ID of the message.
 * @param aEditorOrViewerMode Flags the new Bio control
 * as editor or viewer.
 * @return The newly created object.
 */
IMPORT_C static CMsgBioControl* NewL(
MMsgBioControlObserver& aObserver,
```

```
CMsvSession* aSession, TMsvId aId,
TMsgBioMode aEditorOrViewerMode,
const TFileName& aFileName);
```

There are a number of abstract methods from `MMsgBioControl` that should be implemented. Many of these are related to layout management and command handling.

The `SetMenuCommandSetL()` method allows menu options to be added dynamically. The framework calls it every time that the Options menu is opened. In our example, we have the play option in the received message view. The `IsEditor()` method is used for asking whether the control was launched in the sending or received mode:

```
void CRpsBioControl::SetMenuCommandSetL(
  CEikMenuPane& aMenuPane)
     {
     const TInt KMenuPos = 1;     // The inserting position
                                  // of menu item.
     if (IsEditor())     // is sending view?
         {
         // no special commands when sending
         }
     else
         {
         // view of received msg
         AddMenuItemL(aMenuPane, R_RPS_PLAY,
            KMenuCommandPlay, KMenuPos);
         }
     }
```

The `HandleBioCommandL()` function is the command handler, which you should implement if you have added your own commands in `SetMenuCommandSetL()`:

```
TBool CRpsBioControl::HandleBioCommandL(TInt aCommand)
    {
    // Get the real command id.
    TInt commandId = aCommand -
      iBioControlObserver.FirstFreeCommand();
    if (commandId == KMenuCommandPlay)
        {
        PlayGameL();
        return ETrue;     // the command was handled
        }
    return EFalse;     // the command was not handled
    }
```

There are a number of specialized methods inherited from `MMsg-BioControl` which are used for scrolling by the framework. Some of them have a default implementation, which will do if your view is small enough.

CRichBio is a text control which is tailored for the plug-in viewer scrolling framework. It has the same look and feel as the built in smart message viewers, for example the Business Card smart message viewer. Note that the CRichBio cannot be fully constructed until there is a parent with a window. It means that CRpsBioControl should override SetContainerWindowL(), and construct the CRichBio there after calling SetContainerWindowL from CCoeControl:

```
void CRpsBioControl::SetContainerWindowL(
const CCoeControl& aContainer)
    {
    CCoeControl::SetContainerWindowL(aContainer);
    // constructed here because parent with window needed
    iViewer = CRichBio::NewL(this, ERichBioModeEditorBase);
    AddFieldsToViewerL(*iViewer);
    }
```

The CRichBio has a number of layout methods which also exist in MMsgBioControl or CCoeControl. Most of these calls can be passed on from the CRpsBioControl to the CRichBio instance, for instance the SetAndGetSizeL():

```
void CRpsBioControl::SetAndGetSizeL(TSize& aSize)
    {
    iViewer->SetAndGetSizeL(aSize);
    SetSizeWithoutNotification(aSize);
    }
```

The other calls which are passed straight from CRpsBioControl to the CRichBio instance are listed below:

```
CurrentLineRect()
VirtualHeight()
VirtualVisibleTop()
IsCursorLocation()
OfferKeyEventL()
```

When CRichBio is used, the IsFocusChangePossible() should be implemented like this:

```
TBool CRpsBioControl::IsFocusChangePossible(
    TMsgFocusDirection aDirection) const
    {
    If (aDirection == EMsgFocusUp)
        {
```

```
        return iViewer->IsCursorLocation(EMsgTop);
        }
    return EFalse;
    }
```

Finally, the `SizeChanged()` function should be implemented like this:

```
void CRpsBioControl::SizeChanged()
    {
    iViewer->SetExtent(Position(), iViewer->Size());
    }
```

Note that smart messages can be sent from the device by using either `SendUI` or `SendAs`. The API's have an argument for specifying the smart message type.

17.6 SendUI

SendUI offers a simple API to create and send messages, using different messaging protocols from the applications. SendUI is built on top of the Send As API, which is included in Symbian OS generic technology. The interface is implemented by the `CSendAs` class. SendUI is implemented by the `CSendAppUi` class, and it offers the whole message creation and sending process, including all UI components, such as the menu items shown in the applications and progress dialogs during sending. Both interfaces provide a means to query the sending capabilities of the message type. For example, if a document needs to be sent as an attachment, message types supporting the body text only can be ignored.

`CSendAs` provides a common interface to create a message, using any message type installed in the device. An application using Send As must implement the `MSendAsObserver` interface, which is called by `CSendAs` to gain more information on the required sending capabilities and to render the message in the correct format for sending. Most messaging protocols do not use this interface at all. One exception is fax MTM, which calls the interface to render the material in a suitable format for sending as a fax.

The available message types can be queried at runtime, or the message type can be defined in the application. Although the constructor of Send As does not take `TRequestStatus` as a parameter, and the call to the constructor returns immediately, some initialization is done asynchronously, and the created object is not ready to be used immediately. Thus, it is recommended that the Send As object not be created immediately before it is needed, but, for example, during application launch. The following example sends the file given as a parameter as an attachment to the recipient. The used message type is queried from the user by showing a list of the available message types supporting attachments:

```
#include <sendas.h>          // CSendAs
#include <baflutils.h>       // BaflUtils::CopyFile

void SendMessageL(const TDesC& aRecipientAddress, TFileName aMessage,
     MSendAsObserver& aObserver)
     {
     // create Send As object
     CSendAs* sendAs = CSendAs::NewL(aObserver);
     CleanupStack::PushL(sendAs);

     /* message needs to support attachments, no need to examine
        capabilities further, so 2nd argument EFalse, if 2nd
        argument is ETrue, MSendAsObserver::CapabilityOK()
        is called
     */
     sendAs->AddMtmCapabilityL(KUidMtmQuerySupportAttachments, EFalse);
     // show list of available MTMs to user, who selects preferred
     // message type
     const TInt selected(QueryMtmL(sendAs->AvailableMtms());


     // use selected mtm and create message
     sendAs->SetMtmL(selected);
     sendAs->CreateMessageL();

     // create attachment into entry and copy file into directory
     // returned by CreateAttachmentL
     TMsvId attachmentId;
     TFileName directory;
     sendAs->CreateAttchmentL(attachmentId, directory);
     RFs rfs;
     rfs.Connect();
     CleanupClosePushL(rfs);
     User::LeaveIfError(BaflUtils::CopyFile(rfs, aMessage, directory));
     CleanupStack::PopAndDestroy();      // rfs


     // add recipient to message
     sendAs->AddRecipientL(aRecipientAddress);
     // validate message, particularly recipient syntax for the chosen
     // message type is verified
     if (sendAs->ValidateMessage() != 0)
          {
          // message is not valid, handle error
          }

     // move message to outbox
     sendAs->SaveMessageL();

     CleanupStack::PopAndDestroy(); // sendAs
     }

TInt QueryMtmL(const CDesCArray& aMtmArray)
     {
```

```
    // show selection list to user
    ...
    return selectedMtm;
    }
```

In the above example, it was the responsibility of the user of the API to implement the required user interface. Series 60 includes SendUI, which is built on top of Send As and implements a user interface in conforming to the platform standards. Like `CSendAs`, the constructor of `CSendAppUi` returns immediately, but the created object is not usable immediately. For creating messages, SendUI has two methods,

```
CreateAndSendMessageL()
```

and

```
CreateAndSendMessagePopupQueryL()
```

SendUI offers two overloads of the `CreateAndSendMessageL()` method for creating messages. The first overload is called with a command ID of the SendUI menu item, and the other is called with the UID of the MTM of the wanted message type. `CreateAndSendMessage-PopupQueryL()` shows a query list box of available message types, and the user can select which message type is used. The previous example can be written by using SendUI in shorter and simpler form using the latter method:

```
#include <sendui.h>      // CSendAppUi

void SendMessageL(const TDesC& aRecipientAddress, TFileName aMessage)
    {
    // argument is 0, we're not using UI
    // if UI is used, argument indicates place of the Send menu item
    // in applications menu
    CSendAppUi* sendUi = CSendAppUi::NewLC(0);

    // create array containing attachment(s)
    CDesCArrayFlat* attachments = new (ELeave) CDesCArrayFlat(1);
    CleanupStack::PushL(attachments);
    attachments->AppendL(aMessage);

    // create array containing recipient(s)
    CDesCArrayFlat* recipients = new (ELeave) CDesCArrayFlat(1);
    CleanupStack::PushL(recipients);
    recipients->AppendL(aRecipientAddress);

    // create message, SendUI shows list from where message type is
    // selected, default values are used for some parameters
    // The 1st argument is title shown in popup query
    sendUi->CreateAndSendMessagePopupQueryL(_L("Send message"),
        TSendingCapabilities::ESupportsAttachments,
```

```
        NULL,      // no body text
        attachments,
        KNullUid,      // not a BIO message
        recipients);

    CleanupStack::PopAndDestroy(3);      // sendUi,
                                         // attachments, recipients
    }
```

SendUI also provides menu items to be used in the applications wishing to send information. The client application has to instantiate a `CSendAppUi` object, giving it an argument indicating the location of the menu items of SendUI in the application's menu. There must always be command IDs available after the identification given when creating SendUI; these are used for the menu items shown in the submenu. To show a menu item, `DisplaySendMenuItemL()` is called in the menu initialization method, when the menu is opened:

```
#include <sendnorm.rsg>       // UI item constants
void CExampleAppUi::DynInitMenuPaneL(TInt aMenuId,
  CEikMenuPane*aMenuPane)
    {
    switch (aMenuId)
        {
        case R_EXAMPLEAPP_MENU:
            {
            TInt pos;
            // find position for menu
            aMenuPane->ItemAndPos(ECmdSendUi, pos);
            /* insert menu item, 3rd argument defines
               requirements, if any, using
               TSendingCapabilities object
            */
            iSendAppUi->DisplaySendMenuItemL(*aMenuPane, pos,
              TSendingCapabilities());
            break;
            }
        ...
        }
    ...
    }
```

When the SendUI submenu is opened, `DisplaySendCascadeMenuL()` should be called in the menu initialization method:

```
void CExampleAppUi::DynInitMenuPaneL(TInt aMenuId,
  CEikMenuPane* aMenuPane)
    {
    switch (aMenuId)
        {
        ...
        case R_SENDUI_MENU:      // Defined is sendnorm.rsg
            // show submenu, 2nd argument is an array of MTMs
            // which are not shown in list
```

```
            iSendAppUi->DisplaySendCascadeMenuL(*aMenuPane, NULL);
            break;
        }
    }
```

When a command is selected in the SendUI submenu, the command must be handled and the message created by using `CreateAndSendMessageL()`, giving it the selected menu item ID as an argument. The message is then created and opened embedded in the editor, if the message type uses the editor to create messages. When the editor is closed, control returns to the application which launched it.

17.7 Summary

The messaging architecture of the Series 60 Platform supports e-mail, short messages, multimedia messages, and smart messages. It is possible to implement new message-type (MTMs) modules to support other message types and protocols. Messages are accessed through the Message server, which provides an index of all messages, regardless of their type and other properties. Through the index it is possible to access a message header, but further manipulation typically requires message-type-specific plug-in modules.

Message creation depends on the message type, but most message types contain a header, a body, and attachments. The header contains, at least, the receiver's address and the subject. The body is most often written in the rich-text format. Attachments can be any binary data. In this chapter we have described how to use the services of the messaging architecture to create, send, and receive e-mail messages, short messages, and multimedia messages.

18

Connectivity

A broad definition for **connectivity** is the capability to connect and share information with the external world. Most household devices are still closed systems: stereos, TVs, and even standard phones do have limited data-sharing features. However, the tendency in modern society is to be more and more mobile-centric and information-centric.

With the everincreasing demand to take notes, to get and send information instantly, while on the move around, the actual value of smartphones will be in the data they can store, not in the hardware itself. The data can be either personal or related to business, but given the precious information that smartphones contain, that data must be accessible at any time and in any place. Connectivity and synchronization make sure that the information is always available, secure, safely backed up, up to date, and can be shared with other people, devices, and services.

Communication protocols enable connectivity, and connectivity enables a connection establishment and data change between Symbian OS phones and non-Symbian OS devices. There are two types of connectivity: **local** and **remote**. Local connectivity is short-range, device-to-device connectivity (e.g. to share business cards, or to allow direct desktop synchronization). Remote connectivity relates to data sharing and synchronization over the telephone network and/or the Internet between a smartphone and a server.

SyncML is the industry-approved standard for remote connectivity. This chapter introduces Symbian Connect (a desktop interface from Symbian OS), SyncML, and **synchronization**, and **device management** in general. Specific emphasis is placed on connectivity capabilities used by Series 60 Platform.

18.1 Symbian Connect

Symbian Connect is a component that provides features for desktop data synchronization and connectivity. For end users, Symbian Connect is a standard Windows application, giving direct access to device capabilities and data repositories when a connection is established. For developers, a Symbian Connect software development kit (SDK) provides a powerful tool for utilizing the connectivity services with C++ or Visual Basic and building their own plug-in engines and applications for synchronization, conversion, and task and file management.

The first Series 60 Connectivity release supports, in principle, all Windows versions from Windows 95 onwards, but full interoperability is guaranteed with:

- Windows 98 SE release;
- Windows NT service pack 3; if a Bluetooth connection is used, service pack 5.0 is required;
- Windows 2000;
- Windows ME.

18.1.1 Symbian Connect Features

Symbian Connect provides a large set of functionality, as data operations should be accessed from the desktop side as well as from the smartphone. The feature list can be summarized as follows:

- Backup and restore:
  - backup task driver: this provides the ability to back up a connected Symbian OS phone to the PC. Backed up files can later be restored by using the archive view plug-in.
  - archive application: list and restore operation files are archived by means of the backup facility in Symbian Connect.
- Application installation [SIS files (Symbian installation system files) can contain Symbian OS executables, Java Midlets, audio, etc.].
- Sync engine: this provides underlying sync technology for sync task drivers.
- Contacts and agenda synchronization:
  - Agenda sync task driver: this provides agenda synchronization functionality for appointments, events, to-do lists, and

anniversaries, between Symbian OS Agenda and PC-based PIM (personal information management) applications (Microsoft Schedule+ 7.*x*, Microsoft Outlook 9*x* and 2000, Lotus Organizer 97, 97GS, 5.0 and 6.0).

  - Contacts sync task driver: this provides contact synchronization between Symbian OS Contacts and PC-based PIM applications (Microsoft Schedule+ 7.*x*, Microsoft Outlook 9*x* and 2000, Lotus Organizer 97, 97GS, 5.0 and 6.0).

- Email synchronization:
  - E-mail sync task driver: this provides e-mail synchronization functionality for e-mails between the Symbian OS Message Center and PC-based mail applications (Microsoft Outlook 9*x* and Microsoft Exchange client).

- Machine management:
  - Symbian Connect recognizes a smartphone identifier [ID; such as the International Mobile Equipment Identity (IMEI) code] and can seamlessly manage several devices from a single desktop. Backup data for each machine are stored in separate directory trees, with synchronization details and other settings also being held separately. The appropriate information is automatically invoked on connection; no user input is required, as all the procedures for multiple machine management are totally transparent.

- Capabilities manager: this is an engine module, determining the capabilities of the attached machine.
- Framework user interface: this is a completely new user interface for Symbian Connect, consisting of a frame window that contains the view plug-in functionality.
- View plug-ins: generally, these consist of an engine and a user interface (UI) component, which plug into the framework user interface.
- Control panel: this gives the user access to all the Symbian Connect settings from one place. The individual control panel items are applets that plug into the control panel. These are: connection, log settings, CopyAnywhere, machine manager, and file types.
- Error logger: this is an engine that allows for logging and viewing of errors and information messages within Symbian Connect.
- Task scheduler: this carries out a number of regularly scheduled tasks, including sync and backup. A 'Unify' allows a selection of tasks to be run at a single click or cradle button press.

- Task drivers: these consist of an engine and property pages, and plug into the task scheduler component.
- File operations (supported by Symbian OS but not exposed in Series 60 by default) consist of:
  - onboard conversion;
  - file transfer;
  - Explorer shell extension, allowing one to browse and to move, copy, convert, and delete files on the Symbian OS phone via Windows Explorer;
  - CopyAnywhere, an engine that synchronizes the PC and the phone clipboards.

It is important to note that some Series 60 smartphones (such as Nokia 7650) are designed not to expose their file structure but automatically to handle all supported file types, such as images, music, or contact cards. The reason for this is to minimize the burden of learning and managing complex operations and to let the smartphone do everything automatically. Therefore, document-related operations (i.e. file management, remote print, and clipboard) are not necessarily supported as standard features. However, interfaces should be open for any third-party developer, so the application development is still enabled.

Backup and Restore

A highly desired task provided by Symbian Connect is to back up data residing on the smartphone to a desktop. Users can backup data on demand, automatically each time connected, or at fixed intervals (daily, weekly, or monthly). There are many options, which can be used in conjunction with the backup and restore operations. The following summarizes the provided functionality:

- provision of full or incremental backup or restore. 'Incremental' means that the backup or restore operation is performed only for files that have been changed since the previous operation;
- archiving of updated and deleted files;
- specification of a date range and file extension filters to backup or restore;
- formatting of drives before full restore;
- error detection and handling capabilities;
- automatic closing of applications before backup or restore, and restarting afterwards.

18.1.2 Connectivity Architecture

Symbian Connect has a layered design to allow direct access and the addition of new connectivity and synchronization services to the Symbian OS phone. The connectivity architecture on the PC side is shown in Figure 18.1.

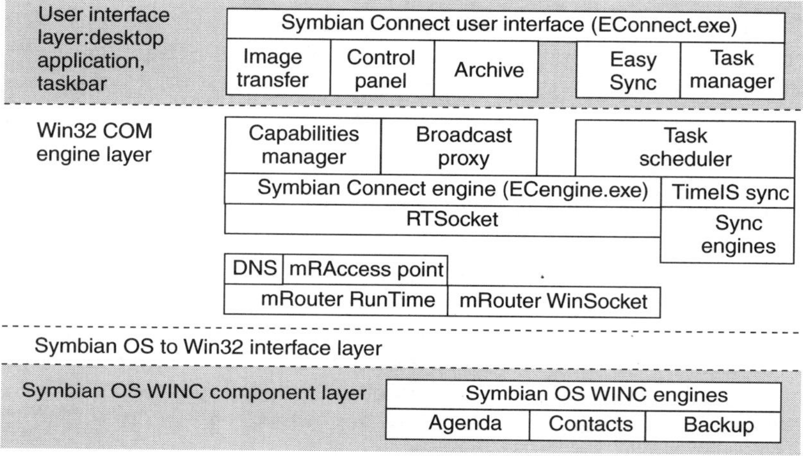

Figure 18.1 High-level connectivity architecture. Note: DNS, domain name service

Symbian Connect architecture can be observed in different layers. The topmost layer is the UI layer, the interface visible to the end-user. It is highly customizable by the device manufacturer and has some plug-in framework for developers.

The Win32 COM layer provides all the engines used by connection management, conversion, and synchronization and backup. The capabilities manager provides all the information on device parameters. Win32 COM engines are common to all Symbian OS phones; hence, it guarantees interoperability between Series 60 smartphones and other platforms. A step towards a non-Windows bind solution is to utilize a TCP/IP (transmission control protocol/Internet protocol) connection with DNS (domain name service) and RAS (remote access server).

Symbian Connect is almost as much a Symbian application as it is a Windows application; a remarkable part of its internal workings is layered on top of the Symbian WINC variant. The WINC component layer provides a utilization of the non-UI parts of Symbian OS, which is available for PCs running an operating system compatible with Microsoft's Win32 application programming interface (API). WINC allows a PC to use exactly the same application engines that run on a smartphone. A PC running Symbian Connect is able to use the facilities offered by two operating systems simultaneously. When a PC is connected to a standalone smartphone, two Symbian machines are actually connected together. The hard work of getting the data from one operating system to another is performed inside Symbian OS to the Win32 components. The interfaces are provided, for instance, to the Symbian Agenda and Contacts models.

Access to the Symbian Connect Engine

PC application can access the Symbian engine to control the file system directly. Once the Symbian Connect engine has been invoked, it can be used in the following operations:

- connecting to a Series 60 smartphone;
- starting and stopping a program on an attached smartphone;
- acquiring information about an attached smartphone;
- synchronizing the machine time on both devices;
- file operations (optional support):
  - reading information about the file systems of both devices, when connected,
  - copying and moving files,
  - renaming and deleting files,
  - changing the attributes of files,
  - parsing filenames and wildcards.

Comparable methods are provided for C++ and Visual Basic in a straightforward and intuitive manner. For example, any program can be started on a smartphone by calling `StartProgram()` with the name and path of the program as a parameter.

Most of the Symbian Connect engine calls are available in both synchronous and asynchronous forms, and multiple commands can be sent to the engine, with each command being held in a queue, until it can be processed.

Backup and Restore Engines

Symbian Connect has separate engines for backing up and restoring data. As shown in the architecture diagram, there is no backup or restore application view in Symbian Connect, but backup is performed by setting backup tasks in Task Manager, and the restore operation is done with an Archive application.

Backup and restore functions are used to detect what machine is connected, to retrieve and use device-specific information, to set backup and archive directories, to analyze file differences, and to add specific files to the list of items to be backed up. Once everything is in place, a `Backup()` function performs all the operations requested.

Backup covers contacts, calendar, documents, and installed components. Device and application settings are also backed up. For data protection and copyright reasons, certain files (such as MIDIs, JARs (Java archives), and Waves) are not saved to backup archive.

Access to Synchronization Engines

The WINC engines are used when synchronization is performed. Data from the Series 60 smartphone enters from the bottom of the diagram (Figure 18.1), via Symbian OS to Win32 interfaces, to the synchronization engines, which convert or synchronize the data and pass the final data back to the relevant PC application. The process works in a similar way in reverse order for data that are to be synchronized or converted from a PC to a smartphone.

The components at the engine layer are UI-independent, so that products with different user interfaces can use the same engine components. Within the Win32 COM layer, there are the converter and synchronizer engines, which are used by the individual converters and synchronizers. These modules are supplied either with the product (by Symbian or the device manufacturer) or added at later stage.

At the Symbian OS to Win32 Interface layer, a series of COM APIs provide access to the different Symbian OS application engines. These interfaces are published so that third parties can implement new versions of applications for different synchronization or conversion services or even operating systems.

Connection Management

The Connection manager supervises connections between a PC (running Symbian Connect) and a Symbian OS smartphone, and it includes both PC-side and phone-side components. The evolution of connection management is described in the following.

Symbian Connect 6.0 This provides the following:

- Abstraction of the hardware connection layer away from the protocol layer; the Connection manager works in the same way over any physical layer it supports. Connections over physical serial links, infrared links, Bluetooth, and USB (universal serial bus) are supported. The minimum configuration for RS232 serial communications is a three-wire mode (using GnD, TxD, and RxD); however, RTS/CTS and DTR/DSR are supported when available.

- Support for multiple client applications on the PC: this is achieved by ensuring that all data transfers are atomic operations. The Connection manager can multiplex or demultiplex data to or from a Symbian OS-side custom server; thus a single instance of a custom server will support a number of client applications on the PC.
- Support for 'unify' functionality: that is, it allows a selection of tasks (synchronization, backup, any user-defined tasks) to be started with a single click or cradle button press. Unify task can be triggered equally from the smartphone side or the PC side.
- Detect ion of unexpected disconnection of a phone and broadcast of the disconnection to all clients on the PC: this ensures that all custom servers open on the phone are shutdown. The broadcast of this information will occur within two seconds of detection.
- No restriction on the amount of data that can be sent to or retrieved from the phone.

Symbian Connect 6.1 This is the first release with Series 60 connectivity:

- The PLP (Psion link protocol) connection protocol is replaced with mRouter, a protocol enabling connection management over a TCP/IP link.
- Access to connection parameters is through a connection icon in the Windows System Tray.
- Windows ME support is provided.
- Bluetooth with Serial Port Profile (serial link emulation) is supported.
- Architecture to add new connectivity protocols is available.

Symbian Connect 7.0 This offers:

- USB support,
- Bluetooth LANP (local area network access profile) support.

mRouter Connection Protocol

The biggest change from Symbian Connect 6.0 release to the Series 60 Platform is the replacement of the PLP (Psion link protocol) with the mRouter protocol.

The PLP was designed for serial cable and infrared use and was somewhat complex in its architecture, APIs were not open, and the protocol was for the most part not documented. In consequence, the only desktop platform supported was Windows, and there was never full connectivity implementation for Macintosh, Unix, or Sun. Some pioneers did reverse-engineer the PLP, but it was obvious that true interplatform solutions, beyond the simple file get and put operations, could not really exist before replacement of the PLP.

Now, connectivity operations are not limited to the Windows or desktop end; they can easily be channeled to a network. The connection manager can now, for example, enable Internet browsing over Bluetooth or SyncML synchronization channeled directly to the server. A scenario where Symbian Connect applications can entirely reside somewhere on the network is now possible. Figure 18.2 illustrates the mRouter connection.

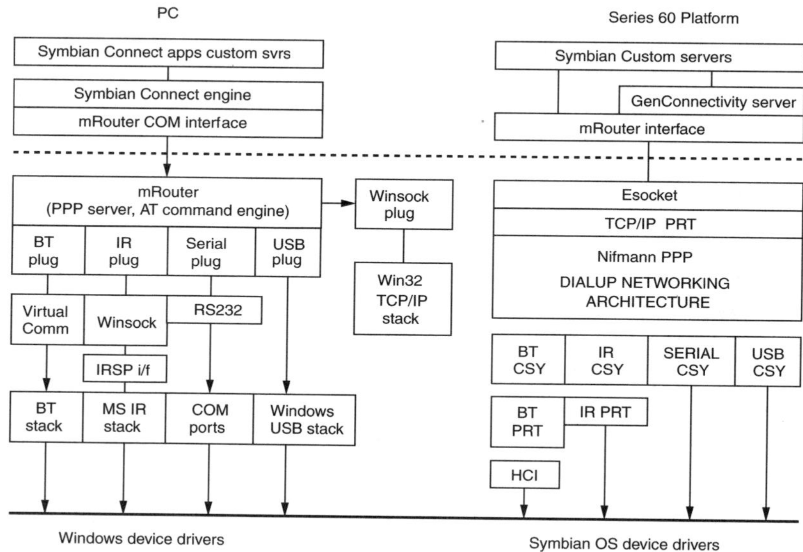

Figure 18.2 Connection over mRouter; for abbreviations, see the glossary

The mRouter protocol consists of components both on the Windows and on the Symbian OS side. Connection from the Series 60 side is established with Esocket, TCP/IP, and dialup networking architecture [Nifman (network interface manager) and PPP (point-to-point protocol)]. Access from the connectivity server to the utilization of infrared, Bluetooth, Serial, or USB is provided through the mRouter interface. Dual homing is used as a connection method. On the Windows side, similar components are provided when connecting the Symbian Connect engine to physical connection layers.

For end-users, the change of connection protocol is transparent, although the connection should be more reliable and will enable trustworthy connection over infrared, Bluetooth, and USB. For the developer, it opens a new world, with standard TCP/IP connections and easy-to-use APIs.

The mRouter protocol provides the following APIs:

- connection/disconnection; permitting the developer to configure the connection, query the current configuration, and start or stop the connection;
- connection monitoring; with a callback function, information on connection progress is provided.

The following parameters can be managed by the protocol:

- connection type [data routed via GPRS/CSD (General Packet Radio Service/circuit-switched data) or mRouter];
- bearer [RS232C, IrComm, RFComm (Bluetooth), USB, Emulator, or unknown];
- Bluetooth profile [SPP (serial port profile), LAP (local area network access profile), or DUNP (dialup networking profile)];
- status of current connection (connecting, connected, disconnecting, disconnected).

18.2 Synchronization

"Data Synchronization is the process of making two sets of data identical." (SyncML initiative 2002)

In the mobile world, prediction of the future is like shooting at a moving target, with a lot of speculation, new trends, and prophecies. Still, there are definite trends that can be identified. A mobile user is a multiple-system user, relying on multiple devices, applications, and data, and shuttling between applications and systems.

Typically, smartphone users create and edit content information through various PIM applications, or any application storing data on the device. Similar information is also kept on their PC or network repository. Rather than force the user to maintain two independent sets of data, synchronization allows the same data to be seamlessly maintained on both devices. Changes at each end must be kept synchronized.

With the everincreasing dependence on constantly updated data, the synchronizing of data becomes a top-priority task. Synchronizing the data between systems is a key element in the new mobile computing reality. Without proper synchronization, disparities in the application data damage user efficiency and reduce users' confidence in the smartphone.

18.2.1 Business Cards, Calendar Entries, and Mail

The assumption built in the concept of synchronization is that information can be categorized into data types, and that the data types are comparable across different applications and different devices. Standardization is the key element as far as interoperability is concerned. Wherever possible, Symbian OS uses the industry standards for personal data interchange (PDI), originally developed by the Versit Consortium, and now maintained by the Internet Mail Consortium (IMC). The two formats used for synchronization are the vCard electronic business card format, and vCalendar, which is an electronic calendar and scheduling format. Version 2.1 of the vCard standard and version 1.0 of vCalendar are supported.

Symbian has defined a vMail format. The full specification is published in the Connectivity SDK (go to www.symbian.com). vMail data can either be mail messages or be memos, including subject and date information. The vMail data may include separate headers and body text, and status and folder information. Although the vMail definition is not maintained or endorsed by IMC, it is based on similar principles of stream and property formatting used within the existing vCalendar and vCard standards. The successor of vCalendar, the iCalendar standard, is not yet widely approved by the smartphone synchronization community, but it can bring certain benefits in implementation (e.g. fully working repeat calendar appointments).

As well as e-mails, the synchronizer engine built in Symbian OS Connect utilizes the following major types of data from the standard Symbian OS engines:

- appointments, which are the basic data type in most calendar systems; the essential data in each entry consist of a start and end time, and some descriptive text;
- events, which are similar to appointments but do not require a start and end time;
- anniversaries, which are, in principle, similar to the events but which are repeated annually;
- tasks, or to-do lists, which are jobs to be done; these can have a deadline and priority information associated with each item;
- contacts, storing names, addresses, phone numbers, and so on.

An example of vCard format is shown below:

BEGIN:VCARD vCard header, a starting sequence of a single vCard record; one file can consist of several vCards

VERSION:2.1 vCard version number; describes the vCard specification version in which the vCard is implemented

REV:20020625T115241Z this property specifies the combination of the calendar date and time of day of the last update to the vCard object

N:Smith;Ira;;; name property; can consist different name attributes (e.g. second name, title, marital status, etc.).

ADR;HOME:;;37 Fulham Park Park Gardens;London;;SW6 4JX;UK address

ORG:MedTwo Inc; organisation, employer

TEL;VOICE:+44 2075632952 telephone number; different tags can be used for mobile, work, and home numbers

EMAIL;INTERNET:irasmith@medtwo.co.uk e-mail address

PHOTO;ENCODING=BASE64: encoded image starts

. . . image data removed

END:VCARD vCard ends

Figure 18.3 illustrates how the same vCard object is displayed in a Series 60 phonebook. Data from the vCard object is first parsed from the Unicode text format and converted to the Contacts database. Same contacts databases can be used in different language versions; only field names will vary.

18.2.2 Synchronization Engine

Symbian Connect provides a default synchronization engine (ISynch, provided by Time Information Services) as well as default tasks for the synchronization. Many users will not need to change these predefined tasks. Others will find that modifications are easy to make, either using the Settings Wizard or directly, making alterations to the task properties.

Although setting up a synchronization is easy to do, specifying what information is to be synchronized is a critical part of the process; it is not really possible to synchronize data from two different applications without prior knowledge of where to find compatible data types. ISynch provides an interface that makes accessing data simple. It provides a `ReadSetup()` function that returns an IStream interface pointer, providing both read and write access to the structures within

Figure 18.3 vCard object displayed in Series 60 Platform phonebook

the file. Elements can be used in the control file to identify a pair of dynamically loadable ISynchApp interfaces.

ISynch provides a single function, `Synchronize()`, that is called to perform the synchronization operation; the function can only be used within Symbian Connect. Since, in some cases, the synchronization operation can take time, ISynch provides an IProgress interface. It is used by the ISynchApp dynamic link libraries (DLLs) for each application to keep the user informed about the progress status and to report if any error occurs.

Each synchronization starts by processing the control file. During this phase, the two ISyncApp application-specific interfaces are instantiated and initialized. The implemented virtual functions do most of the important synchronization work.

Each ISynchApp has an `Open()` function, which is called to establish and validate its supported database. Following this, the `CheckSource()` function is used to verify the contents. The bulk of the synchronization is then performed by the two functions `ReadSource()` and `WriteSource()`. These functions convert the individual entries found in the database to and from the supported data types (vCard, vCalendar, or vMail).

18.3 SyncML

SyncML is an initiative for device synchronization standardization and will be supported by most of the synchronization service providers and handheld, smartphone, and traditional mass-market phone manufacturers. The SyncML initiative was formed in Paris, in February 2000, and the sponsor members are (September 2002) Ericsson, IBM, Lotus Development Corporation, Motorola, Nokia, Openwave Systems Inc., Matsushita Communication Industrial (Panasonic), Starfish Software, and Symbian. The sponsor members primarily fund the forum, and, on an equal basis, manage and steer the standard specification and interoperability processes. Currently, there are about 50 client and 30 server SyncML compliant products.

In June 2002, the announcement of a new global organization – the Open Mobile Alliance (OMA) – was made. The alliance combines the forces of the WAP Forum, the Open Mobile Architecture Initiative, the SyncML initiative, Wireless Village initiative, the Location Interoperability Forum (LIF), and the MMS Interoperability Group (MMS-IOP).

SyncML enables synchronization not only between a Symbian OS device and a PC but also between any device or service using SyncML. SyncML can bind to several bearer protocols. The supported bearers include HTTP (hypertext transfer protocol), WSP [wireless session protocol], OBEX [object exchange; IrDA (infrared data association), Bluetooth, and other local connectivity interfaces], pure TCP/IP networks, and proprietary wireless communications protocols. The SyncML protocol architecture is shown in Figure 18.4. The first Series 60 SyncML version supports only a HTTP bearer. A fundamental of this figure is that equal synchronization layers are presented on both the smartphone side and the server side, and physical connection of different bearers can be used.

SyncML supports arbitrary networked data. In order to guarantee interoperability, the protocol describes how common data formats are represented and transferred over the network. The introduction and plug-in capability of the new formats will ensure extensibility. Developers are also allowed to use their own experimental protocol primitives. The common data characteristics that a protocol should support include (SyncML Initiative, 2002):

- PIM data formats, such as vCard for contact information, vCalendar and iCalendar for calendar, to-do lists, and journal information;
- objects such as e-mail and Internet news;
- relational data;
- XML (extensible markup language) and HTML (hypertext markup language) documents;
- binary data, binary objects.

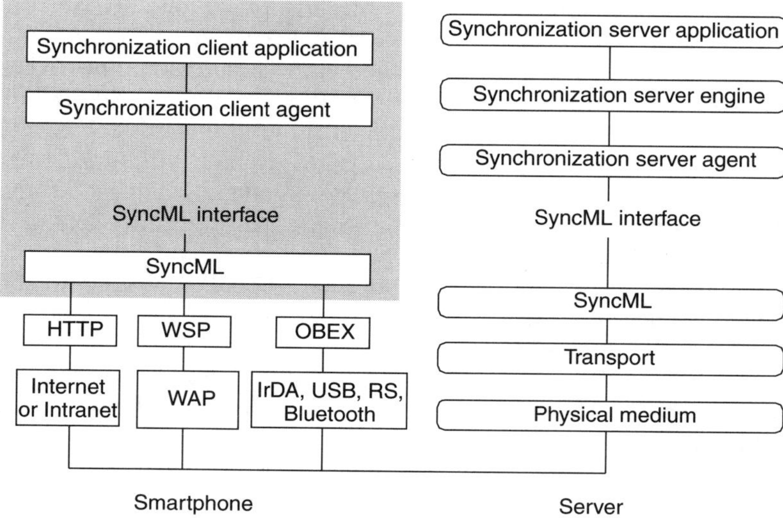

Figure 18.4 High-level SyncML protocol architecture. Note: HTTP, hypertext transfer protocol; IrDA, infrared data association; OBEX, object exchange; RS, RS-232; USB, universal serial bus; WAP, wireless application protocol; WSP, wireless session protocol

By default, the Series 60 Platform supports vCard and vCalendar synchronization only, but in future the plug-in architecture will permit extension to the propriety data types. SyncML contains two protocols: one for device management and one for synchronization. There are several ways in which the synchronization operation can be performed (SyncML Initiative, 2002):

- two-way sync: a standard synchronization type, where a client and a server exchange information about modified data;
- slow sync: a form of two-way synchronization, where a client sends all items in the database to the server and the server does the synchronization analysis on a field-by-field basis; usually, this operation is performed only when synchronizing databases for the first time;
- one-way synchronization from the client only: the client sends its modifications to the server, but the server does not send its modifications back;
- refresh synchronization from the client only: a client sends all its data from a database to the server, which replaces any existing data with that sent by the client;
- one-way synchronization from the server only: the client gets all modifications from the server but the client does not send its modification to the server;

- refresh synchronization from the server only: the server replaces all the data in the client database;
- server alerted synchronization: the server alerts the client to start a specific type of a synchronization with it; this can be done via smart or BIO (bearer-independent object) messaging or by using the WAP PUSH operation.

18.3.1 Device Management

Device management is a term used for the technology enabling third parties to carry out remote configuration of mobile devices on behalf of the end-user. Typical third parties are wireless operators, service providers, and IT (information technology) personnel inside the corporation. Device management enables an external party to install software, set parameters remotely, or take care of troubleshooting of applications. The device management protocol (DMP) is part of the SyncML standard and supports device configuration (read or modify device parameters, installed software, hardware configuration), software maintenance, and diagnostics.

Initially, device management is implemented in Series 60 through BIO messaging [i.e. OTA (Over The Air) smartmessaging standard for the transmission and reception of application-related information in a wireless communications system), but the next version is to support true SyncML-based device management].

All the managed objects in a device supporting SyncML device management make up a management tree. Each object in the tree is identified by a unique identifier uniform resource (URI). The list of properties consists of the following elements:

- Access control list (ACL);(1) allowing hierarchical access to access rights based on the device management server IDs;
- Format; specifying how the object values should be interpreted;
- Name; the name of the object in the tree;
- Size; the object size in bytes (mandatory only for leaf objects);
- Title; the name of the object in text format (optional property);
- Tstamp; giving the time stamp, data, and time of the last change (optional property);
- Type; the MIME (multipurpose Internet mail extension) type of the object (optional for interior objects);
- VerNo; the version number, automatically incremented at each modification (optional).

(1) The use of the abbreviation ACL here is not to be confused with 'asynchronous connectionless'.

Properties may be read and only a few of them (ACL, Name, and Title) may be replaced by the user. The device automatically updates the values of the other properties.

The importance of device management will grow significantly in the near future as mass volumes of smartphones become available. The effort individuals or corporate IT personnel are currently using for maintaining and configuring desktop PCs will be moving towards remotely operated device management. Operators or corporations need to 'control' a large number of customers who have varying computing skills. It is fundamental that smartphones be accessed and supported remotely with as limited end-user intervention as possible. However, all device management operations should be safe and protected from illegal access to smartphone data.

18.4 Summary

The Series 60 Platform provides powerful connectivity tools for easy and straightforward data-sharing and synchronization with the external world. Predefined functionality ensures safe backup and standard synchronization. The flexible plug-in architecture and sound connection management and synchronization permit third-party developers to extend the connectivity experience – and to make future innovations 'true'. The Series 60 connectivity solution is a stepping-stone towards transition from PC-centric data storage and synchronization operations to where most is performed at the multiple mobile Internet locations. A smartphone user's personal data must be always available, always connectable, always safe, and always synchronized.

Part 4

Programming in Java

19

Programming in Java for Smartphones

In addition to smartphones, ordinary mobile phones are starting to support Java. This chapter defines the principles of Java 2 Micro Edition, CLDC (connected limited device configuration) configuration, and MIDP (mobile information device profile) commonly used in mobile phones. In the next chapter, we go through the user interface (UI) programming framework of the MIDP application. The Series 60 Platform supports MIDP and provides some extensions for better device control and performance.

Java is an object-oriented programming language developed by Sun Microsystems. It is designed to work on many different platforms. This is realized by execution of the same compiled source code, called **byte code**, in a virtual machine developed for a specific platform. In the Series 60 Platform, K virtual machine (KVM) is used. Originally, Java was developed for server and desktop computers, but it has evolved to survive in a low-resource mobile device environment (Allin, 2001). The strengths of the Java language in mobile device environment are portability, possibility to secure execution environment, and over the air deliverability to user devices.

19.1 Java 2 Micro Edition

Java 2 Micro Edition (J2ME) is the youngest child of the Java 2 family. The purpose of J2ME is to provide a standard environment for the application developer to create applications for mobile devices and set-top boxes. Since mobile devices have limited memory and execution resources, J2ME is designed to minimize resource usage compared with other Java 2 editions. Still, J2ME has managed to conserve the major language features and the most popular software interfaces of the other Java 2 editions (go to java.sun.com/j2me/docs).

J2ME is easy to learn for first-time programmers, and it can easily be adopted by the experienced Java user.

Java is already known for application and service development for desktop and server devices. The Java 2 Platform is separated into three editions to enable differentiation of the large variation in industry needs. Java 2 Standard Edition and Java 2 Enterprise Edition serve the traditional computer markets. For the emerging wireless market, Sun has added a third edition, the J2ME. Figure 19.1 shows the architecture of the Java 2 platform.

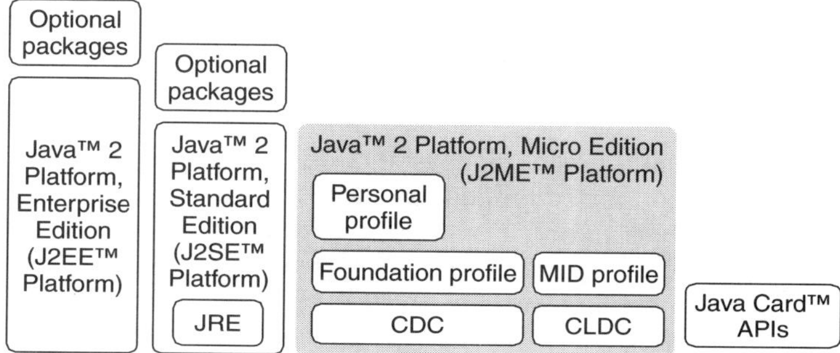

Figure 19.1 Java 2 Architecture. Note: API, application programming interface; CDC, connected device configuration; CLDC, connected limited device configuration; JRE, Java runtime environment

The editions of the Java 2 platform are:

- Java 2 Platform, Standard Edition (J2SE), for the desktop computer market;
- Java 2 Platform, Enterprise Edition (J2EE), for enterprises with scalable server solutions;
- Java 2 Platform, Micro Edition (J2ME), for the consumer market.

All editions of the Java 2 Platform can be further adjusted to target use with optional packages. Optional packages contain software components and frameworks that enable application programmers to create high-functionality applications without inventing the wheel over and over again. The J2ME consists of configurations and profiles to customize the runtime environment to the needs of the large variety of device types and different usage models.

A **configuration** is the base runtime environment built from common application programming interfaces (APIs) and a **virtual machine** (VM) for a device category. From the application programmer's point of view, Java binary code is always executed in the virtual processor environment, called a virtual machine. The configuration defines the basic capabilities of the underlying VM and device such as memory resources and network connection. The configuration also describes the core set of software APIs. The next section, on connection limited device configuration, gives an example of the an API of the J2ME configuration.

A configuration is extended by a profile for the vertical market. The profile specifies a set of APIs that address this market. Typically, a profile defines APIs for the UI framework and persistent data storage. The section on mobile information device profiles (Section 19.3) presents an example of the J2ME profile used in the Series 60 Platform. An application developed for the Java 2 Platform will run on any device that complies with the same configuration and profile.

The J2ME defines two device categories, specified in two different configurations: the connected device configuration (CDC) and the connected limited device configuration (CLDC). Figure 19.2 shows the components of J2ME architecture.

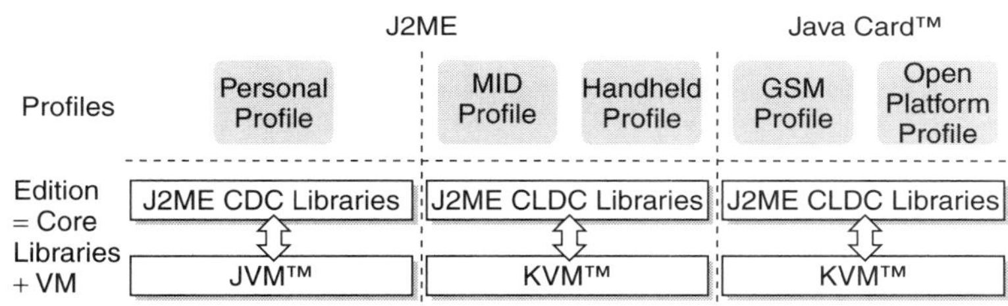

Figure 19.2 Java 2 Micro Edition (J2ME) architecture. Note: CDC, connected device configuration; CLDC, connected limited device configuration; GSM, global system for mobile; JVM, Java Virtual Machine; KVM, K Virtual Machine; MID, mobile information device; VM, Virtual Machine

The CDC is for high-end consumer devices, such as set-top boxes and communicator phones. These devices have a large range of UI capabilities; the typical memory available to a Java application is around 2–4 Mbytes and a high-bandwidth network connection. In this category, 'high-bandwidth network' is quite a broad term, because the bandwidth of a communicator device can be as low as 15–50 Kbit s^{-1}, but a set-top box might provide 2 Mbytes of bandwidth.

The CLDC is for low-end consumer devices, such as mobile phones, two-way pagers, and personal organizers. The user interface is very simple; the minimum memory budget is typically around 128 Kbytes, and only a low-bandwidth network connection is available.

As an extension to the CLDC, the Mobile Information Device Profile (MIDP) defines the APIs for the vertical mobile phone market.

Smartphones using the Series 60 Platform are equipped with the CLDC and the MIDP extension. The configuration and profile will be discussed in detail in the next section.

19.2 Connected Limited Device Configuration

The CLDC specifies the VM and a set of s APIs, which are grouped by functionalities, provided by the API classes. In Java, these class groups are called **packages**. The goal is to support a wide range of devices with limited memory, display, and processing resources. CLDC provides a simple **sandbox security model** to protect the device from hostile applications. The CLDC requires a host operating system, which provides basic multitasking, data storage, communication, and application management functionalities.

19.2.1 Requirements

CLDC defines a number of hardware requirements for the target device. The requirements are defined loosely to keep the door open for a broad range of devices. The memory requirements are:

- at least 128 kbytes of non-volatile memory (typically flash or ROM) for running the VM and the CLDC libraries;
- at least 32 kbytes of volatile memory (typically RAM) during application runtime.

Note: there is no requirement that a device should support dynamic content or application downloading. However, most devices will support this feature.

Additionally, the CLDC defines minimal software requirements. The underlying host operating system must be able to run the KVM and manage the applications on the device. Management of the applications includes selecting, launching, and removing those applications from the application manager.

19.2.2 Security

The design principles of the security model of the J2ME are simplicity and a low memory footprint. The simple sandbox model means that the applications are run in an environment where they can access only predefined APIs and resources of the underlying device and VM.

An application is treated somewhat like a child in a sandbox. The applications can do whatever they come up with within the sandbox, but they cannot harm the environment outside of the sandbox. Every crossing of the sandbox edge is controlled and verified. Only an application from the same application suite can play in the same sandbox, thus helping to prevent applications from harming each other. The following features define a simple sandbox security model:

- all downloaded class files must be verified, because hazardous byte code can break the VM and get access to closed APIs or resources;
- only a predefined set of Java APIs are available to the application programmer; the application programmer cannot open new APIs or in any other way interfere with parts of the device that are advisedly closed because of security reasons;
- there is a class file look-up order; the application programmer cannot overwrite classes in the java.lang.* or javax.microedition.*, packages;
- Java application downloading and management is handled by native implementation, not the JVM; the application programmer cannot interfere with Java byte code downloading.

J2SE features that are eliminated from J2ME MIDP because of security concerns are as follows:

- Java native interface (JNI): the simple sandbox security model provided by CLDC assumes that the set of native functions must be closed;
- user-defined class loaders: the sandbox security model requires elimination of user-defined class loaders, because all classes must be loaded by the system class loader;
- reflection: the CLDC Java applications are not permitted to inspect the number or the content of classes, objects, methods, fields, threads, execution stacks, or other runtime structures inside the virtual machine;
- midlet cannot directly close the VM: the `Exit()` method of the `Runtime` and `System` classes will throw `java.lang.SecurityException`. The midlet can only notify the midlet execution environment that it is ready for shutdown.

The MIDP Java application can access the resources belonging to the same midlet suite. The midlet suite is a collection of MIDP applications and their resource files. Section 19.3, on mobile information device profile, describes the midlet suite in more detail.

Midlet can access resource files, such as images, that are packed inside the JAR (Java archive) file of the midlet suite. For storing the state of the midlet and some other cached data, the midlet can create persistent storage files in nonvolatile memory of the device.

If two midlets belong to the same midlet suite, they can access each other's persistent storage files. To ensure that the removed applications do not leave any trash behind, all persistent storage files of the midlet suite will be erased when the midlet suite is uninstalled.

Class File Verification

Class file verification protects the VM against corrupted code by rejecting this code. This type of low-level security is done by class verification. Since the class file verifier takes up excessive memory for a wireless device, an alternative solution is defined in the CLDC. This alternative is a two-step process:

- Preverification: a preverifier tool processes the class files. Additional attributes are inserted into the class to speed up the verification process. This step is usually done on the development workstation.
- Run-time verification: a runtime verifier runs through each instruction to perform the actual class file verification efficiently.

A class is interpreted, only when the verification process has run successfully.

19.2.3 K Virtual Machine

The Java Virtual Machine is specified in *The Java Language Specification* (Steele *et al.*, 2000) and *The Java Virtual Machine Specification* (Lindholm and Yellin, 1999). The VM specified tries to conform to the specifications as is reasonably possible given the constraints on memory and processing of a mobile device.

The Series 60 Platform provides implementation of the KVM (K Virtual Machine), designed for resource-limited host devices. It is written in C language, and the features are carefully selected to support a low memory footprint. The K in the name of the virtual machine stands for 'kilo', referring to the small footprint of the virtual machine. In comparison, the full-featured VM needs a couple of dozens of kilobytes of memory.

The features listed in Table 19.1 are removed from the KVM because of their memory consumption or the requirement of simple security model of the CLDC.

19.2.4 Packages Overview

The CLDC defines the basic set of packages that a wireless device must support. The classes included in the packages are in the

Table 19.1 Features of the Java Virtual Machine removed for the K Virtual Machine

Compatibility with: Java language specification

Floating point: processors of small devices typically do not have a floating point unit (FPU). Without a FPU, floating point operations are quite expensive in terms of computing time

Finalization: removal of the finalization simplifies the garbage collection process. Because of the limited nature of the CLDC, application programmers do not need to relay on finalization

Error handling: exception handling is supported, but error handling is limited. First, recovery from serious error conditions is highly device-dependent and, second, implementing error handling capabilities fully according to the Java language specification is rather expensive

Compatibility with the: Java Virtual Machine specification

Floating point: Processors of small devices typically do not have an FPU. Without an FPU, floating point operations are quite expensive

Java native interface (JNI): the simple security model of the CLDC assumes that the set of native functions must be closed. Full support of JNI would consume too many resources

Custom class loader: the class loader cannot be replaced, overridden, or modified. The simple security model of the CLDC assumes that applications cannot override the default class loader. The class loader is device-dependent

Reflection: this is removed mainly because of resource limitations

Thread groups: support for threads exists, but not for thread groups. Thread groups are removed because of resource limitations

Finalization: removal of finalization simplifies the garbage collection process. Because of the limited nature of the CLDC, application programmers do not need to relay on finalization

Weak references: removal of weak references simplifies implementation of the virtual machine and, particularly, the garbage collection process

Errors: error handling capabilities are limited and device-dependent. First, recovery from serious error conditions is highly device-dependent and, second, implementing error handling capabilities fully according to the Java language specification is rather expensive

Note: CLDC, connected limited device configuration; Java language and virtual machine specification can be found in Steele *et al.* (2000) and Lindholm and Yellin (1999), respectively.

main derived from J2SE. Some classes have been redesigned; others are newly introduced for the CLDC. The packages are listed in Table 19.2.

The classes in the packages can be grouped into J2SE-derived classes and CDLC-specific classes; these classes are introduced and special aspects highlighted in the following two subsections.

Classes Derived from Java 2, Standard Edition

The J2ME CLDC is largely based on J2SE. In the following we describe which J2SE classes are supported by the J2ME.

System classes Fundamental base classes of the J2ME are listed in Table 19.3.

Table 19.2 Connected limited device configuration (CLDC) packages

Package	Description
java.util	Miscellaneous utility classes (based on J2SE)
java.io	System input and output through datastreams (based on J2SE)
java.lang	Fundamental classes (based on J2SE)
javax.microedition.io	Networking support (CLDC-specific)

Note: J2SE, Java 2, Standard Edition.

Table 19.3 Base classes of Java 2, Micro Edition

Class	Comment
java.lang.Class	Instances of the class `Class` represent classes and interfaces in a running Java application
java.lang.Object	The `Object` class is the root of the class hierarchy
java.lang.Runnable	The `Runnable` interface should be implemented by any class whose instances are intended to be executed by a thread
java.lang.Runtime	Every Java application has a single instance of the `Runtime` class, which allows the application to interface with the environment in which the application is running. Note: the `Exit()` method will always throw the `java.lang.SecurityException`
java.lang.System	The `System` class contains several useful class fields and methods. Note: the `Exit()` method will always throw the `java.lang.SecurityException`
java.lang.Thread	A thread is a unit of execution in a program
java.lang.Throwable	The `Throwable` class is the superclass of all errors and exceptions in the Java language

Data-type classes Table 19.4 describes data types that can be used in J2ME. The table describes object forms of data types and also the corresponding primitive types that can be used. The biggest difference from J2SE primitives is that the J2ME does not have float or double data types.

Collection classes The J2ME supports basic collection types of the J2SE collection framework, as shown in Table 19.5.

Input–output classes The J2ME supports basic Java input–output streams. Both byte and character-encoded streams are supported, as shown in Table 19.6.

Table 19.4 Data types available in Java 2, Micro Edition

Class	Comment
java.lang.Boolean	The `Boolean` class wraps a value of the primitive type `boolean` in an object
java.lang.Byte	The `Byte` class wraps a value of the primitive type `byte` in an object
java.lang.Character	The `Character` class wraps a value of the primitive type `char` in an object
java.lang.Integer	The `Integer` class wraps a value of the primitive type `int` in an object
java.lang.Long	The `Long` class wraps a value of the primitive type `long` in an object
java.lang.Short	The `Short` class wraps a value of the primitive type `short` in an object
java.lang.String	The `String` class represents character strings
java.lang.StringBuffer	A string buffer implements a mutable sequence of characters

Table 19.5 Collection classes

Class	Comment
java.util.Enumeration	An object that implements the `Enumeration` interface generates a series of elements, one at a time
java.util.Hashtable	This class implements a hash table, which maps keys to values
java.util.Stack	The `Stack` class represents a last-in, first-out (LIFO) stack of objects
java.util.Vector	The `Vector` class implements a growable array of objects

Internationalization Classes listed in Table 19.7 provide implementation for character-encoded streams.

Calendar and time classes The J2ME supports date and time handling. The classes listed in Table 19.8 provide ways to handle date and time calculations, which would require many rules and code lines without these classes.

Utility classes Table 19.9 lists the utility classes.

Exception classes Table 19.10 describes exceptions that can be thrown by API methods of the CLDC.

Table 19.6 Input–output classes

Class	Comment
java.io.DataInput	The `DataInput` interface provides an interface for reading bytes from a binary stream and reconstructing from them data in any of the Java primitive types
java.io.DataOutput	The `DataOutput` interface provides an interface for converting data from any of the Java primitive types to a series of bytes and writing these bytes to a binary stream
java.io.ByteArrayInputStream	A `ByteArrayInputStream` contains an internal buffer that contains bytes that may be read from the stream
java.io.ByteArrayOutputStream	This class implements an output stream in which the data are written into a byte array
java.io.DataInputStream	A data input stream lets an application read primitive Java data types from an underlying input stream in a machine-independent way
java.io.DataOutputStream	A data input stream lets an application write primitive Java data types to an output stream in a portable way
java.io.InputStream	This abstract class is the superclass of all classes representing an input stream of bytes
java.io.OutputStream	This abstract class is the superclass of all classes representing an output stream of bytes
java.io.PrintStream	A `PrintStream` adds functionality to another output stream; namely, the ability to print representations of various data values conveniently
java.io.Reader	An abstract class for reading character streams
java.io.Writer	An abstract class for writing to character streams

Table 19.7 Stream-handling classes

Class	Comment
java.io.InputStreamReader	An `InputStreamReader` is a bridge from byte streams to character streams: it reads bytes and translates them into characters according to a specified character encoding
java.io.OutputStreamWriter	An `OutputStreamWriter` is a bridge from character streams to byte streams: characters written to it are translated into bytes according to a specified character encoding

Error classes The J2ME has limited error handling capabilities. The classes listed in Table 19.11 indicate serious error conditions. The VM may throw these errors or silently destroy the application. The applications are not required to recover from these errors.

Classes Specific to the Connected Limited Device Configuration

The CLDC defines new classes for simple network connection support.

Table 19.8 Calendar and time classes

Class	Comment
java.util.Calendar	`Calendar` is an abstract class for getting and setting dates by using a set of integer fields such as `YEAR`, `MONTH`, `DAY`, and so on
java.util.Date	The class `Date` represents a specific instant in time, with millisecond precision
java.util.TimeZone	`TimeZone` represents a time-zone offset, and also works out daylight savings

Table 19.9 Utility classes

Class	Comment
java.lang.Math	The class `Math` contains methods for performing basic numeric operations
java.util.Random	An instance of this class is used to generate a stream of pseudo-random numbers

Network classes and exception classes Classes listed in Tables 19.12(a) and 19.12(b) provide a framework for simple network connections. These connections can be used to make an HTTP (hypertext transfer protocol) request and read files. The CLDC does not provide any concrete implementations (the implementations are provided by profiles).

19.3 Mobile Information Device Profile

As the CLDC defines the common characteristics of a constraint device, the mobile information device profile (MIDP) defines the characteristics for a mobile device on top of the CLDC. The MIDP specification covers the following areas:

- user interface support
- HTTP networking support
- persistent storage support
- miscellaneous classes

In addition, the MIDP specification defines an application execution model, called midlet, in accordance with applets.

Table 19.10 Exception classes

Class	Comment
java.io.EOFException	Signals that an end of file or end of stream has been reached unexpectedly during input
java.io.InterruptedIOException	Signals that an I/O operation has been interrupted
java.io.IOException	Signals that an I/O exception of some sort has occurred
java.io.UnsupportedEncodingException	Character encoding is not supported
java.io.UTFDataFormatException	Signals that a malformed UTF-8 string has been read in a data input stream or by any class that implements the data input interface
java.lang.ArithmeticException	Thrown when an exceptional arithmetic condition has occurred
java.lang.ArrayIndexOutOfBoundsException	Thrown to indicate that an array has been accessed with an illegal index
java.lang.ArrayStoreException	Thrown to indicate that an attempt has been made to store the wrong type of object into an array of objects
java.lang.ClassCastException	Thrown to indicate that the code has attempted to cast an object to a subclass of which it is not an instance
java.lang.ClassNotFoundException	Thrown when an application tries to load in a class through its string name using the `forName()` method in the `Class` class
java.lang.Exception	The class `Exception` and its subclasses are a form of `Throwable` that indicates conditions that a reasonable application might want to catch
java.lang.IllegalAccessException	Thrown when an application tries to load in a class but the currently executing method does not have access to the definition of the specified class because the class is not public or in another package
java.lang.IllegalArgumentException	Thrown to indicate that a method has passed an illegal or inappropriate argument
java.lang.IllegalMonitorStateException	Thrown to indicate that a thread has attempted to wait on an object's monitor or to notify other threads waiting on an object's monitor without owning the specified monitor
java.lang.IllegalStateException	Signals that a method has been invoked at an illegal or inappropriate time
java.lang.IllegalThreadStateException	Thrown to indicate that a thread is not in an appropriate state for the requested operation
java.lang.IndexOutOfBoundsException	Thrown to indicate that an index of some sort (such as to an array, to a string, or to a vector) is out of range
java.lang.InstantiationException	Thrown when an application tries to create an instance of a class using the `newInstance()` method in the `Class` class, but the specified class object cannot be instantiated because it is an interface or an abstract class

Table 19.10 (*continued*)

Class	Comment
java.lang.InterruptedException	Thrown when a thread is waiting, sleeping, or otherwise paused for a long time and another thread interrupts it using the `interrupt()` method in the `Thread` class
java.lang.NegativeArraySizeException	Thrown if an application tries to create an array with negative size
java.lang.NullPointerException	Thrown when an application attempts to use `null` in a case where an object is required
java.lang.NumberFormatException	Thrown to indicate that the application has attempted to convert a string to one of the numeric types but that the string does not have the appropriate format
java.lang.RuntimeException	`RuntimeException` is the superclass of those exceptions that can be thrown during the normal operation of the Java Virtual Machine
java.lang.SecurityException	Thrown by the security manager to indicate a security violation
java.lang.StringIndexOutOfBoundsException	Thrown by the `charAt()` method in the `String` class and by other `String` methods to indicate that an index is either negative or greater than or equal to the size of the string
java.util.EmptyStackException	Thrown by methods in the `Stack` class to indicate that the stack is empty
java.util.NoSuchElementException	Thrown by the `nextElement()` method of an `Enumeration` to indicate that there are no more elements in the enumeration

Note: I/O, input–output.

Table 19.11 Error classes

Class	Comment
java.lang.Error	An `Error` is a subclass of `Throwable` that indicates serious problems that a reasonable application should not try to catch
java.lang.OutOfMemoryError	Thrown when the Java Virtual Machine cannot allocate an object because it is out of memory, and no more memory could be made available by the garbage collector
java.lang.VirtualMachineError	Thrown to indicate that the Java Virtual Machine is broken or has run out of resources necessary for it to continue operating

Table 19.12 (a) Network classes and (b) exception classes

Class	Comment
(a) Network classes	
javax.microedition.io.Connection	This is the most basic type of generic connection
javax.microedition.io.ContentConnection	This interface defines the stream connection over which content is passed
javax.microedition.io.Datagram	This is the generic datagram interface
javax.microedition.io.DatagramConnection	This interface defines the capabilities that a datagram connection must have
javax.microedition.io.InputConnection	This interface defines the capabilities that an input stream connection must have
javax.microedition.io.OutputConnection	This interface defines the capabilities that an output stream connection must have
javax.microedition.io.StreamConnection	This interface defines the capabilities that a stream connection must have
javax.microedition.io.StreamConnectionNotifier	This interface defines the capabilities that a connection notifier must have
javax.microedition.io.Connector	This class is a place holder for the static methods used to create all the connection objects
(b) Exception classes	
javax.microedition.io.ConnectionNotFoundException	This class is used to signal that a connection target cannot be found

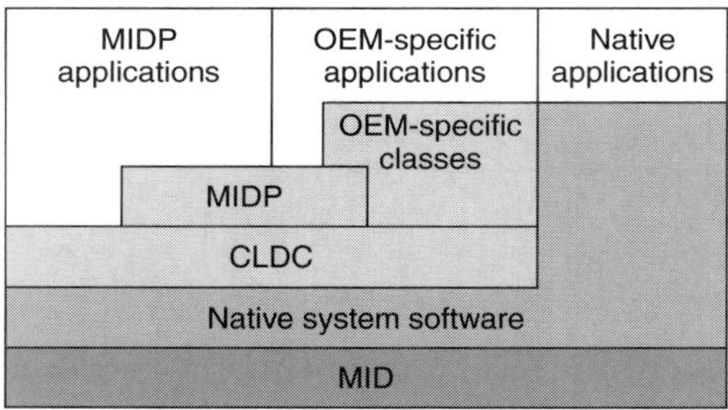

Figure 19.3 Midlet architecture. Note: CLDC, connection limited device configuration; MID, mobile information device; MIDP, mobile information device profile; OEM, original equipment manufacturer

Figure 19.3 presents the architecture of J2ME MIDP. The architecture supports three different kinds of applications. The native applications are run on top of the host operating system and are host-device-specific applications, such as the midlet application manager. MIDP applications are very portable because they can be run in any device that supports MIDP profile.

Applications specific to the original equipment manufacturer (OEM) can typically provide more device-specific functionality than can pure MIDP applications because they can utilize OEM-specific APIs. For example, Nokia UI API allows direct access to the UI resources of a Series 60 device. Utilization of OEM-specific APIs limits the applications to run only in devices of that manufacturer or even in some device group of that manufacturer.

19.3.1 Requirements

The MIDP defines its own requirements on top of the CLDC hardware requirements. It needs:

- 8 kbytes of nonvolatile memory for persistent data;
- at least a 96 × 54 pixel screen;
- availability of user input (through a keypad, keyboard, or touch screen);
- two-way networking with limited bandwidth.

These requirements are satisfied in a typical mobile phone. Thus, mobile phones with Java MIDP capabilities are increasingly appearing on the market. Typical MIDP applications (midlets) used in these devices are simple games.

19.3.2 Package Overview

MIDP specific packages shown in Table 19.13 extend to the CLDC classes with user interface framework, midlet application framework and persistent storage classes. The user interface framework is designed to fit in a small display and meet low resource allocation requirements. The application framework is simple sandbox application environment with persistent storage possibility. The persistent storage framework allows applications to create simple persistent record stores.

Table 19.13 Mobile information device profile (MIDP) packages

Package	Description
javax.microedition.lcdui	Provides the user interfaces for MIDP applications
javax.microedition.midlet	MIDP applications and interactions
javax.microedition.io	Networking support
javax.microedition.rms	Provides a mechanism for midlets persistently to store data and later retrieve those data

Classes Specific to the Mobile Information Device Profile

The MIDP includes all classes described in the CLDC. The MIDP also describes classes for UI framework, persistent data storage and an implementation for network connection framework.

Table 19.14 User interface classes

Class	Comment
javax.microedition.midlet.MIDlet	A `midlet` is an MIDP application
javax.microedition.lcdui.Choice	Choice defines an API for a selection of user interface components from predefined number of choices
javax.microedition.lcdui.CommandListener	This interface is used by applications that need to receive high-level events from the implementation
javax.microedition.lcdui.ItemStateListener	This interface is used by applications that need to receive events that indicate changes in the internal state of the interactive items within a Form screen
javax.microedition.lcdui.Alert	An alert is a screen that shows data to the user and waits for a certain period of time before proceeding to the next screen
javax.microedition.lcdui.AlertType	The `AlertType` provides an indication of the nature of an alert
javax.microedition.lcdui.Canvas	The `Canvas` class is a base class for writing applications that need to handle low-level events and to issue graphics calls for drawing to the display
javax.microedition.lcdui.ChoiceGroup	A `ChoiceGroup` is a group of selectable elements intended to be placed within a `Form`
javax.microedition.lcdui.Command	The `Command` class is a construct that encapsulates the semantic information of an action
javax.microedition.lcdui.DateField	A `DateField` is an editable component for presenting date and time (calendar) information that may be placed into a `Form`
javax.microedition.lcdui.Display	`Display` represents the manager of the display and input devices of the system
javax.microedition.lcdui.Displayable	This is an object that has the capability of being placed on the display
javax.microedition.lcdui.Font	The `Font` class represents fonts and font metrics
javax.microedition.lcdui.Form	A `Form` is a `Screen` that contains an arbitrary mixture of items: images, read-only text fields, editable text fields, editable date fields, gauges, and choice groups
javax.microedition.lcdui.Gauge	The `Gauge` class implements a bar chart display of a value intended for use in a form
javax.microedition.lcdui.Graphics	Provides simple two-dimensional geometric rendering capability
javax.microedition.lcdui.Image	The `Image` class is used to hold graphical image data
javax.microedition.lcdui.ImageItem	This is a class that provides layout control when `Image` objects are added to `Form` or to `Alert`
javax.microedition.lcdui.Item	A superclass for components that can be added to a `Form` and `Alert`
javax.microedition.lcdui.List	The `List` class is a `Screen` containing a list of choices
javax.microedition.lcdui.Screen	The common superclass of all high-level user interface classes
javax.microedition.lcdui.StringItem	An item that can contain a string
javax.microedition.lcdui.TextBox	The `TextBox` class is a `Screen` that allows the user to enter and edit text
javax.microedition.lcdui.TextField	A `TextField` is an editable text component that may be placed into a `Form`
javax.microedition.lcdui.Ticker	Implements a 'ticker-tape', a piece of text that runs continuously across the display

Note: API, application programming interface; MIDP, mobile information device profile.

User interface classes Classes listed in Table 19.14 describe the UI and midlet application frameworks of the MIDP.

Persistent data storage classes The MIDP describes a framework for storing persistent data to the nonvolatile memory of the device. The corresponding classes are described in Table 19.15. Midlets in the same midlet suite can access the data stores of each other.

Table 19.15 Data storage classes

Class	Comment
javax.microedition.rms.RecordComparator	An interface defining a comparator that compares two records (in an implementation-defined manner) to see if they match or to determine their relative sort order
javax.microedition.rms.RecordEnumeration	A class representing a bidirectional record store `Record` enumerator
javax.microedition.rms.RecordFilter	An interface defining a filter that examines a record to see whether it matches (based on an application-defined criteria)
javax.microedition.rms.RecordListener	A listener interface for receiving `Record Changed`, `Added`, or `Deleted` events from a record store
javax.microedition.rms.RecordStore	A class representing a record store

Network classes and exception classes The MIDP provides the HTTP connection to the Network connection framework of the CLDC. Network and exception classes are listed in Table 19.16.

Table 19.16 (a) Network classes and (b) exception classes

Class	Comment
(a) Network classes	
javax.microedition.io.HttpConnection	This interface defines the necessary methods and constants for a hypertext transfer protocol (HTTP) connection
(b) exception classes	
javax.microedition.midlet.MIDletStateChangeException	Signals that a requested `midlet` state change failed

19.3.3 Midlets

The application model supported in the MIDP is the midlet. A midlet runs with related applications in a midlet suite. A suite supports application interaction and data sharing to save memory.

A MIDP application is a class derived from the midlet class. As an introduction to midlet programming, look at the following example:

```
package s60.j2me;

import javax.microedition.midlet.*;
import javax.microedition.lcdui.*;

public class HelloWorld extends midlet implements CommandListener {
```

```
     Command iCmdExit;
     TextBox iText;
public HelloWorld() {
     iCmdExit = new Command("Exit", Command.EXIT, 1);
     iText = new TextBox("Midlet says:", "Hello World!", 20,
TextField.ANY);
     iText.addCommand(iCmdExit);
     iText.setCommandListener(this);
}

  protected void startApp() {
        Display.getDisplay(this).setCurrent(iText);
  }

  protected void pauseApp() {
        // empty
  }

  protected void destroyApp(boolean aIsUnconditional) {
        // empty
  }

  public void commandAction(Command aCmd, Displayable aDisp) {
        if(aCmd == iCmdExit) {
            destroyApp(false);
            notifyDestroyed();
      }
  }
}
```

The HelloWorld midlet is a simple application including only one class, derived from the `midlet` class. This enables the midlet to be run in a midlet environment. The HelloWorld midlet also implements the `CommandListener` interface to get events from the exit command. The HelloWorld midlet defines two member fields that store UI components: `iCmdExit` stores a reference to the exit command, and `iText` stores a reference to the 'Hello World' text object.

The constructor of the HelloWorld midlet creates the text box object and the exit command. The constructor also adds the exit command to the UI framework and adds itself to an event listener for the exit command. The Java application manager calls the `startApp()` method when the midlet is made visible. The `startApp()` method of the HelloWorld midlet simply sets the `iText` object to the current visible object. The `pauseApp()` method is called when the midlet needs to release all unnecessary resources. The HelloWorld midlet simply has nothing to release. The application manager calls the `destroyApp()` method before the application will be destroyed. The midlet should save its state, but the HelloWorld midlet has nothing to save. Finally, the `commandAction()` method implements the `CommandListener` interface. The application manager calls this method when the user activates the exit command of this midlet. The `commandAction()`

method simply calls the `notifyDestroyed()` method to notify the application manager that the midlet can now be destroyed. Because the close of the midlet was initiated by the midlet it self the application manager does not call the `destroyApp()` method, so the `destroyApp()` method is called before the `notifyDestroyed()` method is called.

Midlet Lifecycle

A Midlet takes several states during its lifecycle, with well-defined transitions (see Figure 19.4). It is always in one of the three states: paused, active, or destroyed. The state are listed in Table 19.17.

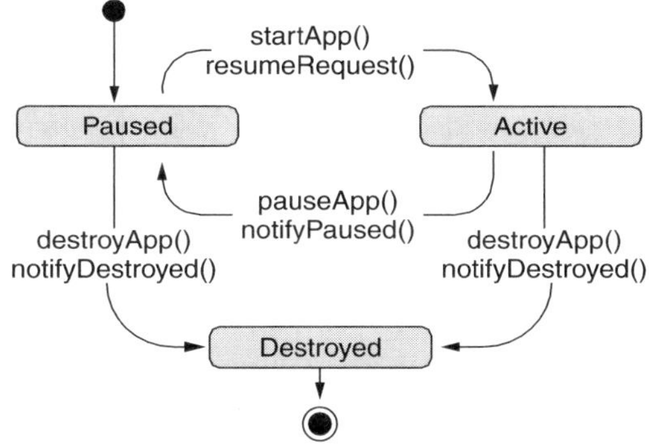

Figure 19.4 Midlet lifecycle

Table 19.17 Midlet states

State	Comment
Paused	The midlet must release shared resources and become quiescent; can still receive asynchronous notifications
Started	The midlet may hold all necessary resources for execution
Destroyed	The midlet must release all resources and save any persistent data: cannot re-enter any other state

The transition between states can be triggered from the midlet itself or externally from the Java application manager (JAM). The application manager is the host OS application that provides tools for the end-user of the device to manage the application installed in the device. At a minimum, the JAM enables the end-user to browse, install, and remove applications to the device. The JAM is described in more detail in Chapter 20.

The application manager calls `startApp`, `pauseApp`, or `destroyApp` to notify the midlet about a state change. From inside the midlet, the `resumeRequest`, `notifyPaused`, and `notifyDestroyed` methods notify the application manager about a state change request. If the constructor of the midlet throws an exception, then the midlet is directly changed into the destroyed state.

State changes from the JAM are as follows:

- `startApp()` is called form the application manager when the midlet is placed into the start state. This can happen at startup of the midlet or when re-entering the state after a pause. If an exception occurs, the midlet can throw `MIDletStateChangeException` for transient failures, otherwise it should call the `destroyApp()` method to cleanup.
- `pauseApp()` is called from the application manager when the system requires suspending or pausing the midlets on the device, typically when the device runs into a low-memory situation.
- `destroyApp()` is called from the application manager when the system requires terminating the midlet. The method's boolean parameter indicates whether the shutdown is unconditional. If the value is true, it means the shutdown is unconditional and the midlet has no choice but to free all resources. Otherwise, the midlet may request to stay alive by throwing a `MIDletStateChangeException`. It is still in the control of the application manager to accept the request.

State changes from within midlet are as follows:

- `resumeRequest()` indicates to the application manager that a paused application requests to re-enter the start state to perform some processing. This is typically called after some timer events.
- `notifyPaused()` signals the application manager that the application wants to enter the pause state at its own request. Typically, the application has set up a timer and waits.
- `notifyDestroyed()` signals the application manager that the application is ready to shutdown. This time, all shared and persistent data should already be stored, because the `destroyApp()` method will not be called.

19.3.4 Midlet Suites

Midlet applications are designed to run on a mobile device and to share data and resources. This is accomplished by placing midlets into a suite. The suites consist of a Java archive (JAR) file and a Java application descriptor (JAD) file. The JAR file contains the packed midlets and additional resources, such as images or text files, whereas the JAD file contains the midlet suite 'advertising' information.

Application Manager

The application manager is responsible for retrieving, installing, launching, removing, and upgrading the midlet suites on the device (Figure 19.5). Single midlets are not handled by the application manager, only the suites. The MIDP specification defines only the minimum operations, and the implementation of the application manager is device-manufacture-dependent.

The retrieval of midlets consists of several steps. First, the medium identification step allows the user to select the transport medium. Typical mediums are serial cable, infrared, or a type of wireless network. The second step is the negotiation step, where the midlet is advertised to the user and the device. A device may not be capable of installing the midlets because of insufficient memory. After the information is verified and the user accepts, the retrieval step starts and the actual midlet is downloaded to the device.

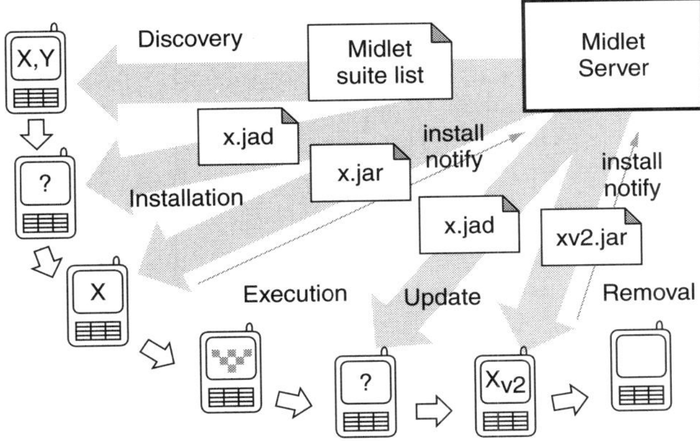

Figure 19.5 Midlet delivery and installation

Installing the suite is the next phase, and it starts with verification that the midlet does not violate any device security policies. After this has been verified, the midlet can be transformed into the device-specific representation, which is, for example, storing the midlet into the persistent memory.

After the application manager has successfully installed the midlet suite, it will notify the midlet server. This will give the application provider the possibility of knowing when the midlet suite is successfully delivered to the user. This information can be used as a basis for application billing.

Now the midlet is ready to be launched. Through an inspection step the user is presented with a list of available midlets. The user can now select a midlet, and the application manager launches the selected midlet.

To upgrade an existing midlet the application manager keeps track of the installed midlets and their version through the identification and version management step. The older version can now be upgraded to a newer version of the suite.

Removing an existing midlet suite consists of inspection of the steps and deletion. The suite and all related resources, including the persistent data, are removed from the device by this operation.

Java Application Descriptor File

The application manager advertises the midlet suite through the JAD file. This file is optional for the midlet suite developer. The JAD file has an extension .jad and is of the MIME (multipurpose Internet mail extension) type text/vnd.sun.j2me.app-descriptor'. The MIME type is relevant for web server to recognize the file. Support of a JAD file allows the application manager to download first the relatively small JAD file for advertising information, before the user really wants to download the midlet suite, packed in the JAR file.

The JAD file provides information about the JAR contents through specific midlet attributes. The information is stored through a name–value pair. The MIDP specification defines mandatory and optional attributes. The mandatory attributes of JAD are described in Table 19.18(a); optimal attributes are described in Table 19.18(b).

Additional own attributes can be defined. All attributes starting with 'midlet-' are reserved; also, attribute names are case-sensitive. An attribute is retrieved through the `MIDlet.getAppProperty()` method.

The following shows an example of a typical JAD file with mandatory attributes:

```
MIDlet-Name: MyMidletSuite
MIDlet-Version: 1.0
MIDlet-Vendor: Company ABC
MIDlet-1: MyMidlet1, /MyMidlet1.png, j2me.MyMidlet1
MIDlet-2: MyMidlet2, /MyMidlet2.png, j2me.MyMidlet2
MicroEdition-Profile: MIDP-1.0
MicroEdition-Configuration: CLDC-1.0
```

Table 19.18 Java application description attributes: (a) mandatory and (b) optional attributes

Attribute	Description
(a) Mandatory attributes	
MIDlet-Name	The name of the midlet suite that identifies the midlets to the user
MIDlet-Version	The version number of the midlet suite
MIDlet-Vendor	The organization that provides the midlet suite
MIDlet-<1...n>	The name, icon, and class of the *n*th midlet in the Java archive file separated by a comma
MicroEdition-Profile	The Java 2 Micro Edition profile required, using the same format and value as the System property microedition.profiles (e.g. 'MIDP-1.0')
MicroEdition-Configuration	The Java 2 Micro Edition configuration required using the same format and value as the system property microedition.configuration (e.g. 'CLDC-1.0')
(b) Optional attributes:	
MIDlet-Icon	The name of a PNG (portable networks graphics) file within the Java archive file used to represent the midlet suite
MIDlet-Description	The description of the midlet suite
MIDlet-Info-URL	An URL for information further describing the midlet suite
MIDlet-JAR-URL	The URL from which the Java archive file can be loaded
MIDlet-JAR-Size	The number of bytes in the Java archive file
MIDlet-Data-Size	The minimum number of bytes of persistent data required by the midlet; the default value is zero

Note: URC, universal resource locator.

The above JAD file defines two midlets – MyMidlet1 and MyMidlet2 – in the midlet suite `MyMidletSuite`.

Java Archive File

A midlet suite JAR file contains the class files, resource files, and a manifest describing the contents of the JAR file. The manifest provides information about the JAR contents through specific midlet attributes. The content of the JAR manifest is the same as that of the JAD file, including mandatory and optional attributes. The `MIDlet-Name`, `MIDlet-Version`, and `MIDlet-Vendor` values must match the information provided in the JAD file. In the case of a mismatch, the application manager will reject the JAR file.

19.4 Summary

Java 2 Micro Edition and MIDP provide application programmers with an easy platform to develop and distribute applications to smartphones. J2ME MIDP is targeted at mobile phones and other similar devices. It provides sufficiently good security and distribution models for device manufacturer to allow third parties for provide applications for a large number of handsets. For application programmers, J2ME provides a well-specified and portable environment that allows the developed applications to run in many different devices. In the next chapter we introduce the creation of Java applications for the Series 60 Platform with MIDP and present in detail the available methods and their use.

20

Midlet User Interface Framework

In this chapter we look at the midlet user interface (UI) framework. All views and items supported by the Series 60 Platform mobile information device profile (MIDP) are covered. Nokia UI application programming interface (API) classes are explained within this chapter (Section 20.2). These classes are the extension classes for default MIDP UI classes. The UI API contains also support for sound.

We also show how networked midlets can be implemented. Like most things in the MIDP, networking is made as easy as possible. Currently, the Series 60 Platform supports only HTTP (hypertext transfer protocol) connections. Thus, it is not possible to use, for example, sockets.

At the end of this chapter, we provide an explanation over-the-air (OTA) provisioning – the way in which midlet suites can be distributed to end-users.

20.1 Defining the Midlet User Interface

The MIDP defines a UI library that is optimized for mobile devices. The user interface on mobile devices is typically key-driven, providing a small screen, and it must be manageable by users who are not necessarily experts in using computers.

The use of UI classes is straight forward, which makes MIDP programming the easiest way to get into mobile programming. The main idea of the MIDP is to provide a way to implement time-to-market applications – applications that can be done with minimum effort and implementation time.

The price of the easy implementation is a quite restricted collection of usable components. These components cannot be modified by developers as might be possible with C++, pure Symbian programming. These constraints exist also because of the MIDPs sandbox framework, as described in the previous chapter (Section 19.3). It must be safe to download and install any available midlet.

The concept that the MIDP expert group introduces is based on displays. A `Display` class is the manager of the screen and the input devices of the system. `Display` is capable of showing and switching between `Displayables`, as shown in Figure 20.1.

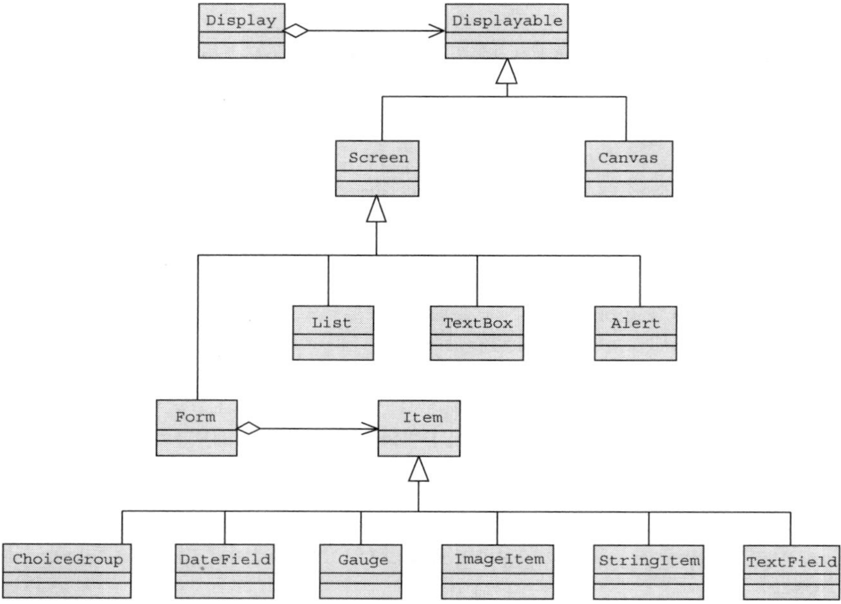

Figure 20.1 Classes in the user interface of the mobile information device profile

`Displayable` is an abstract class, which may also have commands and listeners. Commands are used to retrieve user actions such as pressing a button on the command button array (CBA). `Listener` must be set to `Displayable` to retrieve these commands.

There are two main categories of `Displayables` – the abstract `Canvas` and the `Screen` class. The `Canvas` class is able to handle graphics and low-level events, such as the user having pressed a button. `Screen` is a high-level base class for the UI classes, with additional support for a ticker and title. The ticker is a piece of text that runs across the screen.

Concrete `Screen` classes are `Alert`, `List`, `TextBox`, and `Form`. The `Alert` class shows the date to the user and waits a certain period of time to proceed to the next screen. `List` contains the list of choices. `TextBox` is an editable text screen. `Form` is the container for `Items`.

An abstract Item is a labeled component that can be added to Form. The concrete Items are ChoiceGroup, DateField, Gauge, ImageItem, StringItem, and TextField. ChoiceGroup contains selectable elements. DateField represents a time and/or date editor. Gauge represents a barchart display of a value. ImageItem is a container for Image. StringItem represents text that cannot be modified by the user. TextField is an editable representation of a string.

20.1.1 Example Application

Examples in this chapter are taken from the simple example midlet called Time Tracker. The main idea of this application is to track how long it takes to complete some tasks or projects. This application is useful, for example, for people who work on many different projects and need to estimate the time they spend on those projects per day. Example pictures can be found in later subsections.

The Time Tracker user interface contains three main UI classes: List, Form, and FullCanvas. These classes are basic views that are commonly used, except that Canvas might be more useful than FullCanvas in some cases.

The TTMIDlet class is used only as a startpoint in the application. It is derived from the MIDlet class. TTProjectList is the current displayable after overwritten startApp () method is called.

TTProjectList is derived from the List class and it is the container of all TTProject classes. TTProjectList also shows all projects in a list on the screen.

TTProject is not a UI class; it is only a container for project information such as name, description, and time used.

TTProjectForm is extended from the Form class. It shows information about the project. This information is shown with different types of Items.

TTProjectCanvas is derived from the FullCanvas class. It is used to show the currently used time in the specified project.

Figure 20.2 shows a rough design of the Time Tracker application.

20.1.2 Display

Display represents the manager of the display and input devices of the system. It is capable of showing one Displayable object on the screen and of changing it, when needed. All Displayable objects are responsible for drawing their own context and for handling user actions.

Each midlet has exactly one instance of Display. The application can get the current Display object with the getDisplay()

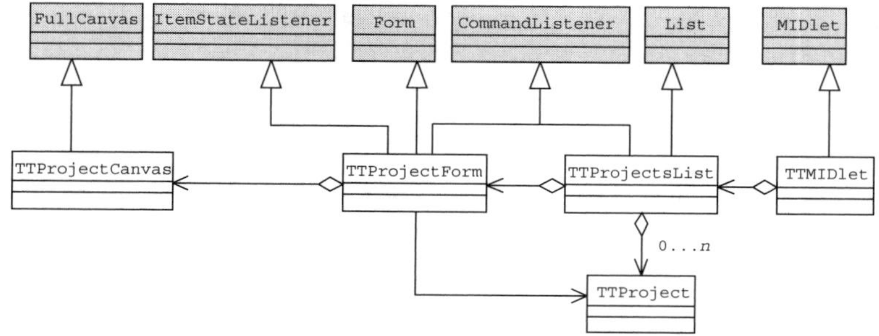

Figure 20.2 Time Tracker design

Table 20.1 Methods of the `Display` class

Method	Comment
`Displayable getCurrent()`	Gets the current `Displayable` object for this midlet
`static Display getDisplay(MIDlet m)`	Gets the `Display` object that is unique to this midlet
`boolean isColor()`	Gets information about the color support of the device
`int numColors()`	Gets the number of colors [if isColor() is true] or gray levels [if isColor() is false] that can be represented on the device
`void setCurrent(Alert alert, Displayable nextDisplayable)`	Shows `Alert` and makes the transition to the next `Displayable`
`void setCurrent(Displayable nextDisplayable)`	Makes the transition to the next `Displayable`
`void callSerially(Runnable r)`	Causes the `Runnable` object to call its `run()` method, serialized with the event stream, soon after completion of the repaint cycle

method. The `Display` for that application will not change during the application lifetime. Methods of the `Display` class are shown in Table 20.1.

In the Time Tracker example, application of the `Display` class is obtained in the `TTMIDlet` class, which extends the `MIDlet` class. The `MIDlet` class is the main class of MIDP applications. When applications are started, a new `MIDlet` object is created and its `startApp()` is called.

In the following code, we get `Display` object in the constructor by using the static `getDisplay()` method of `Display`. The only way to change the current `Displayable` is done with the `setCurrent()` method of `Displayable`. Therefore, the `Display` object must be delivered to all objects if the current `Displayable` is to be changed. In `startApp()` method, we create `TTProjectList` object and give a `Display` to it. So, it can change the current `Displayable` when needed.

In the construction of the midlet, the `Display` does not have any `Displayable` set. It must be set with `setDisplayable()`, as in the `startApp()` method. After that, the new `TTProjectList` `Displayable` is shown in the screen:

```
public class TTMIDlet extends MIDlet
{
    private Display iDisplay;      // The display for this MIDlet
          public TTMIDlet()
    {
          iDisplay = Display.getDisplay( this );
    }

    public void startApp()
    {
          TTProjectsList iProjectsList = new TTProjectsList( this,
             iDisplay );
          // Switching to the TTProjectList view.
          iDisplay.setCurrent( iProjectsList );
    }
    :
    :
}
```

20.1.3 Displayable

The `Displayable` is an object that has the capability of being placed on the display. A `Displayable` object may have commands and listeners associated with it. The `Displayable` content shown on the screen is defined in subclasses derived from `Displayable`. In the MIDP, there are two implemented `Displayables`: `Canvas` and `Screen`. Table 20.2 contains the methods of the `Displayable` class.

Table 20.2 Methods of the `Displayable` class

Method	Comment
`void addCommand(Command cmd)`	Adds a command to the `Displayable`
`boolean isShown()`	Checks if `Displayable` is actually visible on the display
`void removeCommand(Command cmd)`	Removes a command from `Displayable`
`void setCommandListener(CommandListener l)`	Sets a listener for `Commands` to this `Displayable`, replacing any previous `CommandListener`

In the Time Tracker example, we extend the functionality of `List`. `List` extends `Displayable`. There is added a new command for the second CBA button. The place of the command comes with the parameter Command.BACK. The command has now the label 'Exit', and that is displayed in the bottom right-hand corner of the screen. The command is added to `Displayable` with the `addCommand()`

method. A more complete description of the Commands is in the given in Section 20.1.4.

Below, we give an example of CommandListener usage. CommandListener is an interface that can be added to the Displayable object by the setCommandListener() method. When the CommandListener is set, it retrieves high-level user actions. CommandListener must implement one method:

```
public void commandAction(Command c, Displayable s)
```

The Command parameter is used to find out the wanted action (see Section 20.1.4 for further information). The Displayable parameter is the Displayable object, where the command occurred.

There is no rule that CommandListener should be in the same class in which the actual commands exists but, at least in smaller applications, it is convenient to handle commands with views. In the Time Tracker example, we implemented CommandListener in the same class:

```
public class TTProjectsList extends List implements CommandListener
{
     private static final String EXIT_LABEL = "Exit";
     private Command iExitCommand;
     :
     :

     public TTProjectsList( MIDlet aMIDlet, Display aDisplay )
     {
     :
     :
     // Creating new command with exit label, positioning it
     // to the second cba button with priority 1
     iExitCommand = new Command( EXIT_LABEL, Command.BACK, 1 );
     addCommand( iExitCommand );

     // Adding itself as a CommandListener
     setCommandListener( this );
     :
     :
     }

     public void commandAction(Command c, Displayable s)
      {
      :
      :
      }
      :
      :
}
```

20.1.4 Commands

The Command class provides a mechanism for user-controlled input. Displayable and almost all the derived classes support this mechanism through the addCommand() method.

Different implementations of MIDP may present `Commands` in a different way, and the device can map commands with their native style. In the Series 60 Platform, commands can be executed within `Displayables` by means of two softkeys in the lower display area. A single command is mapped to a `Button`, whereas grouped commands are mapped to menus.

Commands support the concept of types and priorities. A type can be Back, Cancel, Exit, Help, Ok, Stop, Item, or Screen. The device uses the type information to place the command to the preferred place. Command types are listed in Table 20.3.

Table 20.3 Command types

Method	Comment
Back	Used to return to previous view
Cancel	Cancels the current action
Exit	Exits the application
Help	User can request for help
Item	Command is associated with `Item`
Ok	Positive answer
Screen	Command is associated with `Screen`
Stop	Will stop some current operation

Priority information is used to describe the importance of the command. Priorities are integer values, where a lower number indicates greater importance. Highest importance is one indicated by a value of, the second most important with a value of two, and so on. Typically, a device sorts commands first by type and then by priority. It is not an error if commands exist with the same type and priority. In that case, device implementation will choose the order of those commands.

The command constructor takes three parameters:

```
Command(String label, int commandType, int priority)
```

The label is the text that is shown to the user to represent the command. The device may override the chosen label and replace it by a more appropriate label for this command on the device. The command methods are described in Table 20.4.

Table 20.4 Methods of the `Command` class

Method	Comment
`int getCommandType()`	Gets the type of the command
`String getLabel()`	Gets the label of the command
`Int getPriority()`	Gets the priority of the command

Command processing is done as follows. First, create a command:

```
Command cmdExit = new Command("Exit", Command.EXIT, 1);
```

Second, add the command to the displayable:

```
addCommand(cmdExit);
```

Third, set the listener to the displayable; the object needs to implement the CommandListener interface:

```
setCommandListener(this);
```

Fourth, fill the listener method with some reasonable code:

```
public void commandAction(Command c, Displayable d)
{
  if(c == cmdExit)
  {
  // do something
  }
}
```

20.1.5 Canvas

The `Canvas` class extends the `Displayable` class. `Canvas` is used in applications that need to control drawing to the display and retrieve low-level user inputs. It can be shown with the `setCurrent()` method of `Display` and acts like other `Displayables`.

`Canvas` is an abstract class, so it must be extended before it can be used. The paint method must be overridden in order to perform custom graphics on the canvas. The paint method argument is a `Graphics` object. The `Graphics` class has standard primitive drawing methods to draw on-screen.

By overwriting the `keyPressed()`, `keyReleased()`, and/or `keyRepeated()` methods, key events can be captured. This is very useful, for instance, when developing games or other applications that need many instant user actions. The key codes are delivered in the method, and the most common keys can be found as constants in the `Canvas` class.

The `Canvas` class provides access for pointer events, but there is no pointer in the Series 60 and therefore these are useless.

The `Canvas` class is notified for showing and hiding through the `showNotify()` and `hideNotify()` methods. By overwriting those methods, notification can be fetched. All methods of the `Canvas` class

Table 20.5 Methods of the `Canvas` class

Method	Comment
`int getHeight()`	Gets height of the displayable area, in pixels
`int getKeyCode(int gameAction)`	Gets a key code that corresponds to the specified game action on the device
`String getKeyName(int keyCode)`	Gets an informative key string for a key
`int getWidth()`	Gets width of the displayable area, in pixels
`boolean hasPointerEvents()`	Checks if the platform supports pointer press and release events
`boolean hasPointerMotionEvents()`	Checks if the platform supports pointer motion events (pointer dragged)
`boolean hasRepeatEvents()`	Checks if the platform can generate repeat events when key is kept down
`protected void hideNotify()`	The implementation calls `hideNotify()` shortly after `Canvas` has been removed from the display
`boolean isDoubleBuffered()`	Checks if `Graphics` is double-buffered by the implementation
`protected void keyPressed(int keyCode)`	Called when a key is pressed
`protected void keyReleased(int keyCode)`	Called when a key is released
`protected void keyRepeated(int keyCode)`	Called when a key is repeated (held down)
`protected abstract void paint(Graphics g)`	Renders `Canvas`
`protected void pointerDragged(int x, int y)`	Called when the pointer is dragged
`protected void pointerPressed(int x, int y)`	Called when the pointer is pressed
`protected void pointerReleased(int x, int y)`	Called when the pointer is released
`void repaint()`	Requests a repaint for the entire `Canvas`
`void repaint(int x, int y, int width, int height)`	Requests a repaint for the specified region of the `Screen`
`void serviceRepaints()`	Forces any pending repaint requests to be serviced immediately
`protected void showNotify()`	The implementation calls `showNotify()` immediately prior to this `Canvas` being made visible on the display
`int getGameAction(int keyCode)`	Gets the game action associated with the given key code of the device
`int getHeight()`	Gets height of the displayable area, in pixels

are described in Table 20.5. An example of `Canvas` usage can be found in Section 20.2.2, in the subsection on `FullCanvas`.

20.1.6 `Screen`

The `Screen` class is a displayable object that can be set to 'current' and displayed in the display. It extends the functionality of `Displayable` with support for the title and ticker. Screen is an abstract class and must be extended to offer custom functionality. Instantiable subclasses of Screen are Alert, Form, List, and TextBox.

In the Time Tracker example, we have used many Screens. The Form screen is used to display project information and to allow the user to edit it. The Alert class is used to inform the user of errors and other events, and the user can select one project from the List object. Each Screen object is described more precisely in other sections of this chapter. The methods of the Screen class are described in Table 20.6.

Table 20.6 Methods of the Screen class

Method	Comment
String getTitle()	Gets the title of Screen
void setTicker(Ticker ticker)	Sets a ticker for use with this Screen, replacing any previous ticker
void setTitle(String s)	Sets the title of Screen
Ticker getTicker()	Gets the ticker used by this Screen

20.1.7 Graphics

The Graphics class provides simple drawing methods and control for color, translation, and clipping. The basic drawing shapes are arcs, lines, and rectangles, as shown in Figure 20.3. It is also possible to draw images and strings.

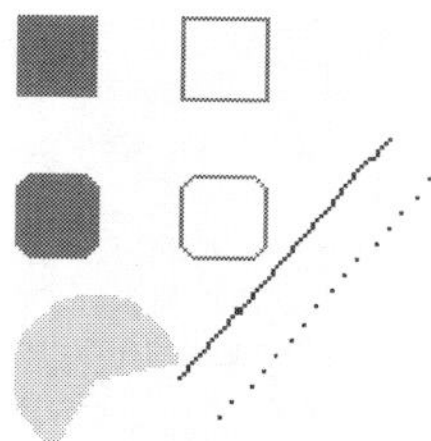

Figure 20.3 Arcs, lines, and rectangles

Graphics can be used with Canvas and Image. With the Canvas object, Graphics is delivered with the paint() method. The Graphics object can be used to draw to an off-screen image. The Graphics object from Image can be obtained with the getGraphics() method. The methods of the Graphics class are described in Table 20.7.

Table 20.7 Methods of the `Graphics` class

Method	Comment
`void clipRect(int x, int y, int width, int height)`	Intersects the current clip with the specified rectangle
`void drawArc(int x, int y, int width, int height, int startAngle, int arcAngle)`	Draws the outline of a circular or elliptical arc covering the specified rectangle, using the current color and stroke style
`void drawChar(char character, int x, int y, int anchor)`	Draws the specified character using the current font and color
`void drawChars(char[] data, int offset, int length, int x, int y, int anchor)`	Draws the specified characters using the current font and color
`void drawImage(Image img, int x, int y, int anchor)`	Draws the specified image by using the anchor point
`void drawLine(int x1, int y1, int x2, int y2)`	Draws a line between the coordinates (x1,y1) and (x2,y2) using the current color and stroke style
`void drawRect(int x, int y, int width, int height)`	Draws the outline of the specified rectangle using the current color and stroke style
`void drawRoundRect(int x, int y, int width, int height, int arcWidth, int arcHeight)`	Draws the outline of the specified rounded corner rectangle using the current color and stroke style
`void drawString(String str, int x, int y, int anchor)`	Draws the specified `String` using the current font and color
`void drawSubstring(String str, int offset, int len, int x, int y, int anchor)`	Draws the specified `String` using the current font and color
`void fillArc(int x, int y, int width, int height, int startAngle, int arcAngle)`	Fills a circular or elliptical arc covering the specified rectangle
`void fillRect(int x, int y, int width, int height)`	Fills the specified rectangle with the current color
`void fillRoundRect(int x, int y, int width, int height, int arcWidth, int arcHeight)`	Fills the specified rounded corner rectangle with the current color
`int getBlueComponent()`	Gets the blue component of the current color
`int getClipHeight()`	Gets the height of the current clipping area
`int getClipWidth()`	Gets the width of the current clipping area
`int getClipX()`	Gets the *X* offset of the current clipping area, relative to the coordinate system origin of this graphics context
`int getClipY()`	Gets the *Y* offset of the current clipping area, relative to the coordinate system origin of this graphics context
`int getColor()`	Gets the current color
`Font getFont()`	Gets the current font
`int getGrayScale()`	Gets the current gray-scale value of the color being used for rendering operations
`int getGreenComponent()`	Gets the green component of the current color
`int getRedComponent()`	Gets the red component of the current color

(continued overleaf)

Table 20.7 *(continued)*

Method	Comment
`int getStrokeStyle()`	Gets the stroke style used for drawing operations
`int getTranslateX()`	Gets the *X* coordinate of the translated origin of this graphics context
`int getTranslateY()`	Gets the *Y* coordinate of the translated origin of this graphics context
`void setClip(int x, int y, int width, int height)`	Sets the current clip to the rectangle specified by the given coordinates
`void setColor(int RGB)`	Sets the current color to the specified RGB (red–green–blue) values
`void setColor(int red, int green, int blue)`	Sets the current color to the specified RGB values
`void setFont(Font font)`	Sets the font for all subsequent text rendering operations
`void setGrayScale(int value)`	Sets the current gray scale to be used for all subsequent rendering operations
`void setStrokeStyle(int style)`	Sets the stroke style used for drawing lines, arcs, rectangles, and rounded rectangles
`void translate(int x, int y)`	Translates the origin of the graphics context to the point (*x*, *y*) in the current coordinate system

Strings can be drawn with specified font and color, and there is a special method for drawing substrings, (see Section 20.1.14 for further information on `Font` usage). Characters and character arrays can be also drawn without converting these to `Strings`.

20.1.8 `Form`

The `Form` class is a graphical container for `Item` objects. Items can be appended on, added to, or inserted in the `Form`, or removed from it. `Item` objects are controlled within the `Form` by their index numbers. Same instance of `Item` can be only in a one `Form`. If the application attempts to append `Item`, which is already appended to some `Form`, `IllegalStateException` is thrown.

Items are placed in `Form`, one upon the other. `Form` takes care of the layout of the `Items` and scrolls all `Items` when needed. The `Item` objects may not have their own scrolling functionality. Methods of the `Form` class are described in Table 20.8.

In the Time Tracker example, we have extended `Form` and added some `Items` to it, as shown in Figure 20.4. `TTProjectForm` is used to view and edit project information. Each project has a name, duration, and description. These properties are presented with `Items`, which are added to the `Form` by the append method, in the `setupFormFields()` method.

Table 20.8 Methods of the `Form` class

Method	Comment
`int append(Image img)`	Adds an item consisting of one `Image` to the form
`int append(Item item)`	Adds an item into `Form`
`int append(String str)`	Adds an item consisting of one `String` into the form
`void delete(int itemNum)`	Deletes the Item referenced by `itemNum`
`Item get(int itemNum)`	Gets the item at a given position
`void insert(int itemNum, Item item)`	Inserts an item into the `Form` just before the item specified
`void set(int itemNum, Item item)`	Sets the item referenced by `itemNum` to the specified `Item`, replacing the previous item
`Void setItemStateListener(ItemStateListener iListener)`	Sets the `ItemStateListener` for the `Form`, replacing any previous `ItemStateListener`
`int size()`	Gets the number of items in `Form`

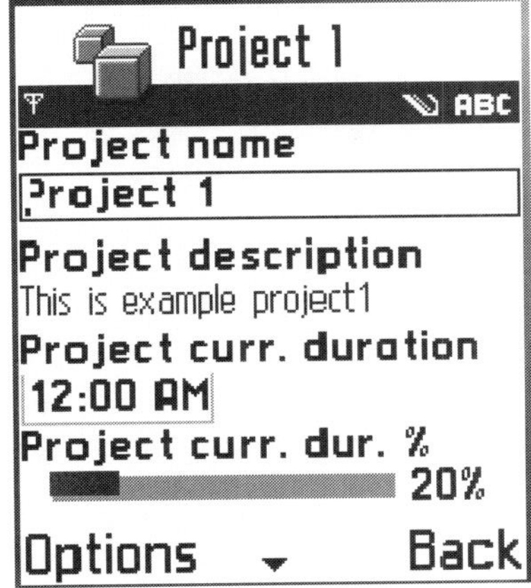

Figure 20.4 `Form` in Time Tracker with `TextField`, `StringItem`, `DateField`, and `Gauge`

We added `TextField`, `StringItem`, `DateField`, and `Gauge` items to `Form`. `TextField` is used to edit the project title. `StringItem` represents the project description, which cannot be modified in this view. `DateField` is used to show the current duration of the project.

The `Gauge` object shows what percentage of time this project has taken compared with other projects. It is constructed with a maximum value of 100, and the initial value is zero. The value of the `Gauge` object is changed with the `setValue()` method, when a new project is set to this `Form`.

TTProjectForm implements ItemStateListener, which can be used to observe when the Items state changes. The ItemStateListener must be set to Form with the setItemStateListener() method to make it function, which is done in this example by the setupFormFields() method:

```
public class TTProjectForm extends Form implements
  CommandListener, ItemStateListener
    {
    :
    private TTProjectsList iProjectsList;
    private TTProject iProject;

    // Project name TextField
    private static final String PROJECT_NAME = "Project name";
    private static final int PROJECT_NAME_MAX_LENGTH = 20;
    private TextField iNameItem = new TextField(
      PROJECT_NAME, "", PROJECT_NAME_MAX_LENGTH,
      TextField.ANY );

    // Project description StringItem
    private static final String PROJECT_DESC =
      "Project description";
    private StringItem iDescItem = new StringItem(
      PROJECT_DESC, "" );

    //Project current duration
    private static final String PROJECT_TIME =
      "Project curr. duration";
    private DateField iTimeItem = new DateField(
      PROJECT_TIME, DateField.TIME );

    //Project duration percents in all projects
    private static final String PROJECT_TIME_PERCENTS =
     "Project curr. dur. %";
    private Gauge iPercentGauge = new Gauge(
      PROJECT_TIME_PERCENTS, false, 100, 0);
    public void setProject( TTProject aProject )
        {
        iProject = aProject;
        // Setting form header to projects name
        setTitle( aProject.getName() );
        // Setting project name to the TextField also
        iNameItem.setString( aProject.getName() );
        // Setting project description to the StringItem
        iDescItem.setText( aProject.getDescription() );
        // Setting used time to the DateField
        iTimeItem.setDate( aProject.getTime() );
        // Setting all percents to the Gauge
        iPercentGauge.setValue( (int)(
          (aProject.getTimeInMillis()*100) /
```

```
        iProjectsList.getProjectsDuration() ) );
    }

private void setupFormFields()
    {
    append( iNameItem );
    setItemStateListener( this );
    append( iDescItem );
    append( iTimeItem );
    append( iPercentGauge );
    }
    :
public void itemStateChanged( Item item )
{
:
}
:
}
```

20.1.9 Image

The `Image` class is used to contain graphical image data. The `Image` objects do not depend on the display of the device and will not be drawn unless they are especially told to. Images can be created from the data array, from another `Image`, or from image files.

`Images` can be categorized into two groups: mutable and immutable. An immutable `Image` cannot be modified after it is created. Methods of the `Image` class are described in Table 20.9. An example for creating `Image` from the byte array can be found in Section 20.3.

Table 20.9 Methods of the `Image` class

Method	Comment
`static Image createImage(byte[] imageData, int imageOffset, int imageLength)`	Creates an immutable image that is decoded from the data stored in the specified byte array at the specified offset and length.
`static Image createImage(Image source)`	Creates an immutable image from a source image
`static Image createImage(int width, int height)`	Creates a new, mutable image for off-screen drawing
`static Image createImage(String name)`	Creates an immutable image from decoded image data obtained from the named resource
`Graphics getGraphics()`	Creates a new `Graphics` object that renders to this image
`int getHeight()`	Gets the height of the image in pixels
`int getWidth()`	Gets the width of the image in pixels
`Boolean isMutable()`	Checks if this image is mutable

20.1.10 Ticker

The `Ticker` class provides scrolling text across the display. The direction and speed of the scrolling are determined by the implementation. In the Series 60 Platform the text scrolls from right to left in the upper part of the display, as shown in Figure 20.5.

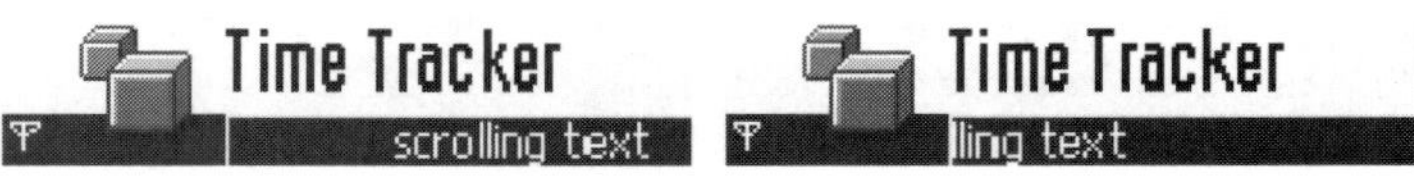

Figure 20.5 Scrolling ticker

A `Ticker` object can be used with `Screen` objects. `Ticker` is set to the `Screen` object with the `setTicker()` method, and it is shown when `Screen` is displayed. The same `Ticker` object can be set to several `Screen` objects. A typical use is for an application to place the same ticker on all of its screens. When the application switches between two screens that have the same ticker, a desirable effect is for the ticker to be displayed at the same location on the display and to continue scrolling its contents at the same position. This gives the illusion of the ticker being attached to the display instead of to each screen. The methods of the `Ticker` class are shown in Table 20.10.

Table 20.10 Methods of the `Screen` class

Method	Comment
`String getString()`	Gets the string currently being scrolled by the ticker
`void setString(String str)`	Sets the string to be displayed by this ticker

While animating, the ticker string scrolls continuously. That is, when the string finishes scrolling off the display, the ticker starts over at the beginning of the string. The following `Screen` method can show `Ticker`:

```
iScreen.setTicker( new Ticker( "scrolling text" ) );
```

20.1.11 TextBox

The `TextBox` class is a `Screen` for text editing and text insertion. It can have constraints to restrict the user's input capabilities. The input constraints are defined in the `TextField` class. (see Section 20.1.20 for further information on constraints).

`TextBox` constructor is as follows:

```
TextBox(String title, String text, int maxSize,
  int constraints)
```

The title is the text shown in the `Screen` title. The text parameter is the text placed initially in the `TextBox` before the user or the application edits the contents. The `maxSize` parameter defines the maximum number of characters that can be stored to `TextBox`. The maximum must be defined, and the implementation may also need to downscale the capacity. The real capacity of `TextBox` can be obtained with the `getMaxSize()` method. Applications should check the capacity with this method and not exceed the capacity.

After construction, the maximum size of the `TextBox` can be changed by the `setMaxSize()` method. If the current number of characters in the `TextBox` is too great, the contents will be truncated to fit to the new size. The method returns the new size of the `TextBox`. The size can be smaller than the requested size. The maximum size is not the number of characters that is shown in the screen. If the `TextBox` has more characters that can be shown on the screen at the same time, the `TextBox` will allow the user to scroll the text and edit it at the selected point.

The content of `TextBox` can be modified and obtained by several methods. The application can obtain the context of `TextBox` with the `getChars()` or `getString()` methods. The method `getChars()` copies the context of `TextBox` into a character array, and the method `getString()` returns the `String` object containing the context. There are also similar methods for setting the context and partially modifying the context. Table 20.11 contains the methods of the `TextBox` class.

20.1.12 Choice

`Choice` is an interface for the UI classes that implement selection from a predefined number of choices. The methods of `Choice` are listed in Table 20.12. In MIDP, there are two implemented `Choice` classes: `List` and `ChoiceGroup`. The use of both classes is basically the same. There are only a few differences, such as supported selection modes.

ChoiceGroup

`ChoiceGroup` is an implementation of `Choice` and extends the `Item` class. Therefore, `ChoiceGroup` is used within the `Form` class. `ChoiceGroup` only supports exclusive-choice (shown in Figure 20.6) and multiple-choice types. The implementation must provide a different graphical presentation for each mode. The exclusive mode might be represented, for example, with radio buttons, and multiple modes with check boxes.

Table 20.11 Methods of the `TextBox` class

Method	Comment
`void delete(int offset, int length)`	Deletes characters from `TextBox`
`int getCaretPosition()`	Gets the current input position
`int getChars(char[] data)`	Copies the contents of `TextBox` into a character array, starting at index zero
`int getConstraints()`	Gets the current input constraints of `TextBox`
`int getMaxSize()`	Returns the maximum size (number of characters) that can be stored in this `TextBox`
`String getString()`	Gets the contents of `TextBox` as a string value
`void insert(char[] data, int offset, int length, int position)`	Inserts a subrange of an array of characters into the contents of `TextBox`
`void insert(String src, int position)`	Inserts a string into the contents of `TextBox`
`void setChars(char[] data, int offset, int length)`	Sets the contents of `TextBox` from a character array, replacing the previous contents
`void setConstraints(int constraints)`	Sets the input constraints of `TextBox`
`int setMaxSize(int maxSize)`	Sets the maximum size (number of characters) that can be contained in this `TextBox`
`void setString(String text)`	Sets the contents of `TextBox` as a string value, replacing the previous contents
`int size()`	Gets the number of characters that are currently stored in this `TextBox`

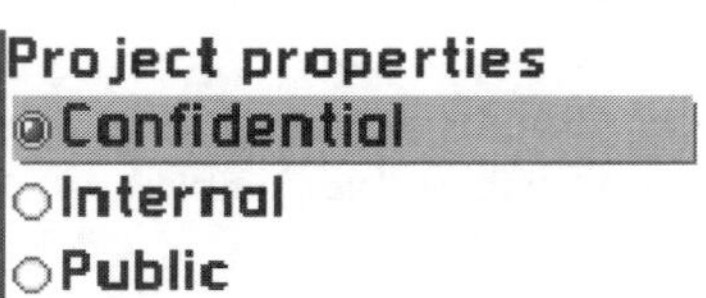

Figure 20.6 Exclusive `ChoiceGroup`

Constructors are as follows:

```
public ChoiceGroup(String label, int choiceType)
public ChoiceGroup(String label, int choiceType,
    String[] stringElements,
    Image[] imageElements)
```

The `String` parameter label is text shown in the `Item` label. The Choice type cannot be implicit, but all the other choice types defined in `Choice` can. The group can be constructed with the `String` element array. Each element in the array forms one choice. The `Image` array is optional. If the image element array is not null, the length of the array must be at least the number of string elements.

The state of `ChoiceGroup` can be observed with `ItemStateListener`. The listener is called every time the user changes the selections in choice group. When the listener is called the choice group can be

Table 20.12 Methods of the `Choice` interface

Method	Comment
`String getString(int elementNum)`	Gets the `String` part of the element referenced by `elementNum`
`Image getImage(int elementNum)`	Gets the `Image` part of the element referenced by `elementNum`
`int append(String stringPart, Image imagePart)`	Appends an element to `Choice`
`void insert(int elementNum, String stringPart, Image imagePart)`	Inserts an element into `Choice` just prior to the element specified
`void delete(int elementNum)`	Deletes the element referenced by `elementNum`
`void set(int elementNum, String stringPart, Image imagePart)`	Sets the element referenced by `elementNum` to the specified element, replacing the previous contents of the element
`boolean isSelected(int elementNum)`	Gets a `boolean` value indicating whether this element is selected
`int getSelectedIndex()`	Returns the index number of an element in `Choice` that is selected
`int getSelectedFlags(boolean[] selectedArray_return)`	Queries the state of a `Choice` and returns the state of all elements in the `boolean` array `selectedArray_return`
`void setSelectedIndex(int elementNum, boolean selected)`	For `MULTIPLE`, this simply sets an individual element's selected state
`void setSelectedFlags(boolean[] selectedArray)`	Attempts to set the selected state of every element in `Choice`
`int size()`	Gets the number of elements present

queried for selected elements. ItemStateListener must implement only one method:

```
public void itemStateChanged( Item item )
```

A parameter item is the choice group where the event occurred. To get events, the listener must be added to Form.

List

The List class is a Screen that provides a list of choices, as shown in Figure 20.7. As List extends Screen, it can have a ticker and contain commands. List supports all selection modes defined in Choice.

Figure 20.7 Project list from Time Tracker

Constructors are as follows:

```
public List(String title, int listType)
public List(String title, int listType, String[] stringElements,
  Image[] imageElements)
```

The list can be observed for changes only in the implicit selection mode. It can be observed with registered CommandListener. The list has a special select Command, which will be delivered to CommandListeners when the state of the list changes:

```
public class TTProjectsList extends List implements CommandListener
    {
    static private final String LIST_NAME = "Time Tracker";
    private Vector iProjects = new Vector();
    :
    :
```

```
public TTProjectsList( MIDlet aMIDlet, Display aDisplay )
    {
    // Creating implicit list
    super( LIST_NAME, IMPLICIT );
    :
    }

private void addProject( TTProject aProject )
    {
    // Appending project to list by its name, no picture
    append( aProject.getName(), null );
    iProjects.addElement( aProject );
    }

public void commandAction(Command c, Displayable s)
    {
    :
    else if( c == List.SELECT_COMMAND )
        {
        // User has selected one of the projects from the list
        // getting which one with getSelectedIndex()
        iProjectForm.setProject( (TTProject)iProjects.elementAt
          ( getSelectedIndex()) );
        :
        }
    }
    :
}
```

20.1.13 Alert

The Alert class is a user interface informing the user about errors and other events: Alert extends the Screen class, but does not accept any application-defined Commands. Alert has a label and possibly an image. Alert stays activated for a certain amount of time or until the user dismisses it.

The method constructs a new Alert object with a given title and content String (alertText):

```
public Alert(String title, String alertText, Image alertImage,
  AlertType alerttType)
```

The image is optional and the parameter can be null if no image is needed. All images in alerts must be immutable.

The AlertType class is a class presenting the type of the alert. There are five predefined alert types: ALARM, CONFIRMATION, ERROR, INFO (information), and WARNING. Alert types are listed in Table 20.13.

The AlertType class has a method to enable sounds in the alerts:

```
public boolean playSound(Display display)
```

Table 20.13 `Alert` types

`AlertType`	Comment
`ALARM`	An ALARM is a hint to alert the user to an event for which the user has previously requested notification.
`CONFIRMATION`	A CONFIRMATION `AlertType` is a hint to confirm user actions
`ERROR`	An ERROR `AlertType` is a hint to alert the user to erroneous operation
`INFO`	An INFO `AlertType` typically provides 'nonthreatening' information to the user
`WARNING`	A WARNING `AlertType` is a hint to warn the user of a potentially dangerous operation

The user is alerted when the sound for this `AlertType` is played. The `AlertType` instance is used as a hint by the device to generate an appropriate sound. Instances other than those predefined above may be ignored. The implementation may determine the sound to be played. The implementation may also ignore the request, use the same sound for several `AlertTypes`, or use any other means suitable to alert the user.

Alerts can be shown with the following `Display` method:

```
public void setCurrent(Alert alert,
  Displayable nextDisplayable)
```

This method sets the alert to the current `Displayable`. When the alert dismisses the nextDisplayable `Displayable` is set current. Display object will not wait for the alert to dismiss, and returns immediately. The next displayable must be defined and cannot be the alert object. Methods of the Alert Clas are listed in Table 20.14.

There is an alert of the `CONFIRMATION type`, as shown in Figure 20.8.

20.1.14 `Font`

The `Font` class represents the fonts and font metrics. The applications do not create the fonts by themselves. Instead, the fonts must be queried with font attributes, and the system will attempt to provide a font that matches the requested attributes as closely as possible. Since only static system fonts are used, it eliminates the garbage creation normally associated with the use of fonts. Methods of the `Font` class are listed in Table 20.15.

All fonts are queried with the following method:

```
static Font getFont(int face, int style, int size)
```

Table 20.14 Methods of the `Alert` class

Method	Comment
`void addCommand(Command cmd)`	Commands are not allowed on an `Alert`, so this method will always throw `IllegalStateException` whenever it is called
`int getDefaultTimeout()`	Gets the default time for showing `Alert`
`Image getImage()`	Gets the `Image` used in `Alert`
`String getString()`	Gets the text string used in `Alert`
`int getTimeout()`	Gets the time this `Alert` will be shown
`AlertType getType()`	Gets the type of `Alert`
`Void setCommandListener(CommandListener l)`	Listeners are not allowed on an `Alert`, so this method will always throw `IllegalStateException` whenever it is called
`void setImage(Image img)`	Sets the `Image` used in `Alert`
`void setString(String str)`	Sets the text string used in `Alert`
`void setTimeout(int time)`	Set the time for which `Alert` is to be shown
`void setType(AlertType type)`	Sets the type of `Alert`

Figure 20.8 Alert in the Time Tracker

The Fonts attributes are `style`, `size`, and `face`. Values for attributes must be specified in terms of symbolic constants:

```
Face:FACE_MONOSPACE, FACE_PROPORTIONAL, FACE_SYSTEM
Style:STYLE_BOLD, STYLE_ITALIC, STYLE_PLAIN,
  STYLE_UNDERLINED
Size:SIZE_LARGE, SIZE_MEDIUM, SIZE_SMALL
```

The values for the `style` attribute may be combined by using the logical OR operator:

The following example shows how a new `Font` can be set to `Graphics`, and how its measurement methods are used:

Table 20.15 Methods of the `Font` class

Method	Comment
`int charsWidth(char[] ch, int offset, int length)`	Returns the advance width of the characters in 'ch', starting at the specified offset and for the specified number of characters (length)
`int charWidth(char ch)`	Gets the advance width of the specified character in this font.
`int getBaselinePosition()`	Gets the distance in pixels from the top of the text to the baseline of the text
`static Font getDefaultFont()`	Gets the default font of the system
`int getFace()`	Gets the face of the font
`static Font getFont(int face, int style, int size)`	Obtains an object representing a font having the specified face, style, and size
`int getHeight()`	Gets the standard height of a line of text in this font
`int getSize()`	Gets the size of the font
`int getStyle()`	Gets the style of the font
`boolean isBold()`	Returns true if the font is bold
`boolean isItalic()`	Returns true if the font is italic
`boolean isPlain()`	Returns true if the font is plain
`boolean isUnderlined()`	Returns true if the font is underlined
`int stringWidth(String str)`	Gets the total advance width for showing the specified `String` in this `Font`
`int substringWidth(String str, int offset, int len)`	Gets the total advance width for showing the specified substring in this font

```
public class TTProjectCanvas extends FullCanvas
    {
    :
    private static final String STOP_TXT = "Stop";
    :
public void paint( Graphics aG )
    {
    :
    //Getting a new font
    Font fnt = Font.getFont(Font.FACE_PROPORTIONAL,
      Font.SIZE_MEDIUM, Font.STYLE_PLAIN );
    //Setting it to Graphics object
    aG.setFont( fnt );
    //Drawing stop text to exactly bottom right corner
    aG.drawString( STOP_TXT, getWidth()-fnt.stringWidth(
      STOP_TXT ), getHeight()-fnt.getHeight(), aG.TOP | aG.LEFT );
    :
    }
    :
    }
```

At first, the desired font is obtained with the `getFont()` method, which takes attributes describing the `Font` features as a parameter. The `Font` is set to the current `Font` of the `Graphic` object. Text 'stop' is drawn to the bottom right-hand corner of the display. The drawing position is calculated by using `Canvas` size and `Font` dimensions.

20.1.15 Item

The `Item` class is an abstract base class for the UI components that can be added to `Form`. All `Item` objects have a label, which is a string that is attached to the `Item`. The label is typically displayed near the component when it is displayed within a screen. In the Series 60 Platform, the label is above the `Item` object and scrolls with the item. Item methods are listed in Table 20.16

Table 20.16 Methods of the `Item` class

Method	Comment
`String getLabel()`	Gets the label of this `Item` object
`void setLabel(String label)`	Sets the label of `Item`

Items must be added to a `Form` to make them visible. `Item` is shown when `Form` is the current `Displayable` and the `Form` focus is above the `Item`.

20.1.16 DateField

`DateField` is an editable component for presenting the calendar date and time information that may be placed into a `Form`. An instance of a `DateField` can be configured to accept the date or time information or both. Input modes and types in the `DateField` class are defined in Tables 20.17(a) 20.17(b), respectively. There is an example of `DateField` use in the `Form` example (Section 20.1.8).

Table 20.17 The `DateField` class: (a) input modes and (b) input types

	Comment
(a) Input mode	
`DATE`	Date information (day, month, and year)
`TIME`	Time information (hours and minutes)
`DATE_TIME`	Date and time information
(b) Input type	
`Date getDate()`	Returns date value of this field
`int getInputMode()`	Gets input mode for this date field
`void setDate(Date date)`	Sets a new value for this field
`void setInputMode(int mode)`	Set input mode for this date field

20.1.17 Gauge

The `Gauge` class implements a barchart display of a value intended for use in a Form:

```
Gauge(String label, boolean interactive, int maxValue, int initialValue)
```

The Gauge accepts values from zero to a maximum value established by the application. The application is expected to normalize its values into this range. The device is expected to normalize this range into a smaller set of values for display purposes. In doing so, it will not change the actual value contained within the object. The range of values specified by the application may be larger than the number of distinct visual states possible on the device, so more than one value may have the same visual representation. The methods of the Cauge class are listed in Table 20.18.

Table 20.18 Methods in the Gauge class

Method	Comment
int getMaxValue()	Gets the maximum value of this Gauge object
int getValue()	Gets the current value of this Gauge object
boolean isInteractive()	Tells whether the user is allowed to change the value of the Gauge
void setMaxValue(int maxValue)	Sets the maximum value of this Gauge object
void setValue(int value)	Sets the current value of this Gauge object

There is an example of the use of Gauge in the Form example (Section 20.1.8).

20.1.18 ImageItem

ImageItem is an Item class that contains an Image and label.
The label is typically shown above the image part.

```
ImageItem(String label, Image img, int layout, String altText)
```

It can be added to a Form, and then it acts like other Items. The label is typically shown above the image part.

The Label parameter is the label of the Item. Parameter img is the image part of the ImageItem. If the img parameter is null, the item uses only the space needed by the label. The whole Item is not shown if there is no image and no label.

The layout parameter specifies how the image is drawn to the screen. Some of the layout constants can be combined, such as in LAYOUT_NEWLINE_BEFORE | LAYOUT_LEFT. The default layout parameter cannot be combined with other layouts. The layout constants are listed in Table 20.19.

An alternative text layout is shown if the image part cannot be shown, which may occur when the image exceeds the capacity of the display device. The methods of the ImageItem class are listed in Table 20.20.

Table 20.19 Layout constants of the `ImageItem` class

Method	Comment
`LAYOUT_CENTER`	Image should be horizontally centered
`LAYOUT_DEFAULT`	Use the default formatting of the 'container' of the image
`LAYOUT_LEFT`	Image should be close to the left edge of the drawing area
`LAYOUT_NEWLINE_AFTER`	A new line should be started after the image is drawn
`LAYOUT_NEWLINE_BEFORE`	A new line should be started before the image is drawn
`LAYOUT_RIGHT`	Image should be close to the right edge of the drawing area

Table 20.20 Methods of the `ImageItem` class

Method	Comment
`String getAltText()`	Gets the text string to be used if the image exceeds the capacity of the device to display it
`Image getImage()`	Gets the image contained within `ImageItem`, or is null if there is no contained image
`int getLayout()`	Gets the layout directives used for placing the image
`void setAltText(String text)`	Sets the alternative text of the `ImageItem`, or is null if no alternate text is provided
`void setImage(Image img)`	Sets the image object contained within the `ImageItem`
`void setLayout(int layout)`	Sets the layout directives

20.1.19 `StringItem`

The `StringItem` class is an item that can contain a string. `StringItem` is display-only; the user cannot edit the contents. Both the label and the textual content of `StringItem` may be modified by the application. The visual representation of the label may differ from that of the textual contents. The methods of the `StringItem` class are listed in Table 20.21. There is an example for `StringItem` in the `Form` example (Section 20.1.8).

Table 20.21 Methods in the `StringItem` class

Method	Comment
`String getText()`	Gets the text contents of `StringItem`, or is null if the `StringItem` class is empty
`void setText(String text)`	Sets the text contents of `StringItem`

20.1.20 `TextField`

The `TextField` class is an `Item` that contains text. The user can edit the content of the `TextField` and the input capabilities can be restricted with constraints. The `TextField` constructor is as follows:

```
TextField(String label, String text, int maxSize,
  int constraints)
```

The `label` parameter is the label of the `Item`, which is shown above the `Item`. The text parameter is the initial text, which is placed in the `TextField` when it is shown. The maximum size parameter (`max Size`) is the number of characters that can be used in the `TextField`.

The user input capabilities may be restricted by the `constraints` parameter. The `constraints` parameter is a predefined constant in the `TextField` class. *Any* constraint can be used to allow the user to enter any text. Other constraints are e-mail address, number, password, and URL (universal resource locator).

Methods in the TextField class are listed in Table 20.22. There is an example the use of TextField in the Form example (Section 20.1.8).

Table 20.22 Methods in the `TextField` class

Method	Comment
`void delete(int offset, int length)`	Deletes characters from `TextField`
`int getCaretPosition()`	Gets the current input position
int getChars(char[] data)	Copies the contents of `TextField` into a character array, starting at index zero
`int getConstraints()`	Gets the current input constraints of `TextField`
`int getMaxSize()`	Returns the maximum size (number of characters) that can be stored in this `TextField`
`String getString()`	Gets the contents of `TextField` as a string value
`void insert(char[] data, int offset, int length, int position)`	Inserts a subrange of an array of characters into the contents of `TextField`
`void insert(String src, int position)`	Inserts a string into the contents of `TextField`
`void setChars(char[] data, int offset, int length)`	Sets the contents of `TextField` from a character array, replacing the previous contents
`void setConstraints(int constraints)`	Sets the input constraints of `TextField`
`int setMaxSize(int maxSize)`	Sets the maximum size (number of characters) that can be contained in this `TextField`
`void setString(String text)`	Sets the contents of `TextField` as a string value, replacing the previous contents
`int size()`	Gets the number of characters that are currently stored in this `TextField`

20.2 Nokia User Interface Classes

Java in mobile devices is used mostly for entertainment, usually games. MIDP is designed to be hardware-independent, but most games may need some low-level access to the device. The Nokia UI API provides access to the control keys and the display more efficiently than can be achieved with the original MIDP classes. The main goal of this extension API is to enhance the developer's capability of implementing high-quality games for Nokia devices. The main features in the Nokia UI API are as follows:

- low-level access to image pixel data
- transparency support
- full-screen drawing
- sound
- vibration and device light control

20.2.1 Package Overview

The Nokia UI API contains two packages – user interface and sound, as shown in Table 20.23. These packages come with the Series 60 Platform MIDP as built-in, so the developer does not need to worry about compatibility between Nokia devices. However, it must be kept in mind that these classes cannot be found elsewhere.

Table 20.23 Packages in the Nokia user interface

Package	Description
`com.nokia.mid.ui`	Provides user interface extensions for MIDP (mobile information device profile) applications
`com.nokia.mid.sound`	Contains a class and an interface for simple sound playback

20.2.2 Classes and Interfaces

`DirectGraphics`

`DirectGraphics` is an extension interface for the original the MIDP `Graphics` class. It contains means of drawing and filling triangles and polygons, drawing rotated or flipped images, alpha channel support (transparency), and access to the raw pixel data of the graphics context.

The `getDirectGraphics()` method of `DirectUtil` can be used to convert the `javax.microedition.lcdui.Graphics` object to Nokia `DirectGraphics` object. The following example shows the syntax:

```
DirectGraphics dg = DirectUtils.getDirectGraphics( g );
```

The operations in DirectGraphics also affect the original graphics object, and vice versa. For instance, setting the color with DirectGraphics also affects the Graphics object, and setting a clipping rectangle to the Graphics affects DirectGraphics. Developers may consider DirectGraphics as a new way to make calls to Graphics. DirectGraphics does not inherit Graphics because of the API dependency.

Some DirectGraphics methods take the color as the ARGB (alpha–red–green–blue) value. This value is in the form of 0xAARRGGBB, and it corresponds to the native format specified by TYPE_INT_8888_ARGB. An alpha value 0xFF indicates opacity, and a value of 0x00 full transparency.

All drawPixels() and getPixels() methods take the format of a parameter. There are six main types that are supported in the Series 60 Platform, and these formats are listed in Table 20.24. Table 20.25 contains the list of supported methods.

Table 20.24 Supported image formats

Image format	Comment
TYPE_BYTE_1_GRAY	1 bit format, 2 distinct color values (on or off), stored as a byte
TYPE_BYTE_1_GRAY_VERTICAL	1 bit format, 2 distinct color values (on or off), stored as a byte
TYPE_USHORT_444_RGB	4 bits for red, green, and blue component in a pixel, stored as a short (0x0RGB)
TYPE_USHORT_4444_ARGB	4 bits for alpha, red, green, and blue component in a pixel, stored as a short (0xARGB)
TYPE_INT_888_RGB	8 bits for red, green, and blue component in a pixel (0x00RRGGBB)
TYPE_INT_8888_ARGB	8 bits for alpha, red, green, and blue component in a pixel (0xAARRGGBB)

Table 20.25 Methods in the DirectGraphics class

Method	Comment
void drawImage (javax.microedition.lcdui.Image img, int x, int y, int anchor, int manipulation)	This method is almost similar to drawImage in Graphics, but it also has a manipulation facility that can be used to rotate and/or flip the image
void drawPixels(byte[] pixels, byte[] transparencyMask, int offset, int scanlength, int x, int y, int width, int height, int manipulation, int format)	This method is used to draw raw pixel data to the graphics context to the specified location. Only the byte-based formats are used (TYPE_BYTE_1_GRAY and TYPE_BYTE_1_GRAY_VERTICAL). The transparency mask is in the same format as pixels
void drawPixels(int[] pixels, boolean transparency, int offset, int scanlength, int x, int y, int width, int height, int manipulation, int format)	This method is used to draw pixels in the int-based formats (TYPE_INT_888_RGB or TYPE_INT_8888_ARGB). If transparency flag is true then transparency information is used (if there is any)

Table 20.25 *(continued)*

Method	Comment
`void drawPixels(short[] pixels, boolean transparency, int offset, int scanlength, int x, int y, int width, int height, int manipulation, int format)`	This method is used to draw pixels in the `short`-based formats (`TYPE_USHORT_444_RGB` or `TYPE_USHORT_4444_RGB`)
`void drawPolygon(int[] xPoints, int xOffset, int[] yPoints, int yOffset, int nPoints, int argbColor)`	*Draws* a closed polygon defined by the arrays of the x- and y-coordinates
`void drawTriangle(int x1, int y1, int x2, int y2, int x3, int y3, int argbColor)`	Draws a closed triangle, defined by the coordinates
`void fillPolygon(int[] xPoints, int xOffset, int[] yPoints, int yOffset, int nPoints, int argbColor)`	Fills a closed polygon, defined by the arrays of the x- and y-coordinates
`void fillTriangle(int x1, int y1, int x2, int y2, int x3, int y3, int argbColor)`	Fills a closed triangle, defined by coordinates
`int getAlphaComponent()`	Gets the alpha component of the current color
`int getNativePixelFormat()`	Returns the native pixel format of an implementation (`TYPE_USHORT_4444_ARGB` in the Series 60 Platform)
`void getPixels(byte[] pixels, byte[] transparencyMask, int offset, int scanlength, int x, int y, int width, int height, int format)`	Copies pixels from the specified location in the graphics context to the byte array. Only the `byte`-based formats are used (`TYPE_BYTE_1_GRAY` and `TYPE_BYTE_1_GRAY_VERTICAL`)
`void getPixels(int[] pixels, int offset, int scanlength, int x, int y, int width, int height, int format)`	Copies pixels from the graphic context to the `int` array in the int-based format (`TYPE_INT_888_RGB` or `TYPE_INT_8888_ARGB`)
`void getPixels(short[] pixels, int offset, int scanlength, int x, int y, int width, int height, int format)`	Copies pixels from the graphic context to the `short` array in the `short`-based format (`TYPE_USHORT_444_RGB` or `TYPE_USHORT_4444_RGB`)
`void setARGBColor(int argbColor)`	Sets the current color to the specified ARGB (alpha–red–green–blue) value

Combining `FLIP_HORIZONTAL`, `FLIP_VERTICAL`, `ROTATE_90`, `ROTATE_180`, or `ROTATE_270` with `OR` operator forms manipulation parameters to `drawPixels()` and `drawImage()`. These combinations are always handled in the same order: first the rotation, then the vertical flip, and, finally, the horizontal flip. The creation of a combination is shown later, in an example.

The following example is an except from the `TTProjectCanvas`. It draws the background image and then a 50% transparent red polygon in the middle of screen. After this, it draws the current time over that polygon.

```
void paint( Graphics aG )
    {
    :
    // Getting direct graphics
    DirectGraphics dg = DirectUtils.getDirectGraphics( aG );

    // drawing background image, no manipulation
    dg.drawImage( iBackGround, 0, 0, aG.TOP | aG.LEFT, 0 );

    // initializing position for polygon
    int[] xPos = {0, getWidth(), getWidth(), 0};
    int[] yPos = {getHeight() / 2 - 20, getHeight() / 2 - 20,
      getHeight() / 2 + 20, getHeight() / 2 + 20};

    // drawing 50% transparent red polygon in the middle of
    // the screen
    dg.fillPolygon( xPos, 0, yPos, 0, xPos.length, 0x77FF0000 );
    // Setting color to white
    aG.setColor( 255, 255, 255 );

    // Getting a new font
    Font fnt = Font.getFont(Font.FACE_PROPORTIONAL,
      Font.SIZE_MEDIUM, Font.STYLE_PLAIN );

    // Setting it to Graphics object
    aG.setFont( fnt );

    // drawing current time string to the middle of the screen
    aG.drawString( iCurrentTimeStr, (getWidth()/2)-
       fnt.stringWidth( iCurrentTimeStr )/2,
       (getHeight()/2)-(fnt.getHeight()/2), aG.TOP );
    :
    }
```

`TTProjectCanvas` is shown in Figure 20.9.

DeviceControl

`DeviceControl` contains a collection of methods (listed in Table 20.26) for controlling the special features of the device, such as vibration and screen backlight. For example, flashing the backlights for 1 sec and can be done as follows:

```
DeviceControl.flashLights( 1000 );
```

DirectUtils

The `DirectUtils` class contains a collection of static utility methods, listed in Table 20.27. These methods are used for converting the

Table 20.26 Methods in the `DeviceControl` class

Method	Comment
`static void flashLights(long duration)`	Temporarily flashes the lights for a specific length of time, in milliseconds
`static void SetLights(int num, int level)`	Turns lights on and off. The first parameter indicates which lights are controlled; 0 is assigned to backlights and the rest is left unspecified. The level controls the brightness; values are between 0–100 and, in most cases, 1–100 indicate on, and 0 is off
`static void startVibra(int freq, long duration)`	Activates vibration for a given length of time in milliseconds and with a frequency between 0–100. A value of 0 is no vibration. If the device does not support different vibration frequencies then the device default is used
`static void stopVibra()`	Stops any vibration

Table 20.27 Methods in the `DirectUtils` class

Method	Comment
static javax.microedition.lcdui.Image createImage(byte[] imageData, int imageOffset, int imageLength)	Creates a mutable image that is decoded from the data stored in the specified byte array at the specified offset and length
static javax.microedition.lcdui.Image createImage(int width, int height, int ARGBcolor)	The method will return a newly created mutable `Image` with the specified dimension and all the pixels of the image defined by the specified ARGB (alpha–red–green–blue) color
static DirectGraphics getDirectGraphics(javax.microedition.lcdui.Graphics g)	Converts standard javax.microedition.lcdui.Graphics to `DirectGraphics`

standard `Graphics` to `DirectGraphics`, creating images with specified ARGB color and creating mutable images from the encoded image byte arrays.

Creating an image from the byte array is done with the same semantics as in `createImage()` of `Image`. Creating a new `Image` with dimensions of 100 × 100 pixels and with red, 50% transparent, color can be done as follows:

```
DirectUtils.createImage( 100, 100, 0x77FF0000 );
```

Converting the standard `Graphics` to `DirectGraphics` is described in the DirectGraphics subsection.

FullCanvas

`FullCanvas` is used in midlets that need a full-screen painting area. The full screen is accessed simply by extending one's own canvas

Figure 20.9 FullCanvas view from the Time Tracker

class in FullCanvas. The biggest difference (after full-screen access) from the original Canvas class is that Commands cannot be added to FullCanvas. Pressing the softkey or other command buttons will result in keystroke events. These can be caught by overwriting the keyEvent(), keyPressed(), and/or keyReleased() methods in the Canvas class. There are also some new constants used for these key codes. All the normal key codes are reported to FullCanvas.

The callback methods such as paint(Graphics g) work with the same semantics as in Canvas. This means that converting some of an existing canvas to FullCanvas can be done by extending it in FullCanvas instead of Canvas (if the implementation does not use any Commands).

Although it is said that in the Nokia UI API status indicators such as networking should be shown, there is little deviation from the Series 60 Platform. The Series 60 UI style guide (Nokia, 2001g) says that applications that take the whole screen should not show any indicators.

This Time Tracker class shows how FullCanvas can be created and how key codes can be captured with it. Using the full screen is done simply by extending TTProjectCanvas from FullCanvas. The key events are captured by overwriting the keyPressed() method:

```
public class TTProjectCanvas extends FullCanvas  {
    :

    public void paint( Graphics aG )
        {
        :
        }
```

```
public void keyPressed( int aKey )
    {
    switch( aKey )
        {
        //Key was right cba key
        case FullCanvas.KEY_SOFTKEY2:
            {
            :
            // Switching away from FullCanvas
            iDisplay.setCurrent( iProjectForm );
            break;
            }
        :
        default:
            {
            :
            break;
            }
        }
    }
                                                    }
```

Sound

This class is used to play simple sound effects such as tones and digitized audio. MIDP 1.0 does not have an API for sounds, and therefore this extension is particularly needed for games. The Series 60 implementation supports WAV, RNG, AMR, MIDI, and all formats supported in the Media Server. WAV, AMR, and MIDI files can be initialized with the type `FORMAT_WAV` and RNG with the type `FORMAT_TONE`.

The sound API (shown in Table 20.28) provides support for playing simple beeps of a given frequency and duration. The following example shows how to play a single beep for one second with a frequency of 2000 Hz:

```
Sound s = new Sound( 2000, 1000 );
s.play( 1 );
```

Sound objects can easily be recycled with the `init()` method. For example, if the object is constructed as a monotone beep it can be turned to, for example, WAV simply by calling the `init` with different parameters. Some runtime memory consumption can be lowered in this way. If, for example, the `Sound` objects are constructed in a loop for generating a tone sequence, it is quite certain that an out-of-memory exception will occur at some point. The MIDP devices have restricted memory and resources, and therefore objects should be always recycled as much as possible.

Table 20.28 Methods in the `Sound` class

Method	Comment
`static int getConcurrent-SoundCount(int type)`	Returns the maximum number of concurrent sounds the device can play for a specific audio type. Because the Media Server of the Series 60 Platform supports only one sound channel this method returns 1 for all types
`int getGain()`	Gets the volume of the sound. This value is between 0 and 255
`int getState()`	Gets the current state of `Sound`. States can be `SOUND_PLAYING`, `SOUND_STOPPED`, or `SOUND_UNINITIALIZED`
`static int[] getSupportedFormats()`	Returns the supported audio formats as an `int` array. In the Series 60 Platform this returns both `FORMAT_WAV` and `FORMAT_TONE`
`void init(byte[] data, int type)`	Initializes `Sound` based on byte array data. Byte array can be in WAV, RNG, AMR, MIDI, or any other format that Media Server supports
`void init(int freq, long duration)`	Initializes (Jep) `Sound` to play a simple beep. The frequency must be in the range 0–13 288 Hz and the duration must be at least 1 milliseconds
`void play(int loop)`	This method is used for starting the playback from the beginning of a sound object. Loop count 0 indicates an endless loop; the maximum value of the loop is 255
`void release()`	Releases audio resources reserved by this, object. After this, the object must be initialized again
`void resume()`	The method will continue playback of the stopped sound object from the position from which it was stopped. Owing to the restrictions of Media Server this is not supported in the Series 60 Platform
`void setGain(int gain)`	Sets the volume for the sound object. The gain must be between 0 and 255
`void setSoundListener(Sound-Listener listener)`	Registers a listener for playback state notifications
`void stop()`	This method will stop sound playback will and change the state to `SOUND_STOPPED`

The Symbian Media Server provides only one channel for sounds, so sounds cannot concurrently play. This means that if another sound is started before the first one is finished, then the first one is stopped and the second one starts immediately.

`Sounds` follows the current profile. This means that if the midlet user has turned the warning and game tones off in the device (or emulator) no sounds will be played in Java. The key sounds are always turned off when the `Sound` class is used. This is done to avoid problems with the single sound channel.

SoundListener

This interface is used by applications that need to monitor states of the `Sound` objects. By implementing this interface (Table 20.29) and adding it to `Sound` by the `setSoundListener()` method the application can receive the `SOUND_PLAYING`, `SOUND_STOPPED`, or `SOUND_UNINITIALIZED` events.

Table 20.29 Method in the `SoundListener` interface

Method	Comment
`void soundStateChanged(Sound sound, int event)`	Called when the playback state of a sound has been changed

20.3 Networked Midlets

20.3.1 Basics

The CLDC 1.0 Specification defines an abstract generic connection framework for all connections. This framework provides support for stream-based connections and datagrams. The configurations should not provide implementation and therefore the CLDC (connected limited device configuration) does not provide any protocol implementation. Profiles that are on top of the configurations do not have to implement all protocols. Therefore, the currently used MIDP 1.0 specification demands implementation for HTTP 1.1 stream-based client connections only. This connection type is the only one supported in the Series 60 Platform.

20.3.2 Creating a Connection

In the CLDC generic connection framework, shown in Figure 20.10, the starting point for all connections is the class `Connector`. The `Connector` class has a static open method that is used to obtain the object that implements the `Connection` class or, more likely, one of its subinterfaces. In the Series 60 Platform and generally in MIDP 1.0 there is only a `HttpConnection` object that can be fetched from the `Connector`.

20.3.3 `HttpConnection`

The `HttpConnection` interface is part of the package `javax.microedition.io`. This interface defines the necessary methods and constants for an HTTP (hypertext transfer protocol) connection. The

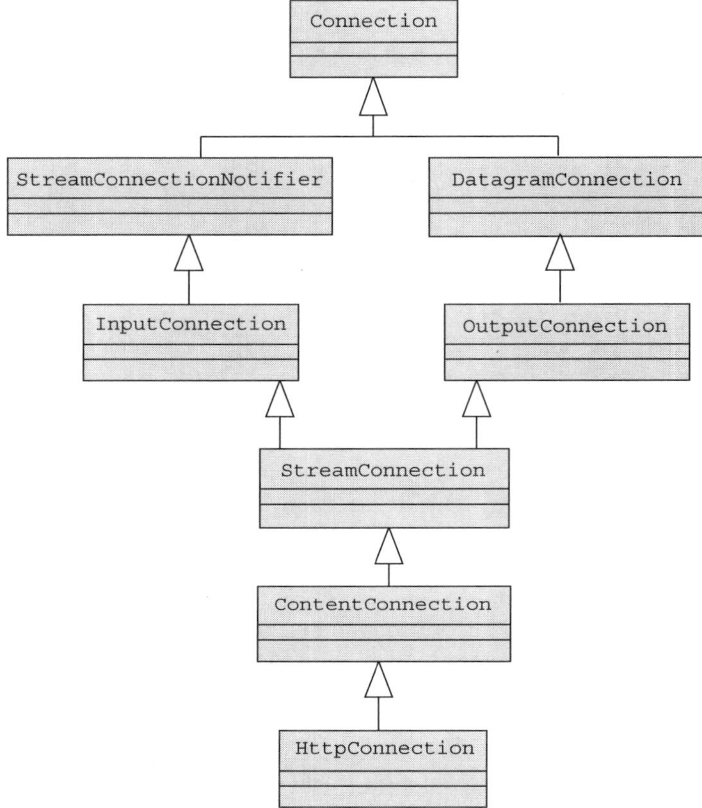

Figure 20.10 Connection framework

HTTP is a request–response application protocol that needs its parameters to be set before sending a request, and therefore it has three states: `Setup` (no connection, setting parameters), `Connected` (request send waiting response), and `Closed` (connection closed). The basic `HttpConnection` is made in the following fashion:

```
HttpConnection conn = (HttpConnection)Connector.open
  ("http://www.digia.com");
```

The connection is then in the setup state and the parameters can be set; for example:

```
conn.setRequestMethod( HttpConnection.POST );
```

The final translation to the connected state is done by one of the following methods:

```
openInputStream(), openOutputStream(), openDataInputStream(),
  openDataOutputStream(), getLength(), getType(), getDate(),
  getExpiration().
```

Closure of the connection is done by the `close()` method in `HttpConnection`. The opened input or output streams must be also closed. The following example shows how an image can be downloaded from the Internet:

```
private Image loadPicture( String aUrl )
    {
    ByteArrayOutputStream baos = new ByteArrayOutputStream();
    try
        {
        //Opening the connection to the logo
        HttpConnection conn = (HttpConnection)Connector.open( aUrl );
        InputStream inputStream = conn.openInputStream();

        byte [] nextByte= new byte[1];
        //Reading the image byte by byte
        while ( ( inputStream.read( nextByte, 0, 1 ) )!=( -1 ) )
            {
            baos.write( nextByte[0] );
            }

        // Closing the connection
        conn.close();
        inputStream.close();
        }
        catch( IOException e )
            {
            System.out.println( "loading the picture failed" );
            }
        //Creating new image from the fetched byte array
        return Image.createImage(baos.toByteArray(), 0, baos.size() );
    }
```

20.4 Over-the-Air Provisioning

Over-the-air (OTA) provisioning is a de facto standard that is provided by Sun Microsystems. The main goal of this recommended practice is to describe how the midlet suites can be deployed in devices. This standard should ensure interoperability between devices from all manufacturers and in that way make it easier for cellular operators to make their own MIDP services.

Midlet suite vendors can expect all devices to have mechanisms that allow users to download suites to the device. Such a download can be done via the resident browser [e.g. (WAP) wireless application protocol] of the device or by some application that is written specifically for suite downloading. During the download and installation process the user should get basic information on the suite that is about to be installed. This information is usually read from the Java application descriptor (JAD). After installation, the device sends notification of successful installation to the server. This gives an opportunity for cellular operators to sell midlet suites and to be sure that customers have successfully installed the suite.

The user should have an opportunity to control the suites in the device once they are installed. This control is usually done by the Java application manager (JAM) application. JAM provides mechanisms for installing, browsing, launching, and removing the midlet suites.

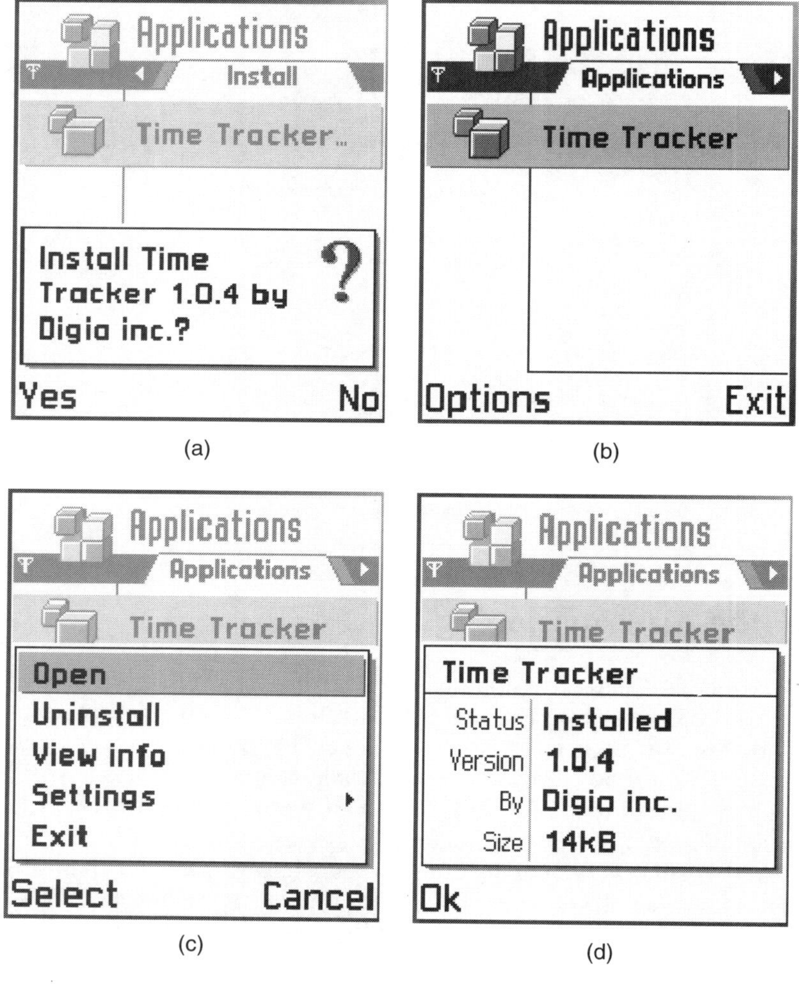

Figure 20.11 Suite download: (a) installation; (b) the installed suite; (c) the Java application manager menu for the suite; (d) midlet suite information

20.4.1 Over-the-Air Provisioning in the Series 60 Platform

In the Series 60 Platform, midlet suite download is normally done via a WAP browser or beamed with IrDa (infrared data association). After download, the Java recognizer starts the JAM installer that installs the suite in the device. The installer displays a verification to the user (Figure 20.11a) that contains the version number and the midlet vendor. This information is read in the JAD file. After installation, the midlet suite can be found in the JAM (Figure 20.11b). After this, the user is able to run the suite, uninstall it, or view its information (Figure 20.11c), which is basically everything that is found in JAD (Figure 20.11d).

20.5 Summary

In this chapter and is Chapter 19 we have described the essentials of Java programming in the Series 60 Platform. In this chapter we have concentrated on use of the midlet UI framework in application development. All UI classes that the Series 60 Platform MIDP supports have been described. In addition, code examples have been used to show how the classes can be applied in UI implementation. The Series 60 Platform supports an extension to standard MIDP UI classes. This extension, which is provided by Nokia, supports also the use of sound in Java applications. The classes in the extension packages have also been described and used in programming examples. In addition to UI programming, communication support between midlets and midlet OTA provisioning have been covered.

Appendix

An example of a User Interface Specification

1 Example of How to Specify the Screen Layout

The main view is displayed after the ImagePlus application is launched. It displays two options: Open image and New image (Figure 1; see also Table 1). Selecting Open image displays a dialog view to the images folder. Selecting New image displays a view with a new image template.

2 Example of How to Specify the Options Menus

1. **Insert object** <**insert_object**> inserts an object.
2. **Edit** <**edit**> opens the edit submenu.
   (a) **Cut** <**object_cut**> submenu function cuts the selected object.
   (b) **Copy** <**object_cut**> submenu function copies the selected object.
   (c) **Paste** <**object_paste**> submenu function pastes the copied object.
3. **Frame** <**frame**> opens the frame submenu.
   (a) **Adjust** <**frame_adjust**> submenu item adjusts the selected frame.
   (b) **Crop** <**frame_crop**> submenu item selects the crop tool.
4. **Image** <**image**> opens the image submenu.
   (a) **Brightness/Contrast** <**image_brightness_contrast**> submenu item opens the Brightness/Contrast Setting View.
   (b) **Color** <**image_color**> submenu item opens the Color Setting View.

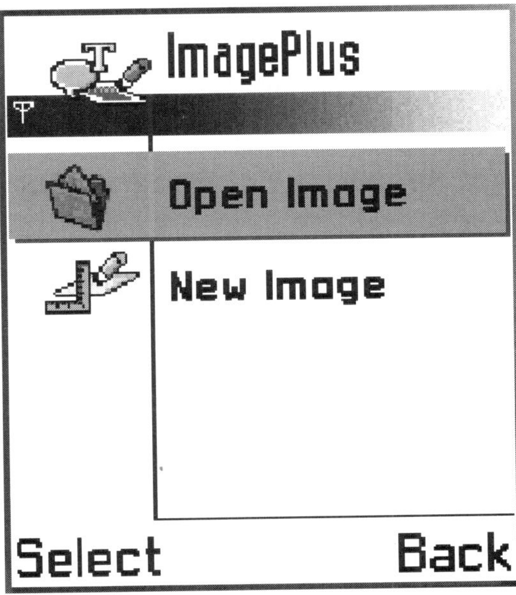

Figure 1 Layout example of the main view

Table 1 The main view layout

Layout area	Description of the layout area
Signal pane	Displays the signal strength
Context pane	Displays the application icon <application_icon>
Title pane	Displays the application name: **ImagePlus** **<application_title>**
Navigation pane	Displays text: **Start up** **<application_startup>**
Battery pane	Displays the battery pane in idle state; the battery pane indicator is not displayed in the applications
Main pane	List type: single list with large icon
AB column, first line	Displays a large icon <open_image>
AB column, second line	Displays a large icon <new_drawing>
C column, first line	**Open image** **<image_open>** opens an existing image
C column, second line	**New image** **<image_new>** opens a new drawing template
	Note: no scrolling indicator arrows are needed
Control pane	
Left softkey	**Select** **<left_softkey_select>** selects the highlighted item
Right softkey	**Back** **<right_softkey_back>** displays the previous view

Note: there is no OK options menu.

(c) **Sharpen <image_color>** submenu item opens the sharpen view.

(d) **Blur <image_blur>** submenu item opens the blur view.

3 Example of How to Specify the Notes and Queries

For details, see Table 2.

Table 2 Notes and queries in main view

Note or query name	Description
Save image note	**Do you want to save image?** **<note_saveimage>** is displayed if the user tries to close an image with unsaved changes. Left softkey: **Yes <note_saveimage_leftsk>** displays a saving wait note, if the image has a file name; if the image does not have a filename, an insert filename query is displayed. Right softkey: **No <note_saveimage_rightsk>**
Saving wait note	This query is displayed when the user chooses Save from the options menu, if the image has a filename. **Saving image, please wait. <note_wait_savingimage>**
Insert filename query	This query is displayed when the user chooses Save from the options menu, if the image does not have a filename. Prompt: **Enter filename <query_filename_prompt>** Left softkey:**Ok <query_filename_leftsk>** Right softkey: **Cancel <query_filename_rightsk>**

4 Example on How to Specify the Keypad

For details, see Table 3.

5 Example on a Use Case

See Table 4.

Table 3 Keypad functions in main view

Key name	Description of the hardware key function
Scroll up	Scrolls the selection up
Scroll down	Scrolls the selection down
Scroll left	Scrolls the selection left
Scroll right	Scrolls the selection right
Select key	Opens the OK options menu or opens the selected list item
Send key	In the idle state opens a list of recently called numbers; in the phone book, it calls the selected number
End key	Ends phone call; in applications, it goes to the idle state
Applications key	Press displays application grid: press and hold displays open applications list
Clear key	Clears characters in text-editing mode
Edit key (ABC key)	Opens the edit menu in the text-editing mode (note: this key can be pressed simultaneously with another key press)
* key	Opens the select symbol (special character table) dialog in text editors; see Chapter 10 for all possible functions of this key
# key	Changes the text input mode: see Chapter 10 for all possible functions of this key
Power key	Powers the device up or down
Voice key	Activates the voice calling function

Table 4 Examples of a use case

Use Case #: 1	
Name	Opening image for editing
Description	User selects an image from the photo album to be edited in ImagePlus application.
Actor	End user who starts using the application
Preconditions	Application Launcher is open
Postconditions	An image has been successfully opened to ImagePlus and user can start adding objects on top of it. The Startup view is displayed if no images exist in the photo album.
Success scenario	1. User selects the ImagePlus application from the application grid. 2. A startup view is displayed, the first item highlighted. 3. The user selects the *Open image* item from the startup list. 4. The Photo album contents are displayed in a pop-up window. 5. The user browses the photo album contents and selects one image. 6. The*Opening image* wait note is displayed while the system downloads the image to ImagePlus. 7. The opened image is displayed in the Main view.
Exceptions	(2) The user exits the application by pressing Exit. ImagePlus application is closed. (5) If no photos exist in the photo album, pressing Cancel will display the Startup view again. (5) If the user selects a folder, the contents of the folder are displayed in a list.
Dependencies	Opening image when an existing image in ImagePlus.

Glossary

Terms in bold in the definitions are also defined within this glossary.

ACK: acknowledgement; a positive confirmation that a requested service has been completed successfully.

ACL: (1) access control list; (2) asynchronous connectionless.

Adapters: where data are in one form and are needed in a different form from how they are stored, adapters convert the data from the stored form to the required form.

ADPCM: adaptive differential pulse code modulation.

AH: authentication header.

AIF: application information file; defines the icons, captions, capabilities, and supported **MIME** types of an application.

API: application programming interface.

APPARC: application architecture; provides a policy on how applications are stored in the file system. It also provides the basic framework for applications.

AppTest: software by Digia Inc designed to help to test applications.

App UI: handles application-wide aspects of the user interface, such as menus, starting and exiting views, and creating and destroying views. All Symbian OS applications should have one App UI.

ARM: advanced risk machine.

ARM4: a 32-bit instruction set and binary interface for **ARM**-based processors. If an application is compiled for ARM4, it can call only functions that have also been compiled for ARM4.

ARMI: a 32-bit instruction set and binary interface for **ARM**-based processors. It can call methods compiled for **ARM4** as well as **THUMB**.

ARP: address resolution potential.

ASSERT macros: macros (__ASSERT_ALWAYS and __ASSERT_DEBUG) used to check the value of a boolean condition. Assert failure usually results in a panic.

Assertion: a condition that must be true for the execution of the application to continue. Used for catching programming errors early.

ASSP: application-specific standard port.

Audio streaming: a method of playing audio where the application can concatenate new audio clips to the currently playing clip, providing a continuous stream of audio to the user.

Avkon: The Series 60 platform standard user interface toolkit built on Symbian Uikon technology.

Base porting: an activity where Symbian OS is ported to a new hardware platform.

BC: binary compatibility; BC determines whether an application is able to use an **API** without recompiling. If BC is kept between revisions of an API, applications compiled for one revision work with all revisions.

BIF: **BIO** information file.

BIO: bearer-independent object; used for sending binary data over various bearers, such as **SMS** or **WDP**.

Black-box testing: testing without knowledge about the implementation of a module. This ensures that the module fulfills the **API** it claims to export. The opposite is **white-box testing**.

BMP: Windows bitmap.

BT: British Telecommunications plc.

C32: serial communications server; handles the serial ports of a device. It provides the applications with a generic interface to serial-like communications.

CBA: command button array on Series 60; consists of the functions bound to the two softkeys.

CCP: compression control panel.

CDC: connected device configuration.

CHAP: challenge handshake authentication protocol.

Checkbox: a special type of list, where the user can choose several items from a list of choices.

CLDC: connected limited device configuration; a Java standard that defines common hardware requirements and a set of **APIs** for low-end consumer devices.

CONE: control environment; provides the framework for creating and accessing user interface controls. Includes all classes starting with the prefix CCoe.

Control: elements that own and handle the drawing on a rectangular area on the screen. They can respond to user actions and they can consist of other controls.

CRE: cyclic redundancy check.

CSD: circuit-switched data; the generic 9600 bps **GSM** data connection.

CSY: Serial communications module; provides an implementation of a serial port to the Serial communications server.

DBMS: database management system on Symbian OS; provides easy-to-use database functionality to the applications.

DCE: data circuit terminating equipment.

DEA: data element alternative.

Delayed function call: interrupt handlers must be very fast and small in order to be able to serve all interrupts. If the handling of an interrupt takes more time, the interrupt handle invokes a delayed function call and exits. The kernel runs the function after the interrupt handle has completed.

DES: (1) data encryption standard; (2) data element sequence.

Descriptors: provide a type-safe and easy-to-use way of accessing, manipulating, and storing strings and binary data.

DFC: delayed function call.

Dialog: used for bringing information to the user's attention or for requesting information from the user. They take up screen space and intercept key commands temporarily until the user acknowledges the dialog. If possible, a setting page should be used instead to provide a user interface with Series 60 feel.

DLL: dynamic link library.

DMP: device management protocol.

DNS: domain name service; provides a mapping between **IP** addresses and domain names.

DTE: data terminal equipment.

DTMF: dual-tone multifrequency; a method of sending numbers and a few special symbols via analog phone system. It is used for selecting the phone number of the remote party at the beginning of a call.

DUNP: dialup networking profile.

EDGE: an evolution of the **GSM**, providing faster wireless data connections.

Enbs: Symbian OS enumeration for **NBS**.

Engine: sometimes called the model of the Model–View–Controller design pattern. It is the part of application that manages the non-user-interface aspects of the program logic. The engine should be transferable between different user interface styles.

ESP: encapsulating security payload.

ETel: the telephony server on Symbian OS; provides access to the telephony functions of a mobile phone, such as **GSM**, **GPRS**, and fax.

FEP: front-end processor; a module that intercepts keyboard events from editor controls. Some events are allowed through, but some events are used, for example, to implement the T9 input method.

FIR: fast infrared; a fast version of the **IrDA** protocol. It provides speeds up to 4 Mbps.

Form: a collection fields, that the user can fill. Dialogs use forms for showing the actual items.

FPU: floating point unit.
GCC: gnu compiler collection.
GDI: graphics device interface.
GIF: graphics interchange format.
GPRS: General Packet Radio System; an extension of **GSM**, which provides packet data service.
Grid: user interface control that provides horizontal and vertical browsing of list elements.
GSM: global system for mobile.
GUI: graphical user interface.
HCI: host controller interface.
HSCSD: high-speed circuit-switched data; a fast **GSM** data connection providing speeds from 14 400 bps upwards.
HTML: hypertext markup language.
HTTP: hypertext transfer protocol.
IAP: Internet access point.
ICMP: Internet control message protocol.
Icons: small graphical elements that convey a piece of information to the user. Icons are almost always embedded inside **MBM** files.
IDE: integrated development environment.
IMAP4: Internet mail access protocol; allows connection to and management of mailboxes on remote hosts. Also permits selective download of messages.
IMC: Internet Mail Consortium; maintains, develops, and promotes standards for e-mail in Internet.
IP: Internet protocol.
IPSec: Internet security protocol.
IrCOMM: emulates the RS-232 protocol on **IrDA**. It allows implementation of IrDA support for legacy applications.
IrDA: Infrared data association; maintains and promotes the infrared protocol stack with the same name. The protocol hides the implementation details of infrared communications from the applications and allows them to communicate with another IrDA-capable device with a simple socket **API**.
IrLAN: infrared local area network.
IrLAP: infrared link access protocol.
IrLMP: infrared link management protocol; establishes and maintains infrared links between applications. It provides services such as device discovery and link multiplexing.
IrMUX: a socket protocol providing an unreliable datagram service through the **IrDA** LM-MUX layer (IrDA Link Management MUX).
IrObex: infrared object exchange.
IrTinyTP: infrared tiny transport protocol.

ISM: industrial, scientific, and medical; a radio band freely usable for industrial, scientific, and medical applications.

ISP: Internet service provider.

ITC: interthread communication.

J2ME: Java 2 Micro Edition; a small version of the Java 2 family intended for mobile phones and set-top boxes.

JAD: Java application descriptor; contains an 'advertisement' on a Java application and directions to download it.

JAM: Java application manager.

JAR: Java archive; contains the class files and resources of a Java application.

JFIF: **JPEG** file interchange format.

JNI: Java native interface.

JPEG: Joint Picture Expert Group.

JRN: Java runtime environment.

KVM: K Virtual Machine; a Java virtual machine for low-footprint applications. It is used in the **MIDP** and **CLDC** platforms.

L2CAP: logical link control and adaptation protocol.

LAF: look and feel; the part of the user interface style that defines the actual appearance of the applications.

LAN: local area network.

LANAP: local area network access profile.

LCP: logical control protocol.

LDD: logical device driver.

LIF: Location Interoperability Forum.

LIFO stack: last-in, first-out stack.

Lists: sometimes called list boxes; lists show the user an array of textual or graphical elements. If the list is interactive, the user can choose one or more elements from the list.

LMP: link management protocol.

LOC: localization file; contains the strings of an application that are displayed to the user. There is one LOC file for each application and language.

MARM: multiprocess advanced risk machine; a general name for a target build of type **ARMI**, **ARM4**, or **THUMB**.

MBM: multi-bitmap file; a collection of bitmaps. It is used in Symbian OS because it is an efficient way of storing and retrieving a large number of bitmaps.

MD5: encryption algorithm.

Menu: menus or pop-up menus contain a list of actions the user can perform. The list is brought up only when needed when the user presses, for example, the Options softkey or selection key (context-sensitive menu).

MHTML: defines how **HTML** documents can be embedded inside e-mail messages.

Midlet: Midlet is an application suite written for the **MIDP** platform.

MIDP: mobile information device profile; built on **CLDC** and defines characteristics for a mobile device.

MIME: multipurpose Internet mail extension; extends the Internet mail format to allow different character sets, nontextual, and multipart messages.

MMS: multimedia message service; a standard that allows multimedia messages to be sent between mobile devices.

MMSC: multimedia message service center.

MTM: message-type module; a plug-in module that contains the implementation of a messaging method, such as **SMS**, **IMAP4**, or **POP3**. An MTM is divided into UI MTM, UI Data MTM, Client-side MTM, and Server-side MTM.

MVC: Model-View-Controller.

NACK: negative acknowledgment; see **ACK**.

NBS: narrow-band sockets; a protocol, that is no longer used. NBS port numbers are still in use in **smart messaging** for distinguishing between different smart message types.

NIF: network interface.

NifMan: network interface manager; responsible for network connections, such as dial up features.

Note: a simple dialog-type control. A note can be global to show it on top of any application, or local to show it only in one application, and a tone can be attached to the note to attract the user's attention. Examples of notes are warning note, error note, and progress note.

OBEX: object exchange.

Object provider: this mechanism allows controls to access other controls in the same user interface context. Examples are control scrollbars, which access the scrollbar on **CBA**.

OCP: open-closed principle.

OCR: optical character recognition.

OEM: original equipment manufacturer.

OMA: Open Mobile Alliance.

OOM: out-of-memory.

OSI: open systems interconnection.

OTA: over the air.

Pane: a part of display area dedicated to a specific purpose. The Series 60 platform defines a list of panes, which should be the same for all applications, to show, for example, the application name, signal strength, and **CBA**.

PAP: **PPP** authentication protocol.

PCM: pulse code modulation.

PDD: physical device driver.

PDI: personal data interchange.

PDU: (1) protocol data unit; (2) protocol description unit.

PIM: personal information management; a collection of functions such as calendar, e-mail, and a contacts manager.

PLP: Psion link protocol; a protocol that was used in communication between a PC and a Symbian OS device before the **mRouter** protocol.

PNG: portable network graphics.

Polymorphic interface library: a **DLL** that exports only one method. That method is commonly used for creating an object with a certain M-class interface. All methods accessed from the class are virtual methods and thus need not be exported.

POP3: Post Office Protocol; allows connection to Internet mailboxes on remote hosts and retrieving the mail. Not as advanced as **IMAP4**.

PPM: pulse position modulation.

PPP: pout-to-point protocol.

PRT: protocol module.

PSL: Psion link protocol.

Query: like a **note**, except the user is asked to enter a piece of information.

Radio button: a special type of list, where the user can choose one from a list of choices.

RARP: reverse address resolution protocol.

RAS: remote access server.

RFCOMM: emulates the RS-232 protocol on top of **L2CAP**. It allows implementation of Bluetooth support for legacy applications.

Rich text: text that can contain several types of fonts, styles, and other control information.

RTTI: runtime-type information; allows C++ applications to make decisions based on the runtime type of an object. Not supported in Symbian OS.

SAP: service action point.

SCO: synchronous connection-oriented.

SDP: service discovery protocol; allows applications to register and query Bluetooth services available on the device.

SDP UUID: service discovery protocol universally unique identifier; a 128-bit quantity used by Bluetooth to identify service classes, services, and protocols.

SDU: service data unit.

SendUI: implements the part of the user interface that is needed for sending a message. The application just needs to embed SendUI and implement parts that specify what is sent.

Setting page: like a **dialog**, but it allows modification of only one setting and covers the whole screen area. It gives the user better and more 'Series 60' like visual feedback than does a dialog. A setting page can be embedded inside a **setting view**.

Setting view: a collection of **setting pages**. It shows a list of settings that can be modified, and when the user chooses a specific setting, the corresponding setting page is opened. Most of the settings in default applications on the Series 60 platform are accessed via setting views.

Sinewave tone: sound composed of a single sinewave, providing only one tone of a certain frequency.

SIR: slow infrared; the original **IrDA** protocol. It provides speeds up to 115.2 kbps.

SIS: Symbian Installation System: allows packaging an application into a file, from which the application can be installed on a Symbian OS system.

Smart messaging: a concept where special messages such as ring tones and calendar entries can be sent via **SMS**.

SMIL: synchronized multimedia integration language.

SMSC: short message service center.

SMTP: simple mail transfer protocol; a standard for sending mail from one host to another. It is the backbone of the Internet mail system.

SPP: serial port profile.

SSH: secure shell.

SSL: secure socket layer.

Static interface library: a **DLL** where all the methods and classes available are accessed via exported methods.

SyncML: a standard that allows synchronizing the calendars and contacts of mobile devices with each other, or with a PC host.

Tabs: a complex user interface can be divided into specific views of the user interface. The top part of the screen can contain tabs (i.e. a horizontal list of the views as text or icons).

TCP: transmission control protocol.

TCP/IP: transmission control protocol/Internet protocol.

THUMB: a 16-bit instruction set and binary interface for **ARM**-based processors. Used primarily for **ROM**-based applications If an application is compiled for THUMB, it can call only functions that have also been compiled for THUMB.

TLS: transfer layer security.

TSY: Telephony control module.

UART: universal asynchronous receiver–transmitter; the hardware component that manages serial ports.

UCD: user-centered design.

UDP: user datagram protocol.

UED: user environment design.

UID: unique identifier.

Uikon: generic Symbian user interface toolkit on which all device-specific toolkits build on. Provides generic elements such as **dialogs**, and other **controls**.

UML: user modelling language.

UMTS: Universal Mobile Telecommunications System; one of the major third-generation systems being developed. It will offer data rates up to 2 Mbit s^{-1} with global roaming and other advanced capabilities.

URL: universal resource locator.

USB: universal serial bus.

UTF: unicode transformation format.

UUID: universally unique identifier.

vCalendar: a format for calendar access. Allows sending calendar entries from one device to another and scheduling meetings between participants.

vCard: a format for business contacts. Allows sending contact entries from one device to another.

View: an application displays the user data inside views. One application can contain several views with the means to switch between views linking parts of user data to each other. A view can also be activated from another application.

VM: virtual machine.

vMail: a Symbian-defined format for mail messages. It is based on the same stream and property formatting principle applied in **vCalendar** and **vCard**.

WAE: wireless application environment.

WAP: wireless application protocol.

WDP: wireless datagram protocol.

White-box testing: testing done generally by the implementer of a module. The test cases are designed to cover a large portion of the normal and exceptional flows of control in a module. See also **black-box testing**.

Window: all draw commands are performed ultimately on windows. The window server converts the draw commands to hardware drawing instructions.

WINS: Windows single process.

WML: wireless markup language.

WSP: wireless session protocol.

WTAI: wireless telephony application interface.

WTLS: wireless transition layer security.

WTP: wireless transaction protocol.

XML: extensible markup language.

References

Allin J (2001) *wireless Java for Symbian Devices*. John Wiley, New York.

Bevan N (2002) 'Cost-effective User-centered Design', available at www.usability.serco.com/trump/index.htm.

Beyer H, Holtzblatt K (1998) *Contextual Design: Defining Customer-centered Systems*. Morgan Kaufmann, San Francisco, CA.

Binder R (1999) *Testing Object-oriented Systems: Models, Patterns, and Tools. The Addison-Wesley Object Oriented Technology Series*. Addison-Wesley Longman, Reading MA.

Booch G, Rumbaugh J, Jacobson I (1999) *The Unified Modeling Language*. Addison-Wesley, Reading, MA.

Buschmann F, Meunier R, Rohnert H, Sommerlad P, Stal M (1996) *Pattern-oriented Software Architecture, Volume 1: A System of Patterns*. John Wiley, New York.

Cai J, Goodman D (1997) 'General Packet Radio Service in GSM', *IEEE Communications Magazine* October: 122–131.

CMG, Comverse, Ericsson, Nokia, Logica, Motorola, Siemens, Sony Ericsson (2002) 'MMS Conformance Document, Version 2.0.0', available at www.forum.nokia.com.

Connected, Limited Device Configuration (JSR-30), Version 1.0, Sun Microsystems, 2000. Available as http://jcp.org/aboutJava/commmunityprocess/final/jsr030/index.html

Doraswamy N, Harkins D (1999) *IPSec, The New Security Standard for the Internet, Intranets, and Virtual Private Networks*. Prentice Hall, New Brunswick, NJ.

Douglass B (2001) *Doing Hard Time: Developing Real-time Systems with UML, Objects, Frameworks, and Patterns*. Addison-Wesley, Reading, MA.

Dray S (1998) 'Structured Observation', tutorial presented at CHI-SA (Pretoria, South Africa), CHI 2001 (Seattle, WA), CHI 2000 (The Hague, Netherlands), CHI 1999 (Pittsburgh, PA), CHI 1998 (Los Angeles, CA), CHI 1997 (Atlanta, GA), and CHI 1996 (Vancouver, BC).

Dray S, Mrazek D (1996) 'A Day in the Life of a Family: An International Ethnographic Study', in D Wixon, J Ramey (eds), *Field Methods Casebook for Software Design* John Wiley, New York, 145–156.

ETSI (1998) 'ETSI GSM 02.60: Service Description, Stage 1 (v 7.0.0)', ETSI, www.etsi.org<http://www.etsi.org>.

Fewster M, Graham D (1999) *Software Test Automation: Effective Use of Test Execution Tools*. ACM Press, Addison-Wesley, Reading, MA.

Gamma E, Helm E, Johnson R, Vlissides J (2000) *Design Patterns: Elements of Reusable Object-oriented Software*. Addison-Wesley Longman, Reading MA.

Hoikka K (2001) 'Design Patterns in EPOC Software Development', master's thesis, Department of Information Technology, Lappeenranta University of Technology, Finland.

Holmqvist L (1999) 'Will Baby Faces Ever Grow Up?', available at www.viktoria.se/groups/play/publications/1999/babyfaces.pdf.

Hynninen T, Kinnunen T, Liukkonen-Olmiala T (1999) 'No Pain, No Gain: Applying User-centered Design in Product Concept Development', in S Dray, J Redish (eds), *Interact 99*. The British Computer Society and IFIP TC 13, 201–205.

ISO 13407 (1999) *Human-centered Design Processes for Interactive Systems*. ISO, Geneva.

Jipping M (2002) *Symbian OS Communications Programming*. Symbian Press, John Wiley, New York.

Larmouth J (1999) *ASN.1 Complete*. Morgan Kaufmann, San Francisco, CA.

Lindholm T, Yellin F (1999) *The Java Virtual Machine Specification*, second edition, Addison-Wesley, Reading, MA.

Martin R (1996) 'Open Closed Principle. C++ Report, January', available at www.objectmentor.com/resources/articles/ocp.pdf.

Mery D (2001) Symbian OS version 6.x: 'Detailed Operating System Overview', available at www.symbian.com/technology/symbos-v6x-det.html.

Mayhew D (1999) *The Usability Engineering Lifecycle: A Practitioners Handbook for User Interface Design*. Morgan Kauffmann, San Francisco, CA.

Nelson E (2002) 'Interview of Kent Beck and Alan Cooper: Extreme Programming vs. Interaction Design', available at www.fawcette.com/interviews/beck_cooper/default.asp.

Nielsen J (1992) 'Finding Usability Problems through Heuristic Evaluation', *Conference Proceedings of Human Factors in Computing Systems*. May 3–7, 1992.

Nielsen J (1994) *Usability Engineering*. Morgan Kauffmann, San Francisco, CA.

Nielsen J, Norman D (2002) 'User Experience: Our Definition', available at www.nngroup.com/about/userexperience.html.

Nokia (2000) 'Smart Messaging Specification Revision 3.0.0', Nokia, Finland; available at www.forum.nokia.com.

Nokia (2001a) 'Bluetooth Application Development for the Series 60 Platform, Version 1.0', Nokia, Finland; available at www.forum.nokia.com.

Nokia (2001b) 'Nokia Coding Idioms for Symbian OS', Nokia, Finland; available at www.forum.nokia.com.

Nokia (2001c) 'Designing Applications for Smartphones: Series 60 Platform Overview', Nokia, Finland; available at www.forum.nokia.com.

Nokia (2001d) 'Nokia Multimedia Messaging White Paper', Nokia, Finland; available at www.forum.nokia.com.

Nokia (2001e) 'Nokia Series 60 Games UI Style Guide', Nokia, Finland; available at www.forum.nokia.com.

Nokia (2001f) 'Nokia Series 60 SDK for Symbian OS', Nokia, Finland; available at www.forum.nokia.com.

Nokia (2001g) 'Nokia Series 60 UI Style Guide', Nokia, Finland; available at www.forum.nokia.com.

Nokia (2001h) 'WAP Service Developer's Guide for the Series 60 UI Category Phones, Version 1.2', Nokia, Finland; available at www.forum.nokia.com.

Nokia (2002) 'How to Create MMS Services, Version 3.0', Nokia, Finland; available at www.forum.nokia.com.

Norman D (1999) *The Invisible Computer: Why Good Products Can Fail, the Personal Computer is so Complex, and Information Appliances are the Solution.* MIT Press, Cambridge, MA.

Perry W (2000) 'Effective Methods for Software Testing', Second Edition, 2000, John Wiley & Sons, Inc. Printed in USA. in R Elliot (ed.), Canada.

Pora H (2002) 'Application Size Optimization in Mobile Environment', master's thesis, Department of Information Technology, Lappeenranta University of Technology, Finland.

Pressman R (1999) *Software Engineering: A Practitioner's Approach.* McGraw-Hill International Editions, New York.

Raskin J (2000) *The Humane Interface.* ACM Press, Addison-Wesley, Reading, MA.

Rose M (1990) *The Open Book: A Practical Perspective on OSI.* Prentice Hall, New Brunswick, NJ.

Ruuska S, Väänänen-Vainio-Mattila K (2000) 'Designing Mobile Phones and Communicators for Consumers' Needs at Nokia', in E Bergman (ed.), *Information Appliances and Beyond; Interaction Design for Consumer Products,* Morgan Kaufmann, San Francisco, CA, 169–204.

Schneier B (1995) *Applied Cryptography: Protocol, Algorithms, and Source Code in C,* second edition, John Wiley, New York.

Snyder C (2001) 'Paper Prototyping: Article on IBM DeveloperWorks', available at www-106.ibm.com/developerworks/library/us-paper/?dwzone=usability.

Stallings W (1999) *Data and Computer Communications,* sixth edition, Prentice Hall, New Brunswick, NJ.

Steele G, Gosling J, Bracha G (2000) *The Java Language Specification,* second edition, ed. B Joy, Addison-Wesley, Reading, MA.

Symbian (2002a) 'UIQ for Symbian OS v7.0', available at www.symbian.com/technology/ui/uiq.html.

Symbian (2002b) ''Symbian OS User Interfaces'', www.symbian.com/technology/devices.html.

SyncML Initiative (2002c) 'SyncML Sync Protocol 1.1 White Paper', available at www.syncml.org.

Tasker M, Allin J, Dixon J, Forrest J, Heath M, Richardson T, Shackman M (2000) *Professional Symbian Programming. Mobile Solutions on the EPOC Platform.* Wrox Press, Birmingham, UK.

WAP Forum (2001) 'WAP MMS Architecture Overview', available at www.wapforum.org.

WAP Forum (2002a) 'WAP MMS Client Transactions Specification', available at www.wapforum.org.

WAP Forum (2002b) 'WAP MMS Encapsulation Protocol', available at www.wapforum.org.

Index